Praise for *Electronic Troubleshooting*

"The fourth edition of *Electronic Troubleshooting* has been thoroughly rewritten and updated. The addition of coverage on computer networking and network devices along with expanded subject matter in many other areas has improved a book that has already proven to be very successful. As a former co-author on editions 1–3, I recommend this book and congratulate my friend Dan Tomal and his new co-author on a job well done."

—Neal S. Widmer
Professor, Electrical & Computer Engineering Technology
Purdue University

"If it's electronic, and there is troubleshooting to be done, then this is the book to reach for!"
—Dr. Simon Monk
Author, 30 Arduino Projects for the Evil Genius

"This book approaches the troubleshooting of circuits and systems by first providing an understanding of the theory of operation, so troubleshooting can proceed in a very logical manner. The writing style and organization are very readable and understandable as well as replete with examples."
—Arthur B. Williams
Author, The Analog Filter and Circuit Design Handbook

"The authors did an excellent job explaining enough information for beginners and experienced troubleshooters alike … well worth the value in your collection!"

—Carl D. Vizza
Director, Office of Professional Development
Illinois Institute of Technology

"The authors have provided an outstanding book on electronic troubleshooting with clear, concise, and concrete examples that anyone can relate to."

—James Karagiannes, Ph.D. Physics
Associate Dean of Engineering and Information Sciences
DeVry University, Chicago

"Drs. Tomal and Agajanian provide a practical and understandable approach to theory and troubleshooting electronic devices. I highly recommend this book."

—A.Galip Ulsoy
C. D. Mote, Jr , Distinguished University Professor
and William Clay Ford Professor of Manufacturing
Dept. of Mechanical Engineering, University of Michigan, Ann Arbor, MI

"One of the best and most comprehensive books since the early digital circuits and microprocessors textbooks of the early 1980s. The authors have an up-to-date approach to diagnostic and troubleshooting fundamentals for the modern technologies."

—Hovhannes Mardirossian
Retired Member of Technical Staff at Bell Laboratories
Chairman, Technology and Education
AESA NY NJ Section

"This is a thoughtful and comprehensive book on electronic troubleshooting. The authors provide an excellent basis for understanding the electronic devices and troubleshooting with clear, concise, and concrete examples. I am confident that this book will provide a positive contribution to the electrical and electronic field."

rasteh, Ph.D.
Department,
n Rouge, LA

Electronic Troubleshooting by Dan Tomal and Aram S. Agajanian is a "must read" book for everyone interested in that field. Its concrete examples, concise style, and clear language give a practical direction to it. I highly appreciate the authors' efforts."

—*Levon Lajikyan*
Professor, Yerevan State University
Yerevan, Armenia

"The authors reduce the gap which exists between industry and an educational environment in electronic theory and troubleshooting; therefore, this book should prove of considerable value to engineers, technologists, and students. A job well done!"

—*Garbis Kerestecioglu, P.Eng., M.Sc.*
Richmond Hill, ON

"Drs. Tomal and Agajanian provide a practical approach to troubleshooting electronic devices, with clear and concise examples most people can relate to. I highly recommend this book to all electronics technicians; they should have it available as a reference in their professional workday, to brush up on theories that they have studied during their years as a student."

—*Masis Michael Mirijanian*
Senior Technical Specialist, Automation
Siemens Corp., Los Angeles, CA

"The text has good technical information, and good, easy-to-grade student self-examinations, questions, and problems for each chapter. I highly recommend it."

—*Dudley Outcalt, Ph.D., P.E.*
Milwaukee School of Engineering (MSOE)

"This book provides a comprehensive and excellent coverage of electronic troubleshooting and is a great book for anyone interested in technology and electronics."

—*Daniel J. Fitzgibbons*
President & CEO
Gibson Electric & Technology Solutions
Downers Grove, IL

"This book can help instill the teachings of in-depth troubleshooting, and would be a fantastic book for multiple theories of electronics."

—*Brooks Jacobsen*
Robotic/Electronic System Technologies
Lake Area Technical Institute, South Dakota

"I recommend this book to anyone that needs to troubleshoot electronic devices. I wish that this book was written 30 years ago since its well written directions would have saved me an enormous amount of time."
—*Andrew Kumiega, Ph.D. I.E.*
Chicago, IL

"The book provides both students and professionals with comprehensive and practical knowledge for troubleshooting problem residential and commercial electrical systems. Definitely a valuable resource for those practicing and who are interested in pursuing a career in the electrical and electronics industry."
—*Dr. Thomas M. Korman, P.E., PLS, M.ASCE*
Associate Professor
Cal Poly State University, San Luis Obispo, CA

Electronic Troubleshooting

Daniel R. Tomal, Ph.D.
Aram S. Agajanian, Ph.D.

Fourth Edition

New York   Chicago   San Francisco
Athens   London   Madrid
Mexico City   Milan   New Delhi
Singapore   Sydney   Toronto

Electronic Troubleshooting, Fourth Edition

1 2 3 4 5 6 7 8 9 0   DOC/DOC   1 2 0 9 8 7 6 5 4

ISBN 978-0-07-181990-9
MHID 0-07-181990-8

This book is printed on acid-free paper.

Sponsoring Editor Michael McCabe	**Acquisitions Coordinator** Bridget L. Thoreson	**Proofreader** Mary E. Kanable
Editorial Supervisor Stephen M. Smith	**Project Manager** Sandhya Gola, Cenveo® Publisher Services	**Art Director, Cover** Jeff Weeks
Production Supervisor Lynn M. Messina	**Copy Editor** Cenveo Publisher Services	**Composition** Cenveo Publisher Services

About the Authors

DANIEL R. TOMAL, Ph.D., is a professor, consultant, and award-winning author. He has authored 15 books and over 200 research studies and articles on the topics of electronics, technology, research, and leadership. He is a widely sought-after speaker and lives in Wheaton, Illinois.

ARAM S. AGAJANIAN, Ph.D., is a professor in the Electronics and Biomedical Engineering Technology and Network Communications programs at DeVry University. He has published multiple papers on educational leadership; the papers concentrate on encouraging female enrollment in science, technology, engineering, and math (STEM) programs. He lives in River Forest, Illinois.

Contents

viii   Contents

x   Contents

Foreword

According to the 2012 Bureau of Labor Statistics, over the next 10 years some of the fastest-growing career opportunities will be in technology-related fields. It is critical that technical professionals have the needed skills to compete in this ever-growing and challenging technological society.

This book written by Dr. Tomal and Dr. Agajanian provides valuable knowledge for the careers of both today and tomorrow. It is written in a practical and straightforward manner, and is one of the best and only books that provides a comprehensive array of multiple technologies and electronic devices. This is an outstanding resource for students in the classroom, practicing technicians and electrical engineers, apprentice technicians, service technicians, hobbyists, and electronic troubleshooters.

Candace Goodwin
Chicago Metro President
DeVry University

Acknowledgments

One day I received an e-mail from Dr. Daniel Tomal asking if I would be interested in co-authoring a book called *Electronic Troubleshooting* and of course I said "Yes." I realized that this was not a dream but a reality when I was looking at a contract with McGraw-Hill. I am still at a loss for words on how to thank Dr. Tomal; he gave me the opportunity to fulfill my desire to be a co-author of a book. I have to thank him for assisting, advising, and walking me through this amazing journey of book writing. From the first day we met he has always encouraged and advised me. It has been an honor to be his co-author.

Credit must be given to Professor Patrick O'Connor on so many levels. He was my Dean who hired me in 1994 at DeVry-Chicago. Through the years he mentored, advised, and encouraged me to pursue a Ph.D. and even danced at my wedding. Pat retired early, but when I contacted him with questions about writing Chaps. 9 and 12, he showed no hesitation to help. The material covered in these chapters is his specialty, and his knowledge and skill go far beyond the content that is included in this book. I credit both chapters to him. Because of his knowledge, experience, and help, I was able to include these two important chapters in the book. Pat has also edited and reviewed other chapters, and he drew most of the figures in Chap. 8.

I must recognize and thank Dr. Ahmed Khan for all of his contributions— not only for this book but for the professional papers through the years we have co-authored. His vast number of published books and papers, and his integrity, knowledge, experience, positive encouragement, advice, and friendship are major reasons I undertook this project and was able to complete my chapters.

Credit for the figures and tables in Chaps. 7, 10, and 11 goes to Mike Breckenridge. His knowledge of programs such as Multisim, Eclipse, and Visio as well as his attention to details has contributed to these professional drawings. I thank him for his positive encouragement, help, and friendship.

I would like to thank the beautiful models Stephanie and Isabella Girjikian who very patiently and professionally posed for the pictures taken by their father Hacik Girjikian.

I would also like to acknowledge my friends and colleagues at DeVry-Chicago for all their advice and encouragement through the years.

In addition, I extend my deep appreciation to Joseph Arslan, a retired civil engineer, who is not only my uncle but my mentor and advisor. Since my earliest memory he has always encouraged me to become an engineer. Without his advice, care, and love, I would not have been able to accomplish any of my goals.

Last and most important, I have to thank my wife for her infinite love, support, encouragement, wisdom, and resilience, and for never losing faith in me.

Aram S. Agajanian, Ph.D.

The authors wish to thank the many organizations and people who contributed to the development of this book. We express our appreciation to the many companies that provided information and illustrations such as Zenith Video Tech Corporation, Winegard Company, KB Electronics, Inc., Reliance Electric Company, Lexsco Inc., B & B Motor and Control Corporation, Tektronix Inc., AVO International, Simpson Electric Company, John Fluke Mfg. Co., Inc., Hewlett-Packard Company, Bodine Electric Company, Superior Electric, Allen-Bradley Company, Square D Company, Chapman Electrical Works, Etcon Corp., Globe Products, Inc., Leader Instruments Corporation, Franklin Electric Company, Marathon Electric Company, General Electric Company, IBM, Inc., Texas Instruments, Inc., Howard W. Sams & Co., Electric Power and Power Electronics Center at the Illinois Institute of Technology, NetGain Technologies, LLC, Maxwell Technologies, Megger, Panasonic Corporation of North America, Rich-Mar Corporation, Agilent Technologies, Intel Corporation, Zilog Corporation, and Cisco Corporation. Appreciation is extended to professor Jeremy Hajek and the Embedded Systems program at the Illinois Institute of Technology. Dan Tomal appreciates the many years of working with his brothers Mike, Tony, and Eric Tomal, who had a knack for fixing anything. Lastly, we thank Frank Gurtz of Gurtz Electric Company, Dan Fitzgibbons of Gibson Electric & Technology Solutions, and the Electrical Contractors' Association of the City of Chicago for their support.

Introduction

The fourth edition of this book has been updated to include the latest electrical and electronic devices and troubleshooting methods. It continues to be very suitable for anyone—associate's, bachelor's, or master's degree electrical and electronic engineering or technology students, service technicians, hobbyists, apprentice technicians, trade, industrial, and vocational students—who wants to understand electrical and electronic troubleshooting. This book is ideal for adoption in electrical and electronic troubleshooting courses or as a required supplemental text for technical, vocational, or engineering courses. For optimum textbook use, a basic knowledge of electricity and electronics theory is advised.

The book can also be used by any technician or engineer as a handy reference guide in troubleshooting a wide array of electrical and electronic products and devices.

The book's unique strength is that it consolidates fundamentals and troubleshooting information of a wide range of electrical and electronic devices, thereby eliminating the need for several books. The authors have written the book in a practical and down-to-earth fashion and have included numerous tables, charts, illustrations, graphs, and troubleshooting flowcharts. The authors have included many "rules of thumb" and time-saving "tricks of the trade" acquired through several years of hands-on experience as electrical and electronic troubleshooters and consultants.

This book has also been written to incorporate sound principles of learning and methodology utilizing the cognitive, affective, and psychomotor domains of learning to enable the reader to comprehend and understand the material.

The book blends traditional electrical theory and equipment (motors, controls, and wiring) with modern networking and electronic technology (computers, microprocessors, embedded microprocessor systems); the cutting-edge technology of biomedical equipment is also included. The reference guides at the back of the book will also allow the reader to quickly focus on the problem at hand, instead of requiring her or him to read the entire book for just one problem.

The organization of the book follows a progressive approach of understanding basic troubleshooting analysis and traditional products to more complex and advanced technology. Chapter 1 covers the essential principles and methods of electrical and electronic troubleshooting and provides the basic theory and testing of components such as capacitors, diodes, silicon-controlled rectifiers, integrated circuits, resistors, ultracapacitors, and inductors. This chapter also includes a practical problem-solving approach and presents the various circuit faults commonly found in electronics today.

Chapter 2 provides a valuable overview of modern-day test equipment used in troubleshooting electrical and electronic products and devices. Commonly used test instruments such as the VOM, FET multimeter, DMM, and oscilloscope are covered, as well as specialized test instruments such as the megohmmeter, digital logic probe, optical time-domain reflectometer, and network, spectrum, and waveform analyzer.

Chapter 3 explains the basic theory and repair of electric motors and generators—the heart of industrial machinery. Some of the types of motors covered include shaded-pole, capacitor, three-phase, repulsion, universal, synchronous, and the increasingly popular stepper motor as used in digital applications.

Chapter 4 explains the essential principles, applications, and servicing of industrial controls. Several controllers are presented, such as a manual starter, magnetic overload relay, pneumatic timing relay, bimetallic thermal overload relay, drum switch, electronic drive, and programmable controller.

Chapter 5 covers troubleshooting of residential and industrial wiring. The basic theory and fundamentals of wiring are explained, followed by troubleshooting of such circuits as distribution panels, three-way and four-way switching controls, wye and delta wiring, polyphase industrial wiring, fiber optics, and television and antenna distribution systems.

Chapter 6 covers the troubleshooting of radio and television. It begins by describing the theory and component stages of AM, FM, FM multiplex, and HD, UHD, and smart television circuitry. This chapter then presents the typical problems encountered, along with methods for testing and repairing these products.

Chapter 7 is an introduction to digital systems and troubleshooting. Topics include analog and digital circuits, basic logic gates, digital technologies, safety, open and short digital circuits, troubleshooting examples, and troubleshooting equipment such as oscilloscopes, logic analyzers, logic pulsers, and logic probes.

Chapter 8 explains the building blocks of modern combinational and sequential circuits. It starts with flip-flops, decoders, and encoders. Next, it covers shift registers, counters, and digital-to-analog and analog-to-digital converters that are constructed from flip-flops. Troubleshooting synchronous logic is covered as well.

Chapter 9 introduces microprocessors, microprocessor-based systems, and troubleshooting. Memory, input/output, troubleshooting microprocessor-based systems using in-circuit emulator (ICE), microprocessor chips such as Intel's LGA2011 and Zilog's Z-80180, memory upgrade, and disassembled and decompiled memory dumps are the major topics covered in this chapter.

Chapter 10 presents the cutting-edge technology of biomedical equipment and troubleshooting. Electrical safety is emphasized. Biomedical engineering technology concepts related to equipment such as defibrillators, and ECG, EEG, hemodialysis,

MRI, ultrasound, and X-ray machines; and troubleshooting, and preventive maintenance are introduced.

Chapter 11 provides a modern overview of computer networking. Topics include configuring network devices such as routers and switches using Cisco IOS commands, IP addressing, subnetting, virtual local area networks, network troubleshooting through OSI layers, a network troubleshooting example using a flowchart, and a troubleshooting example for routers.

Chapter 12 covers the exciting field of embedded microprocessor systems. Embedded systems are explained briefly, followed by the troubleshooting of microcontroller-based systems; timing diagrams for microprocessor and microcontroller; Zilog's Z16FMC microcontroller; recurrent-sweep and triggered-sweep oscilloscopes; logic analyzers; programmable devices such as PALs, CPLDs, FPGAs, and FPLAs; and Raspberry Pi.

The authors believe that this book will be valuable and insightful to you in advancing your or your students' knowledge and skills in the exciting and important field of electrical and electronic troubleshooting.

Trademarks

The following trademarks are listed in alphabetical order, not in the order in which they appear in the book.

- ADC0804 Analog-to-Digital Converter is a registered trademark of Texas Instruments® Inc.
- Circuit Trace is a registered trademark of H. W. Sams & Co.
- DAC0808 Digital-to-Analog Converter Circuit is a registered trademark of Texas Instruments® Inc.
- LGA2011 and 4004 Microprocessors are registered trademarks of Intel® Corp.
- Logic Analyzer is a registered trademark of Agilent Technologies.
- The Megger® is a registered trademark.
- Packet Tracer printouts of router running configurations are registered trademarks of Cisco Corp.
- Router and switch IOS commands are registered trademarks of Cisco Corp.
- Windows is a registered trademark of Microsoft, Inc.
- Z16FMC Microcontroller is a registered trademark of Zilog Corp.
- Zilog Z-80180 8-bit Microprocessor is a registered trademark of Zilog Corp.

Chapter 1

Principles of Troubleshooting

A career in electrical and electronic troubleshooting can be a very financially rewarding and personally satisfying one. The expert troubleshooter has a unique blend of understanding electronic theory, problem-solving techniques, and hands-on skills. Most electronic products and devices contain similar components, such as resistors, capacitors, diodes, transistors, contacts, connectors, and wire. Understanding common faults of these components and how to test them is a prerequisite for the troubleshooter. In this chapter, you will learn basic problem-solving analysis, common circuit faults, various troubleshooting methods, and testing procedures for the most common electrical or electronic components.

Problem-Solving Analysis

Before attempting to service a device, you must first develop an understanding of problem solving and how this concept applies to overall troubleshooting and repair. Think of servicing a device as having three phases: (1) situational analysis, (2) problem solving, and (3) decision making. You must proceed in this logical manner; otherwise mistakes, accidents, and wasted time and expense may result. For example, many troubleshooters, upon the discovery of a blown fuse, just replace the fuse, instead of first determining the source of the problem, perhaps only to have another fuse blow.

Therefore, situational analysis is the first step in servicing a device. It involves critical scrutiny and analysis of a problem situation. It allows the troubleshooter to gain insight into an unacceptable condition. It is defined as simply looking at the overall condition of the device and determining whether a problem even exists.

Begin this step by asking questions and making observations as follows:

1. Discuss the defect with the owner or operator.
2. Compare the problem with others from your past experiences.
3. Perhaps there is no problem at all, and it is an operating error.
4. Identify the existing state of operation with the desired state.
5. Make an overall observation of the situation, noting symptoms and relevant changes.

Problem solving is the second phase and is completed when it has been determined through situational analysis that a problem exists that needs further investigation. Problem solving is triggered by a deviation from a standard or a desired state. Examples include a malfunctioning or inoperable device. Troubleshooting is the process of problem solving. It is in this phase that the cause of a problem is identified.

The first step in problem solving is to get organized. Begin by obtaining necessary schematics, manufacturer's specifications and servicing manuals, and tools and equipment. Do not shortchange this step by jumping in and wasting a lot of time attempting to repair a device, when simply reading the servicing guide could have easily solved the problem. In other words, those who fail to plan plan to fail. Once you are organized, begin by doing the following:

1. Describe the problem.
2. Compare the problem situation with known operating conditions prior to the breakdown.
3. Describe all known differences such as the symptoms, noises, and smells noticed when defect occurred.
4. Compare the "what is" with the "what is not." Which components are okay and which are not, and to what degree are the components defective?
5. Analyze differences through testing by paying close attention to obscure and indirect relationships. For example, slight tolerance changes in components or physical color changes can signal causation.

Once you have determined the actual cause of a problem, you are ready to proceed to the final phase, called *decision making*. Decision making is defined as examining various solutions or repair alternatives and selecting the best option. For example, if it is determined that an electric motor is the cause of the problem, there could be several alternatives in deciding how to repair the overall system. The motor could be repaired, it could be replaced by the same model, or an entirely new upgraded model could be substituted depending on the operating conditions of the overall system. The troubleshooter might decide that upgrading the motor is more cost-effective because the likelihood of future premature breakdowns is reduced.

When deciding which alternative to utilize, you must consider all the advantages and disadvantages for each alternative along with contingency planning. Contingency planning takes into account future changes in the overall system, such as expected life, operating conditions, and model changes. For example, it may not be wise to replace the motor with a new one when you suspect the entire system may soon be obsolete and replaced anyway.

Remember to always follow the three phases: situational analysis, problem solving (troubleshooting), and decision making (repair). Following these basic stages and understanding the importance of this sequence are essential to becoming a skilled expert.

Circuit Faults

Ideally, most people would like electrical or electronic products and devices to be "breakdown-proof," but unfortunately this is not possible. Most breakdowns are probably—directly or indirectly—a result of operating abuse or lack of maintenance.

Electrical or electronic breakdowns can be categorized by some very basic causes as follows:

1. Heat
2. Moisture
3. Dirt and contaminants
4. Abnormal or excessive movement
5. Poor installation
6. Manufacturing defect
7. Animals and rodents

Whenever too much heat is applied to electrical or electronic devices, problems occur. Heat increases the resistance of circuits, which in turn increases the current. Heat will cause the materials to expand, dry out, crack, blister, and wear down much more quickly; sooner or later, the device will break down.

Moisture will also cause circuits to draw more current and eventually break down. Moisture (water and other liquids) causes expansion, warping, quicker wear, and abnormal current flow (short circuits).

Dirt and other contaminants, such as fumes, vapors, abrasives, soot, grease, and oils, are materials that cause electrical and electronic devices to "clog" or "gum" up and operate abnormally until they finally break down.

Abnormal or excessive movement can lead to breakdowns. Vibrations and physical abuse are the leading causes of these types of breakdowns.

Poor installation is often the work of an unqualified installer or one who is careless or in a hurry. Failure to tighten a bolt or properly solder a connection results in an electrical or electronic device's breaking down prematurely.

Manufacturing defects are also very common. For example, it is not uncommon to find a loose circuit board after delivery and installation. The shipping and transporting can also loosen or damage circuit boards and components.

Animals and rodents can also be the cause of electrical or electronic breakdowns. A rat or other small rodent may have chewed on an electric wire or found its way into a motor.

It is essential that every troubleshooter understand the four most common causes of circuit faults:

1. Short circuit
2. Open circuit
3. Ground
4. Mechanical fault

Basically, a short circuit results when the current takes a direct path across its source. For example, a short circuit in an electric motor is caused by a defect in the motor in which two wires of the circuit touch and cause a bypassing of the normal current flow.

Short circuits draw more current because the resistance in the circuit decreases; as a result, the voltage decreases. Typical signs of short circuits are the following:

1. Blown fuses
2. Increased heat
3. Low voltage
4. High amperage
5. Smoke

An open circuit results from an incomplete circuit. For example, an electric motor with an open circuit can be caused by a break in the motor circuit, which prevents the current from flowing in a complete path.

An open circuit will have infinite (unlimited) resistance and zero current since its path has been broken. Typical signs of an open circuit are (1) infinite resistance, (2) zero amperage, and (3) completely dead (inoperable) device.

A ground results when a defect in the insulation or placement of a wire or component causes the current to take an incorrect (abnormal) route in the circuit. For example, an electric motor with a grounded circuit results when part of the windings make electrical contact with the iron "frame" of the motor. A ground is theoretically similar to a short circuit; however, it has distinct characteristics. Generally, the short circuit causes the device to stop operating and trips a circuit breaker due to the direct bypass. However, in the grounded circuit, the device often keeps operating due to the indirect circuit bypass, but it operates poorly and draws abnormal currents and voltages. The grounded circuit also can be the most dangerous, since the device often keeps functioning; the operator can experience shocks, especially without proper ground-fault interrupters.

Common grounds result from wires with poor insulation, pinched wires, or misplaced components. Shocks are obtained from a grounded motor because the frame of the motor and the operator have become part of the electric circuit. Typical signs of a ground are as follows:

1. Abnormal amperage reading
2. Abnormal voltage reading
3. Abnormal resistance reading
4. Shocks
5. Abnormal circuit performance
6. Tripped ground-fault interrupters
7. Periodic blown fuses or circuit breakers

Mechanical problems are a result of too much friction, wear, abuse, or vibration, where the physical part of an electrical or electronic device causes the breakdown.

Broken belts, worn bearings, loose bolts, worn contacts, damaged chassis, and broken controls are common examples of mechanical problems. Typical signs of mechanical problems are as follows:

1. Noisy operation
2. Abnormal operation
3. Visual clues
4. Circuit failure

The most important tool or instrument a troubleshooter can use is his or her own senses. Most troubleshooting problems can be found by the use of one or all of the main senses: sight, smell, touch, and hearing.

Before any sophisticated attempt is made to analyze a problem, first visually look for an obvious cause. A cracked circuit board, broken wire, burned or charred component, or any type of damaged item can quickly lead the troubleshooter to the problem.

To the troubleshooter, there is not a more identifiable smell than that of a burned transformer. A good troubleshooter should easily be able to recognize this smell. Also, burned cables, insulation, wires, and components can give obvious clues to circuit faults and help isolate the main cause.

Many troubleshooters rely on their sense of touch to locate component faults. Integrated circuits (ICs) should never be hot when touched. A hot IC would indicate a short circuit in the IC. Likewise, a hot, smoky motor is a common sign of a short circuit. On the other hand, a 10-watt (W) line resistor should not be cold when touched; it should feel warm or hot. This cold resistor would indicate an "open" component. Through experience, troubleshooters learn that different components tend to have degrees of temperatures unique to them and the specific operating environment. When you learn to recognize these differences, troubleshooting defective components will become easier.

Troubleshooting Methods

There are basic techniques used by all troubleshooters when servicing electrical or electronic devices. Which techniques the troubleshooter uses depends on the type of defect or symptom that exists.

The troubleshooting techniques that will be introduced, and later explained in detail throughout this book, are as follows:

1. Voltage measurements
2. Amperage measurements
3. Resistance measurements
4. Substitution
5. Bridging
6. Heat

7. Freezing
8. Signal tracing and injection
9. Component testers, test lamps
10. Resoldering, adjusting, etc.
11. Bypassing
12. Logic analysis and network injection

The voltage measurement of a circuit is usually taken by using a voltmeter or an oscilloscope. A zero-voltage reading may identify an open circuit, while a low-voltage reading may indicate a short-circuited component. Remember, always connect a voltmeter *in parallel* with the circuit when you measure voltage (Fig. 1.1).

The amperage measurement of a circuit is usually taken by using an ammeter or a "clamp-on" ammeter. The ammeter indicates and locates common circuit faults, such as short circuits, open circuits, and grounds. Remember, always connect the ammeter *in series* with the circuit when you measure current (Fig. 1.2).

An ohmmeter is used to measure the "continuity," resistance of a circuit, or resistance of a component. This technique is very valuable in locating short circuits, grounds, and open circuits. Remember, always *shut off the power* before you measure resistance (Fig. 1.3).

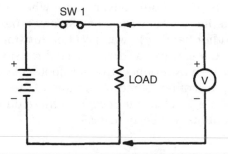

FIGURE 1.1   Always connect a voltmeter in parallel with the circuit.

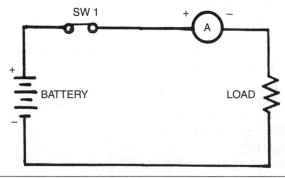

FIGURE 1.2   Always connect an ammeter in series with the circuit.

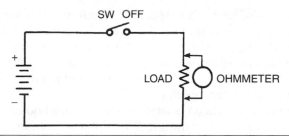

FIGURE 1.3   Always turn off the power in the circuit before measuring resistance.

The substitution technique simply means replacing a suspected faulty component with a known good component. This method can save valuable time and frustration for the troubleshooter. However, there are some risks involved. If a circuit board is replaced with a new one and the underlying problem is not the circuit board, this replacement part may be damaged. Also, once a replacement part has been used in a circuit, many parts distributors may not allow returns since the quality of the used part is questionable. However, substitution is still an important and valuable technique if it is not abused.

When a troubleshooter suspects a component (usually a capacitor) to be faulty, she or he "jumps," or places a known good component across, the suspected faulty component in the circuit. If the circuit starts operating correctly with the new component, the problem is isolated. This is called *bridging*. The troubleshooter can save valuable time and frustration by bridging (Fig. 1.4). Remember, however, this technique is generally limited to open components, not short-circuited components. Bridging a short-circuited component may have no effect or may cause damage to the new component.

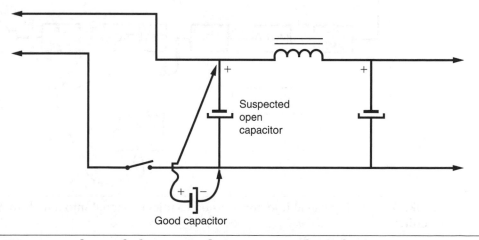

Suspected open capacitor

Good capacitor

FIGURE 1.4   Bridging of a known good component with another.

Application of heat to a suspected "intermittent" component is another troubleshooting technique. This thermally intermittent component breaks down under heat. By applying heat to this suspected intermittent component—usually by using a hot blower—the troubleshooter can determine the quality of the component. Do not use too much heat; otherwise damage to nearby components, particularly plastic components, can result.

The freezing technique is used by the troubleshooter to temporarily restore a component to normal operation. The freezing technique gets its name from the use of cold air from a fan or a chemical coolant. The freezing technique cools the suspected thermally intermittent component, thus temporarily restoring the component to normal operation. Both heat and cold applications can be very useful in identifying micro circuit board cracks and connections. The heat and cold cause expansion and contraction, respectively, which can temporarily trigger a circuit to operate, allowing the troubleshooter to isolate the problem.

Signal tracing or injection is most often used in servicing radio receivers. The technician injects a signal into the malfunctioning receiver in order to locate the specific inoperable (dead) stage (Fig. 1.5). A signal is injected into the various points preceding each stage. A tone is heard at the speaker if the stage is operating. The defective stage will not allow the signal to pass through, and the signal will not be heard at the speaker.

Component testers are instruments used to test the quality of the component. Component testers include insulation meters, megohmmeters, capacitor checkers, test lamps, transistor or diode testers, cathode-ray tube (CRT) checkers, integrated-circuit (IC) testers, and others.

Resoldering, adjusting, and aligning are all techniques used by a troubleshooter on suspected problems. Often, a troubleshooter will use these techniques because

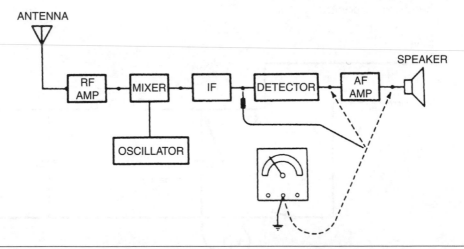

FIGURE 1.5   Using signal injector method to inject a signal into a malfunctioning circuit.

intuition and past experience may indicate that a problem exists. If a specific product's past performance has a high frequency for having a poor electric solder connection (called a *cold solder joint*), a quick touch (resolder) with a solder gun or pencil can rectify the condition.

Bypassing is a technique that a troubleshooter may use to locate a suspected problem. This technique requires unplugging one of several circuits. For example, by "shutting off" a transistor, its effect on the total performance of the circuit can be observed. In other cases, a complete circuit board might be disconnected in order to recheck voltage and other measurements, as well as observe the effect on the overall operating system. For example, a shorted circuit board can put a draw on other circuits. By omitting the shorted circuit board, normal operating values may be restored, thereby isolating the problem.

Logic analysis and network injection are techniques that allow a troubleshooter to debug complex circuit problems. Logic analyzers use super high-speed signals to process data from computers and other digital devices. For example, an optical time domain reflectometer (OTDR) is used to transmit fiber-optic pulses through the fiber cable that scatters the light signal back. The instrument measures the rate of the pulse return that provides detection of the fiber-optic cable length, attenuation, and potential breakpoints.

In diagnosing electrical and electronic troubles, it is important that the troubleshooter follow a logical, systematic procedure to eliminate unnecessary time spent, tests, and replacement of parts. Time is money, and a good troubleshooter needs a good "cookbook" approach to troubleshooting. For example, most troubleshooting procedures can be greatly aided by use of diagrams, schematics, and blueprints.

Schematic diagrams consist of an electrical or electronic layout of the circuit design. These diagrams present specific component values and data. Schematic diagrams often specify normal operating voltages and current, cold resistance values, signal waveforms, and other information.

Basic line drawings and blueprints show layout placement of wire or cable and controls. Line drawings are usually used in residential and industrial wiring and controls to aid in installing, locating, and tracing circuits.

A pictorial diagram can be useful in providing a layout of the location and placement of specific component parts. Often, a schematic diagram will be followed by a pictorial diagram. A pictorial diagram usually shows a "picture" of a circuit. Also, there are many diagrams available on the Internet, which can be located by conducting searches using the manufacturer's web site or by product name.

Often, the success of troubleshooting a device depends on the availability of servicing diagrams. Some unpopular foreign products and equipment are difficult to service when service literature is not available. Often, a troubleshooter will find servicing these products too frustrating and time-consuming and will refuse to work on them.

Regardless of the trouble or situation, a good troubleshooter will always record a mental or written note of a circuit problem that he or she has repaired and will use this information in the future.

Safety Considerations

In working with electrical and electronic devices, safety can never be overly emphasized. No matter how experienced or competent you are as a troubleshooter, you must always respect the power of electricity and follow proper safety considerations. Electricity can cause significant damage to the human body ranging from small burns to the skin, neurological and muscle damage, to death. The combination of voltage and amperage that actually flows through your body is what causes damage. It is possible that a high amount of voltage can zap your hand away from the power source and cause little harm, but a small amount of voltage with a high amount of current can actually kill you. The possible extent of damage by a particular combination of voltage and amperage also depends on the amount of resistance of the body.

For example, when current flows through a person's body the heart and lungs can be interrupted and cause temporary or permanent effects. Suffocation can result when a person is unable to breathe. When the heart is impacted, it can stop pumping blood and the natural rhythm of the heart can be disrupted. In this case, death may be imminent. One common condition is called *ventricular fibrillation*, which causes a disruption of the heart's natural muscle rhythm and makes the muscle contract rapidly and randomly. This condition is typical of victims who experience household deaths as a result of the person's heart triggering into fibrillation.

Electricity can cause physiological effects to the body with as little as 100 milliamperes (mA) of current when applied directly to a person's heart. Body injury can result to muscles due to involuntary muscle contractions with as little as 5 to 10 mA of current. Current above 100 mA can cause muscle paralysis, breathing disruption, skin burns, and possible ventricular fibrillation and death.

Therefore, before servicing any electrical device, it is critical that any troubleshooter carefully read the service manual and supplements. Some general considerations include the following:

1. Ensure that the electrical power has been removed from the device.
2. Always ensure the correct polarity of any installation of any component, especially electrolytic capacitors. The incorrect installation can actually cause the device to explode.
3. Caution should be taken not to use any spray chemicals near the electronic device.
4. Follow the service manual recommendations for cleaning electrical contacts by using only solutions suitable for the device. For example, 90 percent isopropyl alcohol is often recommended for electronic contacts or acetone with application using a pipe clearer or a cotton swab or comparable nonabrasive applicator. However, caution should be taken since this material can be flammable.
5. Never test a high voltage by drawing an arch.
6. Always use a suitable high-voltage test meter such as a digital voltmeter (DVM) or Fetvom with a high-voltage probe attachment.
7. When servicing devices, use proper ground lead from the chassis ground and remove the test receiver ground lead first when completed.
8. When soldering, use a grounded-tip low-wattage soldering iron.

9. Always keep the soldering iron tip clean and well tinned.
10. Avoid overheating electronic components when soldering.
11. Develop a skill of moving quickly when touching the iron tip to the junction of a component lead or printed circuit board and only as long as needed for the solder to flow.

A special caution should be advised for electrostatically sensitive devices. Most semiconductor devices can be easily damaged due to electrostaticity. Typical examples of these components include Fet transistors, Mosfet transistors, and ICs. When unpacking semiconductor components, be sure to avoid any electrostatic charges to your body by maintaining a connection to earth ground. You should generally wear a commercially developed grounding strap that helps prevent electrostatic buildup in the body.

Moreover when possible, use antistatic solder and minimize any motion when unpackaging electronic components that might harm the devices by generating electrostaticity. Also, use antistatic mats on the floor and other commercially produced antistatic products. Also, you should be careful not to work on electronic devices on carpet since this can produce an electrostatic charge. Some popular antistatic devices include a wrist strap and ground bracelet. These devices help keep you grounded and prevent the development of electrostatic charges within the body. These antistatic devices allow for voltage charges to leak through to the ground and prevent static buildup.

Besides the wrist and bracelet straps, the use of ankle and heal antistatic straps are common and operate in the same manner. These straps are particularly useful when people need to be mobile in a work area in which the buildup of voltage can be discharged through a special floor material. More expensive wireless straps are available. Besides damage to electronic devices, proper antigrounding devices should always be used where combustible, flammable, and explosive materials are present.

For example, when a novice works on a personal computer, he or she avoids using antistatic straps when replacing the *random access memory* (RAM) module, which is a common mistake. Three thousand or more volts can actually be produced through static electricity before you can feel it, especially in low humidity environments. Therefore, replacing a RAM module or other IC in a computer could cause damage by as little as 300 to 400 volts (V).

Moreover, when a RAM module is damaged, it is often hard to diagnose, especially if you just inserted a brand new one. The last thing the novice troubleshooter may think is that it was damaged because of static electricity and may be more apt to blame the manufacturer. Therefore, never be penny-wise and pound-foolish.

Keep in mind that just because you have utilized a ground to the electrical ground in a home or office does not mean that the ground is satisfactorily working. It is possible, especially in older homes, that the outlets are not even properly grounded or grounded at all. Therefore, the best recommendation is to use an antistatic wrist strap and antistatic floor pad. Remember to also keep any unneeded devices, such as telephones, away from the computer since the touching of them to the computer may cause static electricity as well.

Testing Basic Components

Several components are common to most electrical and electronic devices. Understanding how to test the most common components is essential for the troubleshooter.

Resistors are manufactured in various shapes, sizes, and values. The main purpose of the resistor is to limit current flow and/or reduce voltage. Most resistors are made of carbon or wire and are manufactured in prescribed ohmic values. For example, a 1000-ohm (Ω) resistor at 10 percent tolerance is color-coded brown, black, red, and silver. Therefore, on an ohmmeter the resistor should measure between 900 and 1100 Ω. An open resistor will have infinite resistance, and a defective resistor could have any value below 900 Ω or above 1100 Ω.

Resistors are rated in watts, which determines the ability of the resistor to absorb the heat produced within the resistor. The actual physical size of the resistor determines its wattage rating. A larger resistor can simply absorb more heat than a smaller one.

The most common defects of resistors are physical—cracking and charring. When excessive current and heat tend to increase the resistance, the resistor opens. A charred or discolored resistor should be replaced, since the resistor will often check good with an ohmmeter but will break down (open) under voltage in the circuit.

The ohmmeter is one of the most important meters used in servicing components. This meter is used to measure continuity or resistance in resistors or other components. A component having continuity has a resistance near zero. On the other hand, a component having no continuity has infinite resistance.

When testing basic components, the troubleshooter is mostly concerned with the resistance of the component or continuity measurement. For example, when a troubleshooter is checking a fuse, a good fuse will read 0 Ω, but an open (blown) fuse will read infinite resistance (Fig. 1.6).

As for checking a fuse, when you are testing cables, cords, or wiring harnesses, a good wire will have continuity and a broken wire (open) will have no continuity.

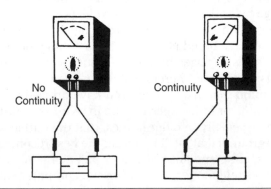

No Continuity

Continuity

FIGURE 1.6   Checking a fuse for continuity by using an ohmmeter.

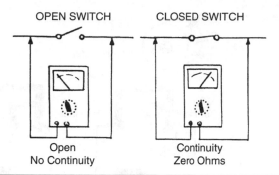

FIGURE 1.7   Checking a switch for continuity by using an ohmmeter.

When checking a wire for a possible defect, while the ohmmeter is connected, gently bend the wire at various points, especially along common faults such as near connections. Since wires often contain intermittent defects, breakdowns occur during movement.

To test a switch, the same procedure is used. A single-pole, single-throw switch should have continuity one way and not the other (Fig. 1.7). As with wires, when you check a switch with an ohmmeter, gently rock the switching mechanism to identify potential intermittent circuit faults. This procedure will also allow you to evaluate the mechanical quality of the switch. Switches should have a smooth, firm operation, not a sloppy action with play and looseness. Some defective switches can easily be repaired by tightening a screw or by cleaning; however, they generally need to be replaced.

Variable resistors are called *potentiometers*, and they can be measured and tested in two simple ways. (1) One way is to use an ohmmeter to measure the value of the potentiometer across the two end terminals. The value should equal the value printed on the potentiometer. Place one lead on the center terminal and the other lead on one of the end terminals. The potentiometer, when turned, should vary the resistance accordingly (Fig. 1.8). (2) Another way to test a potentiometer is to turn the potentiometer while it is in the circuit. If a scratchy, rough sound is heard in the speaker, the potentiometer needs cleaning or replacing. To clean the potentiometer, turn off the power and spray an electronic component cleaner in the sliding contact area while turning the control back and forth.

Batteries can be checked by voltage and amperage testing. The correct voltage output of a battery is very important. An excellent battery should exceed its rated value. For example, a new dry cell rated at 1.5 V direct current (dc) should measure 1.5 to 1.6 V. On the other hand, an old, weak battery will read less than 1.5 V. An automobile battery (lead acid cells) rated at 12 V dc will often exceed 13 V when fully charged. Likewise, worn-out batteries do not have sufficient amperage capacity. When amperage rating is insufficient, the battery needs to be recharged or discarded.

Speakers are common components used in many products such as computers, televisions, and stereo receivers. When you check speakers, first make an

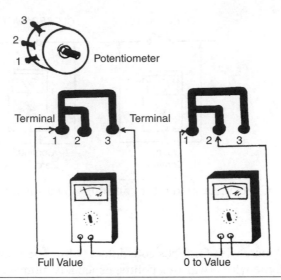

FIGURE 1.8 Checking the potentiometer by using an ohmmeter.

inspection. Rattles and heavy vibration are often signs of a defective speaker. Visually inspect the speaker for cracks, dirt, and other faults. If you are in doubt about the actual quality of a speaker, the best testing technique is substitution. Simply substitute a known good or similar-value speaker in its place. Many speakers will operate intermittently; by gently pressing on the speaker cone, it may begin operating. This is a good indication that the speaker is defective due to a bad voice coil, connection, etc.

When you are replacing a speaker, it is important to replace the speaker with one having the same impedance and power ratings. Primarily, the voice coil of the speaker determines the impedance and power rating of the speaker. The power rating, measured in watts, is the maximum power at which the speaker should be operated. The impedance (in ohms) of the speaker is used to electrically match the speaker input to the output of the receiver. The impedance of a speaker can be roughly approximated by measuring the resistance of the voice coil with an ohmmeter and then multiplying this value by 1.25. Some common speaker values are 3.2, 4, 8, 10, 16, and 20 Ω.

Another method used to check a speaker is to place an ohmmeter across the voice coil terminals. Listen and watch for a small "pop" and movement of the speaker cone. A defective speaker will show no movement or popping sound. Also, this method can be helpful in phasing two or more speakers together. Place the ohmmeter on the voice coil terminals, and note whether the cone moves in or out. Reversing the polarity of the ohmmeter will reverse the movement of the cone. Next, mark the positive side of each cone when it is in the out position (Fig. 1.9). Then connect each speaker to the audio amplifier while observing the correct polarity. Sound reproduction should be improved since the speaker cones will move in and out together. The in-and-out movements of the speaker cones keep the

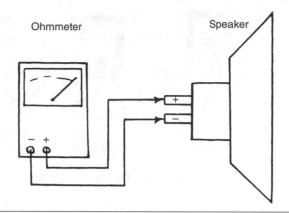

FIGURE 1.9 Phasing speakers with an ohmmeter.

speakers in phase, or in step with each other. If they are out of step, or opposing each other, at certain frequencies, the sound waves will be canceled.

Capacitors are used for hundreds of different purposes, such as filtering, voltage regulating, bypassing, power phase correction, source of power, and frequency controlling. They come in various sizes, shapes, types, and values. Basically, a capacitor is a device that has the ability to store an electric charge. Capacitors consist of two conducting plates, separated by an insulating dielectric material. Types of capacitors include mica, paper, ceramic, plastic, aluminum, and tantalum. The value of a capacitor is expressed by the unit farad (F), but most capacitors are rated in microfarads (abbreviated μF), which represent the amount of electric charge that the capacitor can store.

There are several techniques that can be used to test capacitors:

1. Resistance measurement (ohmmeter)
2. Capacitance measurement (capacitor checker)
3. Spark test
4. Bridging
5. Substitution

A capacitor below 0.25 μF should not show a reading on the ohmmeter since it is too small to advance the meter movement. A near-zero reading indicates a short-circuited capacitor. All capacitors above 0.25 μF should register on the ohmmeter.

When you check a capacitor, place the ohmmeter on a high scale such as 10,000 Ω and place the leads of the ohmmeter across the leads of the capacitor. First, make sure you have discharged the capacitor by short-circuiting the leads with a piece of wire or a screwdriver. When the leads of the meter have been placed across the capacitor, the needle should deflect upward and then slowly drop back down to near zero. Failure to deflect the needle indicates an open capacitor, and failure of the needle to drop down indicates a shorted capacitor (Fig. 1.10).

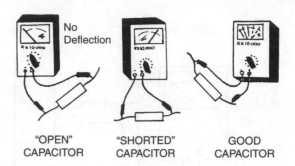

"OPEN"
CAPACITOR

"SHORTED"
CAPACITOR

GOOD
CAPACITOR

FIGURE 1.10   Checking a capacitor with an ohmmeter.

Another method used to check larger capacitors is the spark test method. Momentarily, connect the capacitor across a voltage source. Generally, just 1 second (s) is enough to charge the capacitor. Never leave the voltage applied for too long, or else damage or injury may result. Also, make sure the voltage you have applied does not exceed the voltage rating specified on the capacitor (Fig. 1.11). After the capacitor has been charged, short its terminals together, using a screwdriver or a similar device with an isolating handle so you will not get shocked. A good capacitor will show a spark. Absence of a spark means it is defective.

A capacitor checker is a useful device for testing the performance of a capacitor. Besides, checking its specific capacitance rating, it can also test for other characteristics of a capacitor, such as leakage and opens. Some capacitors can be checked while in the circuit; however, it is generally necessary to test them while they are out of the circuit.

The bridging method is also a good way to check the performance of a capacitor. Usually, filter capacitors that are suspected of being open will be bridged before being replaced. This is a simple method in which the suspected defective capacitor is bridged (or jumped) with a known good capacitor within +10 percent of the rated value. A noticeable difference in the quality of performance of the product or device (such as a radio or TV) under test should result upon bridging the capacitor. For example, defective filter capacitors often produce a noticeable hum in a radio. By

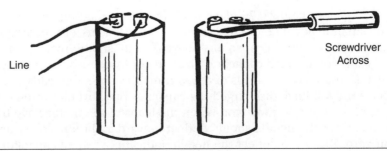

Line

Screwdriver
Across

FIGURE 1.11   Checking a capacitor by the spark test.

bridging a defective capacitor with a good one, the proper circuit operation will be restored and the hum will disappear.

The substitution technique, like the bridging method, determines the quality of the capacitor by means of another capacitor. When using the substitution method, you simply replace the suspected defective capacitor with a known good capacitor of similar value and rating. The performance of the product or device indicates the effect of the new capacitor. Remember, never exceed the voltage rating of a capacitor. A capacitor rated at 100 V should only be replaced with one at 100 V or more; otherwise, it may be destroyed. Substitution boxes are handy aids to assist the troubleshooting. These boxes contain the most common-valued capacitors, eliminating the need to obtain individual capacitors. These substitution boxes can easily be made or purchased, and they offer a quick, convenient, and accessible means of obtaining a substitute capacitor.

Semiconductors

The understanding of basic theory of semiconductors can be a real asset for the troubleshooter in testing these components. One of the first known semiconductors was the crystal detector. This consisted of a lump of crystal galena with a thin wire–cat whisker combination under pressure by a spring. This combination rectified current, allowing it to flow in only one direction.

Although the galena crystal was unreliable, it was the first step in semiconductor application. The development of modern diodes and transistors started with the basic theory and development of p- and n-type materials.

To develop a p- or n-type material, a crystal material of germanium or silicon is used. Silicon has an atomic number of 14 with 4 (valence) electrons in its outer shell; germanium has an atomic number of 32 with also 4 (valence) electrons (Fig. 1.12).

To form a p-type material, an impurity, gallium or indium, is added. The impurity is called a *trivalent* because it has a valence of 3. Thus this trivalent has three electrons in its outer shell. When either gallium or indium is added to silicon or germanium (both have a valence of 4), one valence electron is left unfilled. The unfilled gap, called a *hole*, takes on a positive charge and forms a p-type material. The impurity that created a hole is referred to as the *acceptor impurity* (Fig. 1.13).

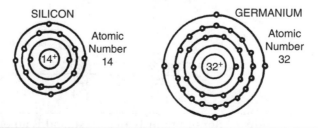

FIGURE 1.12   The atomic structure of silicon and germanium.

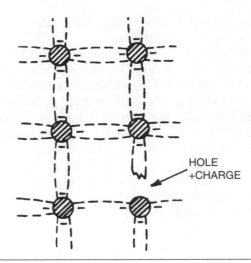

FIGURE 1.13 Adding an acceptor impurity to a crystal leaves a hole, forming a *p*-type material.

To form an *n*-type material, an impurity of arsenic or antimony is added. This impurity, called a *pentavalent* because it has five valence electrons in the outer shell, when added to germanium or silicon, joins with the four valence electrons, adding one free electron. This free electron constitutes a negative charge in the atom, and therefore this impurity is given the name of *donor* (Fig. 1.14).

When a *p*- or *n*-type material is pieced together, a *pn* junction is formed. This *pn* junction is called a *diode* because it allows current to pass in only one direction.

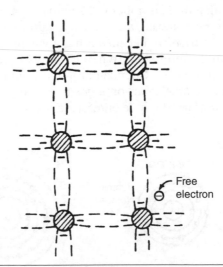

FIGURE 1.14 Adding a donor impurity to a crystal adds an extra electron, forming an *n*-type material.

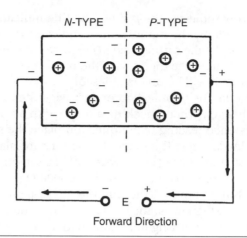

Forward Direction

FIGURE 1.15   Forward-biased diode.

When a battery is hooked up to a diode negative to negative and positive to positive, current flows through the diode. This is called *forward bias* and is shown in Fig. 1.15.

When a diode is forward-biased, the negative charge of the battery repels the negative charges in the diode, forcing them into the positive side of the diode. The positive force of the battery pulls these electrons on through the diode. This completes the circuit.

When the battery is reversed, the diode is reverse-biased, because the positive force of the battery and the positive holes of the diode attract each other, allowing no current flow to take place. A reverse-biased diode is shown in Fig. 1.16.

The *p* (positive) side of a diode is called the *anode*, and the *n* (negative) side is called the *cathode*. It is important for the troubleshooter to know which side is

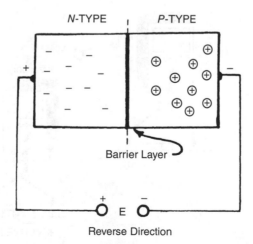

Reverse Direction

FIGURE 1.16   Reverse-biased diode.

which. The arrow indicates the *p* side. The line indicates the *n* side. A line or dot placed on the diode by the manufacturer indicates the cathode side. Notice that the manufacturer marks the cathode side (negative) of the diode with a plus sign or band. When the end of the diode with the band is connected to the positive side of the power source, it will be reverse-biased.

To test a diode, the troubleshooter can use either an ohmmeter DVM or a diode or transistor checker. When checking a diode with an ohmmeter, you can use the low-/high-resistance reading technique. You place the selector switch on R × 100 and you place the leads across the diode. In the forward-biased direction, the ohmmeter should read less than 100 Ω (i.e., low-resistance reading). In the reverse-biased direction, the ohmmeter should read about 5000 Ω (i.e., high-resistance reading). These low/high readings indicate that the diode is probably good. If you obtain a low/low or a high/high resistance reading, the diode is probably defective. Figure 1.17 shows the correct testing of a diode with an ohmmeter.

Most types of diodes can be checked by using an ohmmeter. Keep in mind that the actual resistance value of the diode is not very important as long as you get a low/high reading when the ohmmeter polarity is switched. If any doubt exists after the ohmmeter check, it is recommended that a new diode be substituted. Also, keep in mind that a low resistance reading both ways is common when the diode is tested in the circuit. To be sure that the diode is good, unsolder one lead and check the diode again with the ohmmeter. Keep in mind, when you replace a diode, that a diode will only block a certain amount of voltage in the reverse direction. This is called the *piv rating,* where piv is the acronym for peak inverse (reverse) voltage. Never exceed this voltage rating, or else the diode will be destroyed.

Although there are numerous types of diodes—zener, light-emitting, photoconductive, varactor, and tunnel—each has its own unique characteristics. When you are in doubt about the diode's quality, the best testing method to use is

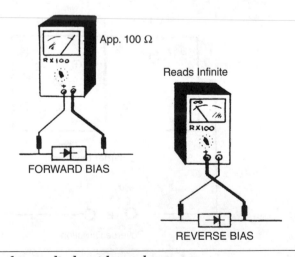

FIGURE 1.17   Checking a diode with an ohmmeter.

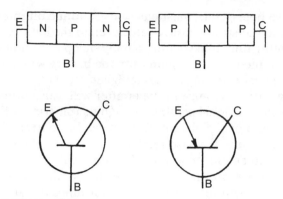

FIGURE 1.18   The three sections of a transistor.

substitution. For example, zener diodes are special diodes that can operate in reverse bias. In reverse bias the zener diode does not conduct until the voltage threshold, or breakdown voltage, is reached, and then it conducts at a relatively constant voltage. This operation allows the zener diode to act as a voltage regulator that can be used for regulated voltage power supplies.

The transistor is actually made up of two diodes back to back. A transistor is either an *npn* or a *pnp* combination. The first section is called the *emitter*. The middle section is called the *base*. The last section is called the *collector*. Figure 1.18 shows examples of both types of transistors.

A troubleshooter should understand the reason why a transistor amplifies. Figure 1.19 shows an *npn* transistor in which the emitter base is forward-biased, so it has low resistance to current flow. Also, the collector base has high resistance to current flow, or is reverse-biased. The negative potential of the battery forces the electrons in the emitter to flow into the base region. Very few of these electrons bond

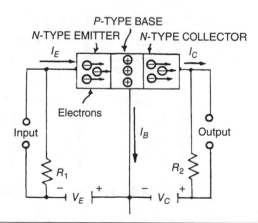

FIGURE 1.19   Electron flow in an *npn* transistor.

with positive holes, since most electrons continue on through to the collector region. This is due to the strong positive attraction of the battery. The electrons complete the circuit by returning to the battery supply. Keep in mind that new positive holes are being pulled into the base region from the battery when electrons fill the holes.

Since the collector region has a higher value of resistance than the emitter region, any current change in the emitter will cause proportionally a greater change in the collector. A signal passing through its transistor will be amplified.

The amount of signal amplification can be controlled by regulating the amount of electron flow into the base region. The amount of electrons supplied to the base region determines the amount of electrons available to the collector region. The regulation of electrons into the base region is called *biasing*. In a transistor, the forward bias (or emitter-to-base bias) determines the amplification of the transistor. The forward biasing of a transistor can be controlled by increasing or decreasing the voltage or resistance at the emitter-base region (see Fig. 1.19).

The basic operation of current flow in the *pnp* transistor is very similar to that of the *npn* except that instead of current flow by electrons, the current flow in the *pnp* transistor is completed by holes. The positive force from the battery forces the positive holes from the emitter through the base-collector region and back to the negative side of the battery. Here again, as with electrons in the *npn* transistor, a small number of holes bond with electrons in the base region, but the majority of these holes continue on into the collector region. Conduction takes place by hole current from emitter to collector. Electron flow is opposite to hole flow (conventional theory). Therefore, electron flow in this circuit is considered to travel in the opposite direction, or from collector to emitter. Do not let this explanation of hole current confuse you; basically, the main function of both transistors in the circuit is the same. Both transistors amplify (Fig. 1.20).

The three basic circuit configurations of transistors are the common base, common emitter, and common collector. Each circuit configuration has its own unique characteristics. Figure 1.21 and Table 1.1 illustrate these basic differences.

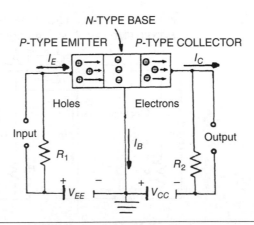

FIGURE 1.20   Electron flow in a *pnp* transistor.

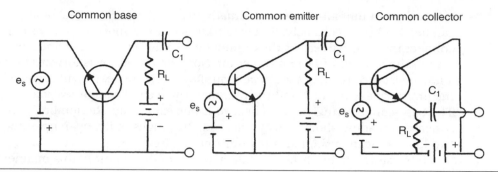

FIGURE 1.21   Three basic transistor circuit configurations.

TABLE 1.1   Three Basic Transistor Circuit Configuration Characteristics

Transistor Function	Common Base	Common Emitter	Common Collector
E gain	High	High	Low
I gain	Low	High	High
Input Z	Low	Moderate	High
Output Z	High	High	Low
Power gain	Moderate	High	Moderate

The circuit operation and troubleshooting tips and pointers will be discussed in greater detail in the succeeding chapters.

Transistors are usually tested either by a transistor checker or by an ohmmeter. Figure 1.22 illustrates how to check a transistor with an ohmmeter.

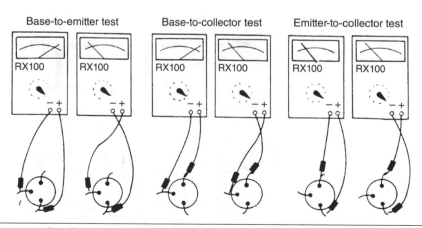

FIGURE 1.22   Checking a transistor for short or open circuits by using an ohmmeter.

Keep in mind that a transistor is actually two diodes back to back and, therefore, can be checked accordingly. To test a transistor for a short or open circuit, simply connect the positive lead of the ohmmeter (R × 100) to the base and the negative lead to the emitter of an *npn* transistor. Now the transistor is forward-biased, and a low resistance reading should be read. Reversing the leads will reverse-bias the emitter-base regions, and the ohmmeter will show a high-resistance reading. The base-collector regions are checked in the same way. Remember, a low-/high-resistance reading should always be seen. Two highs indicate an open transistor; two lows indicate a shorted transistor (out-of-circuit test).

Transistors can be checked while in or out of the circuit in this manner. It is recommended that a transistor that has checked out defective in the circuit be taken out of the circuit and checked again before being replaced.

Using the ohmmeter is another way to help determine the emitter and collector leads and/or the quality of the transistor. First, locate the emitter and collector leads through either the manufacturer's pictorial reference or use of the high/low readings on the ohmmeter. Place one lead of the ohmmeter on the emitter and one on the collector. Some kind of reading should result on the ohmmeter scale. Now short-circuit the base lead to the emitter; the resistance on the meter scale should increase. When the base lead is shorted to the collector, the resistance on the meter scale should decrease (Fig. 1.23).

The *field-effect transistor* (FET) is a special transistor used frequently in electronic circuitry. Although its outward appearance is similar to the bipolar transistor (*npn* and *pnp*), the construction is different. The FET consists of three terminals—source, gate, and drain—which correspond to the emitter, base, and collector of the bipolar transistor (Fig. 1.24).

Current flows between the source and drain, the "resistive" (i.e., semiconductor-substrate channel) parts of the FET. The gate is a diode junction that is reverse-biased rather than forward-biased as in the bipolar transistor. Therefore, the gate has very high resistance, allowing for a very high input impedance desired in many circuits.

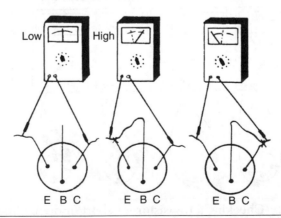

FIGURE 1.23 Checking the quality of a transistor with an ohmmeter.

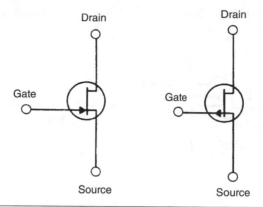

FIGURE 1.24 Schematic symbols for n and p channels for field-effect transistors (FETs).

The FET that is a junction-gate type is called a *junction FET* (JFET). The JFET can be checked with an ohmmeter, as the bipolar transistor can. The ohmmeter (R × 100) shows diode action (high/low resistance) between the source and gate and likewise between the drain and gate. Two high readings indicate an open transistor, and two low readings indicate a short circuit. The ohmmeter reading between the source and drain shows a low resistance in either polarity in a good transistor. Two highs indicate an open circuit (Fig. 1.25).

MOSFET is an acronym for metal-oxide semiconductor FET. This transistor is referred to as an IGFET (insulated-gate FET) type because the gate is electrically insulated from the source-drain channel (i.e., semiconductor substrate) by a thin layer of silicon dioxide (Fig. 1.26). The MOSFET can be a p-channel or n-channel

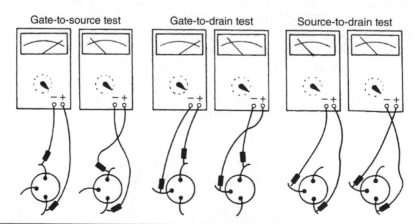

FIGURE 1.25 Checking a junction FET (JFET) for open and short circuits by using an ohmmeter.

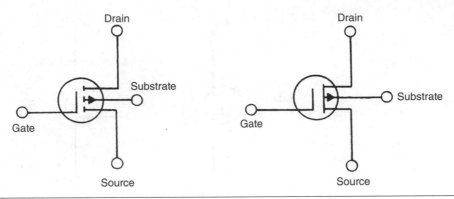

FIGURE 1.26   Schematic symbol for enhancement-type MOSFET. Schematic symbol for depletion-type MOSFET.

type. The current flow in a *p*-channel is reduced by applying a positive voltage and increased by applying a negative voltage. Moreover, three basic types of MOSFETs are the enhancement, depletion, and depletion enhancement. The enhancement type conducts from source to drain when forward-biased and remains "cutoff" (i.e., there is no current flow) at zero bias. The depletion type conducts at zero bias and is reduced with reverse bias. With enough reverse bias, it can be cutoff. The depletion-enhancement type has some conduction with zero bias. Current is reduced with negative bias and increased with positive bias. MOSFETs have high input impedance. They are also sensitive to static electricity and must be carefully handled.

For this reason, the gate is kept short-circuited to the source by twisting their leads together during shipping and handling or by a spiral short-circuiting spring. Dual-gate protected MOSFETs help overcome this problem by reducing the internal load. Figure 1.27 shows the schematic symbol for this type of MOSFET. The dual-gate MOSFET acts as a single-gate MOSFET by connecting the gate leads together. The MOSFET can be checked by using an ohmmeter. There should be zero resistance between the gate and source or drain. A reading on the ohmmeter indicates a short circuit. To check the drain-source condition, place a 15-kW resistor from the gate to drain. With the ohmmeter leads on the source and drain, a change in resistance suggests the MOSFET is okay. Note, however, the best test is by substitution or by use of test instruments (Fig. 1.28).

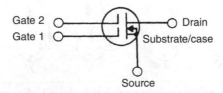

FIGURE 1.27   Schematic symbol for dual-gate *n*-channel depletion MOSFET.

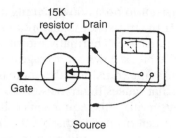

FIGURE 1.28   Checking a MOSFET for open and short circuits by using an ohmmeter.

There are several different testing techniques. Many of these can be used to directly or indirectly test the performance of a transistor. Other testing techniques besides using resistance and component checkers are as follows:

1. Voltage measurements
2. Heating and/or freezing
3. Signal tracing
4. Substitution
5. Transistor cutoff

Voltage readings can be very useful in determining transistor circuit action. For example, the transistor in Fig. 1.29 is given the manufacturer's specified normal operating voltages. If the transistor is open, or not conducting, the transistor will not draw current; and the voltage at the collector will not be 6 V but the full, source voltage of 10 V. If the transistor is short-circuited, the transistor will draw excessive current. This will load down the circuit. Therefore, if the voltage at the collector is low, this could indicate a short-circuited transistor or faulty bias resistor.

Often, transistors can be checked by simply applying a combination of heat and cold. First, apply heat by using a hot blower on the suspected transistor. If the

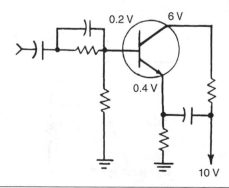

FIGURE 1.29   Typical operating voltages of a transistor.

heat causes the transistor to break down, apply a chemical "freeze" mist coolant or cool air from a fan. If the transistor is restored to normal operation, the transistor can be diagnosed as defective. Thermally intermittent transistors are defective transistors that usually break down after lengthy operation. A rise in temperature increases the current in the transistor; this increased current conduction in turn produces more heat that causes the transistor to draw even more current. Eventually the transistor destroys itself. This sequence is called *thermal runaway.* Remember, do not be too eager or determined to disable the transistor. This technique is used only for suspected thermally intermittent transistors. Never apply too much heat to a transistor, especially specialized or sensitive ones, or else needless damage will result.

Signal tracing can also be used to isolate a particular defective transistor. For example, if a signal is injected into each stage of a malfunctioning transistor receiver, starting at the speaker and working backward, the defective (open) transistor will prevent the signal from reaching the speaker.

The transistor substitution technique can be effective in determining a defective transistor. Keep in mind that when you substitute a transistor, be sure to use a similar transistor. Many troubleshooters prefer to "tack on" the substitution transistor to the copper-clad side of the circuit board first, to see whether the suspected transistor really is bad. This can save the troubleshooter valuable time.

Another technique to determine whether a transistor is operating is to short-circuit the base to the emitter, which cuts off the transistor (Fig. 1.30). A noticeable difference in the overall operation of the equipment should result if the transistor is working. If no noticeable difference is indicated, the transistor is most likely defective. Use caution when you do this test, and make sure you do not short-circuit the base to the collector, since this will cause the transistor to draw excessive current and destroy itself. Also, this method is only useful for certain circuit operations, such as amplifiers or oscillators.

The transistor cutoff technique can be compared to locating a bad spark plug wire on a car. While the automobile is idling, each plug wire is lifted off for a second and the idling operation of the engine is noticed. If the removed spark plug wire causes the engine to idle more roughly, the plug is good; but if the performance of the engine is unchanged, the plug wire is probably defective.

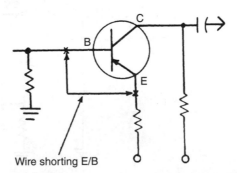

FIGURE 1.30   Cutting off a transistor by short-circuiting the base to emitter.

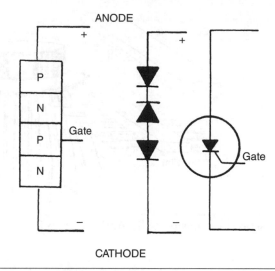

FIGURE 1.31   The construction of a silicon-controlled rectifier (SCR).

It should be mentioned at this point that when transistors are replaced, certain precautions should be taken:

1. Never overheat transistors.
2. Use a heat sink.
3. Use a 35-W, or lower, solder iron.
4. Use an exact or recommended replacement type.
5. Identify E, B, and C positions.

The silicon-controlled rectifier (SCR) is formed when three diodes are arranged back to back (Fig. 1.31). The SCR acts as a rectifier, except that conduction in the forward direction can only take place when sufficient voltage triggers the gate, at which time the SCR conducts as long as sufficient holding current is maintained.

The SCR is a popular device used in burglar alarms and automatic control circuits. It can best be checked by substitution or use of an ohmmeter.

When you check an SCR with an ohmmeter, place the selector knob on the R × 10,000 scale. With the negative lead connected to the cathode and the positive lead to the anode, a good SCR should read over 1 megohm (MΩ). Zero ohm indicates a short-circuited SCR. To see whether the gate junction is operating, short-circuit the gate lead to the anode; the meter should read 0 Ω.

Integrated Circuits

Although the actual construction of an *integrated circuit* (IC) is complicated, the process of checking ICs is much easier to understand.

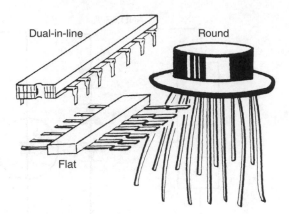

FIGURE 1.32   The three basic configurations of an integrated circuit (IC).

The three basic types of ICs are dual-in-line type, round type, and flat type. Figure 1.32 illustrates these three basic types. The IC basically consists of many micro-size components. One small IC may consist of several resistors, capacitors, diodes, and transistors, all connected into a micro circuit. They are hermetically sealed in a ceramic or plastic package. The two basic methods of IC construction are called monolithic and hybrid. In a *monolithic*, or standard "off-the-shelf," IC, all the components are manufactured on the same substrate. The *hybrid*, however, is custom-made, and this involves manufacturing the different circuits of the IC separately and then assembling them on the substrate. The approach to testing either IC type is the same. Figure 1.33 shows an example of a typical IC diagram.

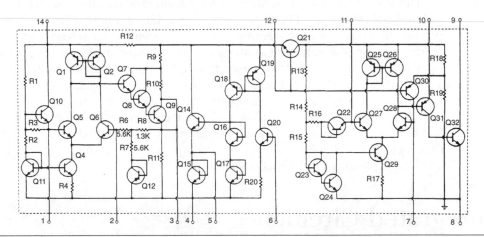

FIGURE 1.33   Typical internal diagram of an IC audio preamp and a 1-W audio output using ECG 1043. (*Courtesy Sylvania, Inc.*)

Although there are various shapes, types, and sizes of ICs, the troubleshooting techniques used are common to all:

1. Use of senses
2. Heating and/or freezing
3. Voltage check
4. Capacitor bypass
5. Substitution
6. Logic pulse probe

The first step in troubleshooting an IC is to use your senses. Look for obvious problems such as corroded, defective, or damaged pins, sockets, or solder connections. Make sure the IC is completely inserted in its socket. Check the component identification number of the IC with those of the manufacturer to make sure the correct IC is in the circuit and is correctly positioned.

Touch is one technique that many service troubleshooters use. While the circuit is in operation, touch the top insulated case of the IC with your finger and note its temperature. A hot IC is a good indication of a defective or short-circuited component. Most IC components should feel cool to warm when touched.

Heating and/or freezing is another technique often used to check for a defective IC. As stated before, a suspected thermally intermittent component can be checked by first heating the component with a hot blower—noting the performance of the circuit—and then cooling or freezing the component. The defective thermally intermittent IC should break down when heated but operate again when cooled off.

Voltage checks can be performed easily with a voltmeter or an oscilloscope. Simply measure the voltage of each pin of the IC and compare it, or the waveform, with the manufacturer's operating voltages or waveform. An incorrect voltage reading probably indicates a faulty IC or surrounding component.

Sometimes a suspected faulty IC can be "jumped" by using a capacitor. The capacitor bypasses the signal around the IC (Fig. 1.34). If the signal increases when the IC is jumped with the capacitor, the IC is probably defective.

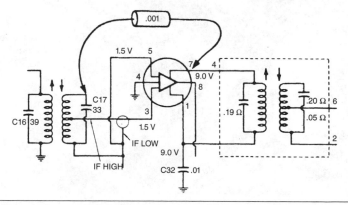

FIGURE 1.34   Bypassing an IC with a capacitor. (*Courtesy Motorola, Inc.*)

Any suspected IC should be replaced by a similar known good IC. This technique of substitution saves valuable time for the service technician. Realistically speaking, troubleshooters do not rely solely on this technique because it would require having a large inventory of ICs on hand, which would be costly. Also keep in mind that if the root of the problem is not a bad IC, replacing a bad IC with a good one could destroy the good one. Many ICs are mounted on circuit boards, and it is often more practical to simply replace the entire board.

IC testers and kits are available to test ICs; however, they may require the IC to be checked while out of the circuit. Also, special multipin clips can be used with a comparator box. These testers and kits can be very handy but are often expensive.

The logic digital IC probe is probably one of the most important test instruments used by troubleshooters. This small, handheld probe is generally used to test logic pulses and levels. The probe contains a complex circuit that identifies, through use of light-emitting diodes (LEDs) (high or low), operating logic-level responses. Like the voltmeter, the logic probe is applied to each IC pin or test point. The response is compared to the manufacturer's data (Fig. 1.35).

The "black box" concept is the standard approach in testing ICs. This system's principle considers the IC as a black box with an input and an output. If you know

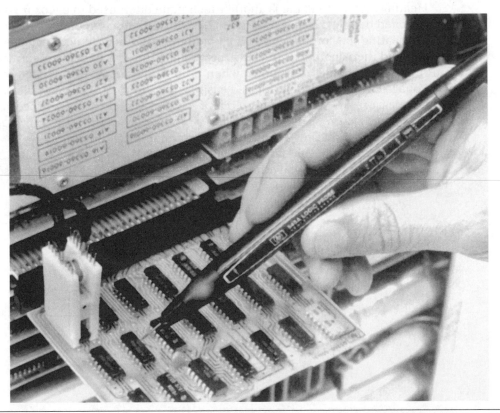

FIGURE 1.35   Using a logic digital IC probe to test an IC. (*Courtesy Hewlett-Packard.*)

what the input and output values of the IC are supposed to be, you can measure these values and generally conclude whether the IC is good. This way of thinking about the IC often eliminates the need to understand the complex internal structure of the IC. For example, you can use an oscilloscope to measure the inputs and outputs and then compare them to the manufacturer's diagram. Digital logic circuits usually have two logic levels: 0 or 1. Connecting negative or positive voltages can determine whether switching occurs.

A signal-tracing method using an oscilloscope is generally preferred to voltage-resistance checks, since true IC operation is dependent on the dynamic operating characteristics of the circuits. The signal-tracing method will be explained in greater detail in later chapters.

When a defective IC has been found, replace it, taking advantage of the following tips:

1. Order the exact replacement.
2. Insert or position the IC exactly as the original IC. It is extremely easy to insert an IC backward! Always identify the number 1 pin of the IC; this is often indicated by the manufacturer's placing a small dot next to this pin.
3. When you insert a dual-in-line 16-pin IC into the socket, it is easy to miss and smash at least one of the 16 pins. Make sure all 16 pins are aligned properly before you press the IC fully down into the socket.
4. Never overheat an IC. If the IC has to be soldered into the circuit, use a small 35-W soldering iron.
5. Do not overuse solder; prevent overflow of solder onto the board. This overflow can cause bridging among adjacent pins and components.
6. Always use desoldering wick or IC suction bulbs to remove excess solder.

Electron Tubes

Since electron tubes are rarely used today, little mention of their theory will be made. Electron tubes are found in some industrial and military applications and in a few guitar amplifiers. Some guitar players prefer the more hollow sound that electron tube amplifiers tend to produce as compared to semiconductor amplifiers. Basically, besides the CRT, the troubleshooter might occasionally run into other tubes such as the diode, triode, tetrode, pentode, gas-filled, and multielement tubes.

The diode tube consists of a negative cathode and a positive plate. The negative cathode, when heated, gives off electrons; and conduction takes place. The process of giving off electrons from the cathode is called *thermionic emission*. When the polarity is reversed, no thermionic emission takes place and no current flows. This valve-like action serves as a one-way gate and is, therefore, used as a rectifier.

The number of electrons that reach the plate from the cathode in the triode tube is controlled by placing a fine-mesh wire called a grid. This control grid is made negative in relation to the cathode. The more negative the grid, the less the current flows; the less negative the grid, the more the current flows. Cutoff is the point where the grid is made too negative and current flow stops. *Saturation* is the point where the grid is at the least negative point and current flow is at a maximum between grid and cathode.

In order to prevent interelectrode capacitance, an undesired effect in the triode, a second grid, called a *screen grid*, is added in the tetrode tube.

When increased performance under certain applications is desired, a third grid is added. It is called a *suppressor grid*. The pentode that contains this grid eliminates secondary emission (uncontrolled acceleration electrons near the plate) by controlling these accelerated electrons.

Power tubes are basically for high-power amplification. Gas tubes are often filled with nitrogen or mercury vapor and are used in high-current applications. The thyratron tube is a common example of a gas tube. Multielement tubes are tubes consisting of two or more tubes enclosed in the same glass envelope. The pentagrid converter is a common example of a multielement tube. It consists of both the local oscillator and mixer stages in a receiver.

The few techniques used to test electron tubes are as follows:

1. Tapping
2. Visual checking
3. Tube checker
4. Substitution

Even though electron tubes are rarely used today, you may still encounter them, especially in servicing old hybrid televisions, computer monitor CRTs, and old industrial and communications equipment. To examine the quality of a tube while the circuit is operating, use the plastic end of a screwdriver. Gently tap the top of each tube while listening or watching the performance of the circuit such as in a radio or television. If any type of noise is heard or anything is seen while tapping on the tube, or if there is any change in the picture, the tube is probably defective. Keep in mind that a loose connection or poor soldering connection in the same area could cause the same problem.

Another way to quickly examine the quality of some tubes is to visually see if the heater is lighted. If the heater is open, no glow will be produced and the tube will not operate. An ohmmeter can also be used to check the heater. A good heater will read near 0 Ω, but an open heater will read infinity.

Although the tube tester can be a helpful instrument for the troubleshooter, it can also be a handicap. For example, there is nothing more comical than years ago to see a "do-it-yourselfer" trying to play troubleshooter by pulling out every tube from a TV set and taking them all to the local drugstore to be tested on a tube tester. Undoubtedly, the tube tester will indicate that many of the tubes need to be replaced but will often fail to pinpoint the one tube that is causing the problem. Valuable time and money can be saved by understanding the function of each tube in a circuit and how to properly use a tube tester. For example, tube testers cannot match the operating conditions of the circuit. They cannot adequately measure interelectrode capacitance. Also, oscillators, limiters, and high-voltage tubes (where characteristic curves are critical) are difficult to test on the tube tester. The best advice to follow is, "When in doubt, replace the tube, but only if you can find a replacement and it is cost-justified."

Substitution of a known tube for a suspected defective tube can be a time saver for the troubleshooter. Remember, if the tube failed because of a circuit problem, the substitution of another new tube will only ruin the new tube. For example, when you are servicing an electronic device, if a rectifier tube is found short-circuited, also look for a short-circuited filter capacitor. Perhaps the filter capacitor caused the rectifier tube to short-circuit. Also, before you replace any tube, it is a good idea to visually inspect the area for charred resistors or any other problem that could have caused the tube to fail.

Unlike the transistor, which could theoretically last forever, the life of an electron tube is limited because the cathode simply wears down and, with time, emits fewer and fewer electrons. Also, mechanical vibrations, excessive heat, and current all contribute to tube breakdowns.

When you are replacing a tube, be sure to use the exact replacement or recommended substitute and make sure the tube socket is clean and free of corrosion. Also, be careful not to bend the tube pins.

Ultracapacitors

Ultracapacitors, also called *double-layer capacitors* (DLCs), are very powerful capacitors that can store over 100 times more electrical energy than conventional capacitors. They operate by using nonreactive, porous plates with an extremely high surface area in an electrolytic solution through the movement of charged ions. Electrical energy is charged electrostatically. The charge separation is also extremely small, which contributes to an ultracapacitor's ability to hold its enormous capacitance. Ultracapacitors offer many benefits over conventional capacitors and batteries, such as quick charging, high energy, low weight, high reliability, long life, and low maintenance.

There are numerous applications for ultracapacitors such as medical devices, computers, children's toys, power tools, wireless transmission, electric hybrid vehicles, and backup power supplies. Figure 1.36 shows an example of an ultracapacitor from Maxwell Technologies that weighs only 6.4 grams (g), but this ultracapacitor delivers approximately 10 F, which is ideal for powering small consumer electronic products. When used in conjunction with batteries, ultracapacitors can also improve efficiency and allow reduced size and weight of batteries by contributing power during peak loads.

One of the more popular applications of ultracapacitors is in the automotive industry. Ultracapacitors are being used in regenerative braking systems, diesel-electric buses, and in conjunction with electrolytic batteries for hybrid vehicles. Ultracapacitors can operate more efficiently and longer than batteries in wide temperature ranges. An ultracapacitor also can operate at any voltage within its voltage rating, and unlike the battery, it can be stored without any charge and quickly recharged.

The use of ultracapacitors in conjunction with batteries can provide excellent power and energy for hybrid vehicle applications. They can extend conventional

FIGURE 1.36 A 10-F ultracapacitor. (*Maxwell Technologies.*)

lead-acid battery life by adding power during peak loads and help provide quick acceleration and regenerative braking. The Maxwell Technologies PC2500 ultracapacitor provides 2700 F that caches 8400 joules (J) of energy at 2.5 V, making it ideal for hybrid vehicle applications (Fig. 1.37). It also has low weight, low current leakage, and excellent cycling reliability, making it suitable for other nonautomotive applications, such as backup power supplies during power outages for industrial and medical facilities.

Like other electrical components, ultracapacitors can experience faults such as internal shorts, open circuits, cell leakage, and material breakdowns, often due to internal stresses from excessive vibration, thermal expansions, damage, or abuse. Common tests for ultracapacitors consist of charge-discharge and measuring

FIGURE 1.37 A 2700-F ultracapacitor. (*Maxwell Technologies.*)

equivalent series resistance (ESR). Parameters such as initial working voltage, discharge current, minimum voltage under loads, voltage after load removal, and the time to discharge from initial charge to a minimum voltage can be measured to test the quality of ultracapacitors. Appendix L gives detailed test procedures from Maxwell Technologies for checking ultracapacitors.

Inductors

An *inductor* is basically an electromagnet that is used in many applications, such as transformers, up and down voltage steppers, filters, oscillators, phase shifters, integrators, and differentiators. The inductor basically opposes any current changes, and this is often referred to as *inductance*. The inductor creates a magnetic field that induces a counter electromotive force (EMF). Inductance L is measured by the unit henry (H). Types of inductors include air core, iron core, ferrite core, fixed, and variable.

One common application of the inductor is in filtering circuits. At a basic level, inductors pass low frequencies and block high frequencies. Capacitors, on the other hand, often pass high frequencies and block low frequencies. Therefore, when both inductors and capacitors are used in conjunction, they can act as a filter. For example, in a sound system, an inductor could be used to block high-frequency music to the woofer speaker, and the capacitor could be used to block low-frequency music to the tweeter speaker. A combination of the inductor and capacitor can be used to provide proper midrange frequency music to the midrange speaker.

Many inductors can be tested using an ohmmeter. While the windings of an inductor are often shorted, open circuits tend to account for the majority of defects. When you use the ohmmeter, an inductor, depending on the size and number of windings, should have resistance from about zero to a few hundred ohms. Larger inductors with many turns of wire generally measure some resistance. A shorted inductor would measure zero resistance. An open inductor would measure infinite resistance. It may be hard to determine whether an inductor is shorted using an ohmmeter since one or more shorted turns of wire may not change the resistance measurement of a normally low-resistance inductor. Therefore, it may be necessary to use an inductor analyzer to check the inductance.

Self-Examination

Select the best answer:

1. Which of the following is not an example of a source of breakdown?
   A. Heat
   B. Moisture
   C. Poor installation
   D. Animals and rodents
   E. None of the above

2. Which of the following is not one of the senses commonly used by service troubleshooters?
   A. Sight
   B. Hearing
   C. Touch
   D. Taste
   E. Smell

3. A hot, smoky product or device is often a sign of
   A. A short circuit
   B. A ground
   C. An open circuit
   D. All the above
   E. None of the above

4. A circuit that has infinite resistance is called
   A. A short circuit
   B. A ground
   C. An open circuit
   D. All the above
   E. (A) or (C)

5. Voltage measurements are often taken by using a voltmeter or
   A. An ammeter
   B. An oscilloscope
   C. An ohmmeter
   D. A wattmeter
   E. None of the above

6. Signal injection or tracing is a method commonly used in troubleshooting
   A. Electric motors
   B. Residential wiring
   C. Industrial wiring
   D. Radio
   E. Any of the above

7. A technique in which a suspected defective component is replaced by a "good" component is called
   A. Bypassing
   B. Substitution
   C. Bridging
   D. Both (B) and (C)

8. A cold solder connection can best be repaired by
   A. Substitution
   B. Bridging
   C. Resoldering
   D. Cooling
   E. Freezing

9. When you take a "step-by-step" analysis approach to troubleshooting, the first step should be
   A. Discussion of defect with customer
   B. Acquisition of service information
   C. Selection of troubleshooting technique
   D. Repairing of the problem
   E. All the above
10. The type of diagram that illustrates the component parts of a product or device is called a
    A. Line drawing
    B. Schematic diagram
    C. Blueprint
    D. Pictorial diagram
    E. Schematic print
11. A component having no continuity would have
    A. Zero resistance
    B. Infinite resistance
    C. Both (A) and (B)
    D. Small resistance
    E. None of the above
12. A good fuse will have
    A. Zero resistance
    B. Infinite resistance
    C. Small resistance
    D. Both (A) and (B)
    E. None of the above
13. The physical size of a resistor, which determines the ability of the resistor to absorb heat, is rated in
    A. Ohms
    B. Volts
    C. Watts
    D. Farads
    E. None of the above
14. The new fully charged lead acid storage battery should measure
    A. Over 12 V
    B. 2 V
    C. 11 V
    D. 12 V
    E. None of the above
15. Capacitors can be tested by
    A. An ohmmeter
    B. A spark test
    C. Bridging
    D. Only (B) and (C)
    E. (A), (B), and (C)

16. In order to make a *p*-type crystal,
    A. A pentavalent of gallium is added
    B. A trivalent of indium is added
    C. A pentavalent of antimony is added
    D. A trivalent of arsenic is added
    E. None of the above

17. The term *acceptor* is referred to in the
    A. Addition of pentavalent to the crystal
    B. Addition of trivalent to the crystal
    C. Both (A) and (B)
    D. None of the above

18. Actually a transistor is
    A. One diode
    B. Two diodes back to back
    C. Three diodes back to back
    D. Four diodes back to back
    E. None of the above

19. High-voltage gain and low-current gain are characteristics of the
    A. Common base circuit
    B. Common emitter circuit
    C. Common collector circuit
    D. Both (A) and (C)
    E. None of the above

20. If the operating voltage at the collector of the transistor were much lower than normal, one could suspect
    A. A defective filter
    B. An open resistor
    C. An open transistor
    D. A short-circuited transistor
    E. None of the above

21. To cutoff a transistor for troubleshooting purposes
    A. Short *E* and *B*
    B. Short *B* and *C*
    C. Short gate to anode
    D. Either (A) or (B)
    E. None of the above

22. The SCR is considered to be _____ diodes back to back consisting of an anode, cathode, and _____.
    A. Two, plate
    B. Three, gate
    C. Four, base
    D. Two, emitter
    E. Three, plate

23. When you check thermally intermittent integrated circuits (ICs), the suggested best troubleshooting technique is
    A. Voltage
    B. Resistance
    C. Heating and/or freezing
    D. Current
    E. Bridging
24. The tube that contains three grids is the
    A. Triode
    B. Tetrode
    C. Pentode
    D. Multielement
    E. Power tube
25. Which of the following techniques is not common in troubleshooting electron tubes?
    A. Tapping
    B. Tube checker
    C. Bridging
    D. Substitution
    E. Both (A) and (C)
26. The gate in the field-effect transistor (FET) is generally
    A. Reverse-biased
    B. Forward-biased
    C. Zero-biased
    D. None of the above
27. The MOSFET is often referred to as
    A. An FET
    B. A bipolar transistor
    C. An IGFET
    D. An SCR
28. The *depletion* MOSFET conducts at
    A. Forward bias
    B. Reverse bias
    C. Zero bias
    D. None of the above
29. Current is reduced in the depletion-enhancement MOSFET with
    A. Positive bias
    B. Negative bias
    C. Zero bias
    D. None of the above
30. Another name for an ultracapacitor is a:
    A. Double-layer capacitor
    B. Dielectric capacitor
    C. Power-cache condenser
    D. Electrostatic condenser

Questions and Problems

1. List and explain seven causes of breakdowns.
2. List and explain four senses commonly used in troubleshooting.
3. What are the four causes of circuit faults? Also list how each one differs from the others.
4. What are the characteristics of a short circuit?
5. What are the characteristics of an open circuit?
6. What are the characteristics of a grounded circuit?
7. What are the characteristics of a mechanical problem in a circuit?
8. Explain the difference between the terms *bridging* and *substitution*.
9. Explain the technique of signal tracing.
10. List and explain different types of service diagrams.
11. Explain the techniques in troubleshooting capacitors.
12. Name several different types of capacitors.
13. Explain the structure of a diode.
14. Explain how to test a diode.
15. What is a diode crystal detector?
16. Explain the structure of a transistor.
17. Explain how to test a transistor.
18. Explain how to test an SCR.
19. Explain the various techniques used to troubleshoot transistors.
20. Explain the various techniques used to troubleshoot ICs.
21. Explain the differences between several types of electron tubes.
22. Explain the few basic ways to check an electron tube.
23. Why should tube testers be used with caution in testing tubes?
24. What is a CRT?
25. What is a thyratron tube?
26. Explain how to check the FET.
27. What is a MOSFET?
28. Explain the different types of MOSFETs.
29. Explain how to check a MOSFET.
30. What special consideration must be taken when shipping or handling the MOSFET?

Chapter 2

Electronic Test Instruments

As electronic products and equipment become more sophisticated, the need for test instruments becomes increasingly important. There are hundreds of different types of test instruments used today. Most electrical and electronic products and devices could not be effectively serviced without proper selection and use of test instruments. The proper use of these instruments improves the troubleshooter's speed and accuracy in isolating and correcting a problem. In this chapter, some of the most popular types of test instruments used by troubleshooters are presented.

General Considerations

The troubleshooter needs to determine the cost/benefit ratio of the financial investment in test instruments. It may not be practical to invest in an instrument that has limited use. However, many troubleshooters have wasted valuable time, obtained inaccurate readings, experienced hours of frustration, and lost business opportunities, all because of ineffective or unavailable test equipment. When you select test instruments, consideration must be given to instrument reliability, traceability, material and international standardization, calibration services, reliability, durability, readout and display features, and overall accuracy and performance.

Always thoroughly read the operating manual before using the test instrument. Many troubleshooters, upon acquisition of a test instrument, fail to take the time to fully understand all the functions and details of the instrument, which results in a surprisingly high incidence of inaccurate reading, misuse, and underutilization. Also, specialized books are available that give further information on test instruments' use and applications.

Multimeter, VOM, FET Multimeter, and DMM

The *volt-ohm-milliamp meter* (VOM) has been a very popular portable instrument for many years (Fig. 2.1). This analog meter is ideal for measuring fluctuating trends and rates, a task which is difficult for digital meters. Many troubleshooters, especially those who work in industrial electrical applications, have favored the observation of needle movement over digital readings. Newer VOMs incorporate a combination of high-energy fusing and a diode network for meter movement protection.

One disadvantage of the VOM is that the impedance of the meter itself can, under certain conditions, "load down" the circuit and affect the measurement of voltage. Therefore, it is not uncommon for the voltage measurement to sometimes be inaccurate. Small, inaccurate measurements of voltage do not usually present a problem for industrial electricians. They do, however, present a problem for the electronic engineer. Small inaccuracies in measuring voltage may greatly affect the electronic diagnosis.

The solid-state field-effect transistor (FET) multimeter overcomes this problem of loading down a circuit by its high input impedance and precision-regulated internal power supply. This meter is a highly versatile, portable meter typically used for field and factory servicing and design testing.

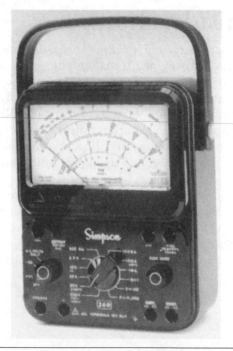

FIGURE 2.1   A typical VOM. (*Courtesy Simpson Electric Co.*)

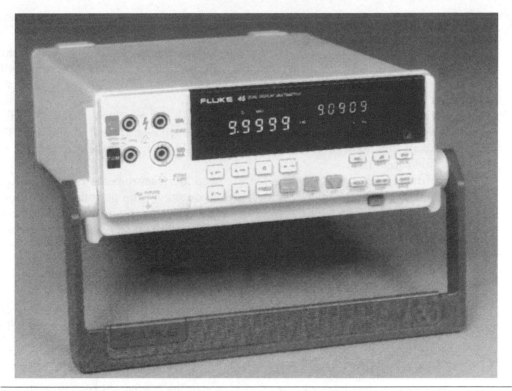

FIGURE 2.2   A dual-display digital multimeter (DMM). (*Courtesy John Fluke Mfg. Co., Inc., reproduced with permission from the John Fluke Mfg. Co., Inc.*)

The *digital multimeter* (DMM) (Fig. 2.2) tends to be the most popular meter for electronic troubleshooters who need extreme accuracy, as in laboratory work and digital equipment testing and servicing. This meter uses circuits that produce numerical readouts using light-emitting diodes (LEDs), and newer ones have LED graphic screens that function similarly to an oscilloscope.

High-performance DMMs feature a five-digit multifunction vacuum fluorescent dual display with selectable reading rates and resolutions. For example, the troubleshooter is able to view two signal parameters from one test point by taking measurements sequentially and simultaneously. This allows the troubleshooter increased versatility in applications requiring two separate examinations of the same signal.

DMMs are generally portable and use standard disposable batteries. Some DMMs have the capability of interfacing with a personal computer for automated recording of measurements. All these advantages of the DMM, along with very accurate readings, make this test instrument very popular for digital equipment bench testing.

Oscilloscope

In the most simplistic sense, the oscilloscope is essentially a visual voltmeter. However, to the novice, this meter, with all its controls and visual screen, induces both fascination and intimidation. The oscilloscope can be one of the most valuable types of equipment in troubleshooting.

The basic advantage of the oscilloscope is that it provides a visible display of the waveform being measured. Most oscilloscopes use electrostatic deflection. The beam sent from the electron gun is deflected vertically or horizontally by pairs of vertical and horizontal plates.

Although the oscilloscope is largely used to measure peak-to-peak voltage, other measurements that can be taken are the frequency, time periods, wave slopes, phase angles, and frequency response. Figure 2.3 illustrates a typical oscilloscope.

The basic operating controls and their functions in an oscilloscope are as follows:

1. Intensity—controls the brightness of the electron beam
2. Focus—adjusts the sharpness of the beam
3. Vertical center control—controls the vertical positioning of the electron beam
4. Horizontal center control—controls the horizontal positioning of the electron beam
5. Vertical gain—adjusts the height of the waveform
6. Horizontal gain—adjusts the width of the waveform
7. Sweep control—adjusts the frequency of the horizontal sweep oscillator
8. Sync selector—permits use of an internal or external synchronization
9. Z axis—varies the intensity of modulation of the trace
10. Calibration scale—permits a scale for measurement of voltage waveforms

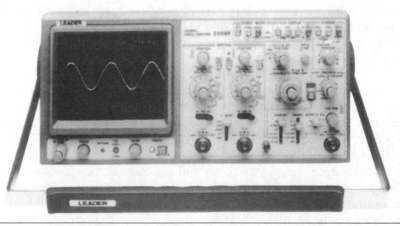

FIGURE 2.3   A typical oscilloscope. (*Courtesy Leader Instruments Corp.*)

The basic setup for an oscilloscope consists of the following:

1. Turn intensity, focus gain, and sync amplitude to minimum.
2. Turn vertical and horizontal controls to midrange.
3. Turn on scope and adjust the intensity control to minimum brightness.
4. Allow 1 to 2 minutes (min) for the scope to warm up and then adjust the focus control for a sharp trace.
5. Center the trace signal by adjusting the vertical and horizontal controls.
6. Connect a 6.3-V (volt) alternating-current (ac) source to the vertical input for calibration.
7. Because 6.3 V root mean square (rms) equals 9 V peak voltage or 18 V peak to peak, it should be adjusted to display 1.8 divisions on the cathode-ray tube (CRT) screen (Fig. 2.4).
8. Adjust the sync control until a stationary pattern displaying three sine waves appears.
9. Now the oscilloscope is "set up" and calibrated.

Each division will now equal 10 V peak. The vertical attenuator can be used to multiply the divisions by 0.1, 1, 10, etc., as desired for proper measurements.

When you calibrate an oscilloscope with an internal calibrator, the scope can be calibrated by adjusting a fixed pattern of 1 V peak to peak.

More sophisticated oscilloscopes have features such as built-in calibrators for calibration checks, separate and independent comprehensive triggering facilities, and beam finders.

When you are troubleshooting with an oscilloscope, three basic accessory probes are commonly used:

1. Low-capacitance probe
2. Demodulation probe or radio-frequency (rf) probe
3. Voltage-divider probe

The low-capacitance probe is generally used to measure high-frequency or high-impedance circuits. The "loading effect" is reduced by using this probe, which increases the accuracy of measurement.

The demodulation probe or rf probe is often used to measure rf signals where the signal must be detected before being displayed on the scope.

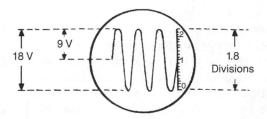

FIGURE 2.4 A calibrated display of an oscilloscope.

The voltage-divider probe is used when the voltage being measured is desired to be "stepped down" (reduced). The usual voltage division ratio is 10:1 or 100:1.

When you select an oscilloscope, it is important to consider the bandwidth or rise time, triggering, and other specialized requirements. Bandwidths can vary from 10 megahertz (MHz) to more than 100 MHz. For example, personal computers used to utilize 5-MHz clocks but are now often seen at 20 MHz. There can be dramatic differences in measurements of the actual waveforms between two different oscilloscopes, especially with digital pulses. An oscilloscope with 10 MHz may be sufficient for automotive requirements but insufficient for video equipment and industrial programming circuitry.

The selection of an analog versus a digital oscilloscope is also important depending on the application. Analog oscilloscopes generally are less costly and are best for measuring analog and high-frequency signals, whereas digital oscilloscopes are used for special digital and storage applications. Also, recent technology now offers analog-digital combinations incorporating digital recording and control with analog familiarity.

Other specialized applications require waveform recording capabilities. The electromyography (EMG) unit, used in biomedical diagnostic testing, is a typical example. The EMG unit uses a built-in oscilloscope to measure the electrical impulses and the nerve conduction velocity that stimulate muscles and allow sensation. In simplistic terms, electrodes record both the electrical activity that travels from one point of the body to another and the muscle or nerve activity at a fixed point. It is important for the practitioner to be able to view more than one signal, freeze a signal, produce an instant hardcopy from a printer, or file a waveform for multiple comparisons. Special triggering requirements, such as delayed sweeps or expansion of full rise time, of a pulse are also available for troubleshooters.

In addition to bench oscilloscopes, there are several types of portable oscilloscopes. These instruments provide sufficient power and high-performance capabilities needed to troubleshoot devices on the job site. For example, some oscilloscopes combine high safety ratings and performance equivalent to many bench oscilloscopes with the extra protection of durable portability. These instruments are useful for plant maintenance engineers and technicians who need a durable test instrument, especially in harsh and hazardous conditions.

Some of the features of portable oscilloscopes include quick sampling rates up to 5 GS/s and 200 pixel shader (ps) resolution, memory capability in excess of 10,000 samples per channel, and waveform with zoom capabilities. They are exceptionally good instruments for three-phase testing in diagnosing industrial systems, power inverters, and converters. Other features include circuit voltage and amperage overloading, signal timing measurements, signal fluctuation, harmonic testing and transience loads, and three-phase power input. One of the more useful functions of portable oscilloscopes includes the paperless recording. This function can be convenient for collecting data by allowing the plotting of values over an extended period of time. In addition, voltages, amperages, temperature, and frequency measurements can be recorded and stored in the unit.

The more advanced bench oscilloscopes provide an extensive range of capabilities. For example, the phosphor oscilloscopes contain advanced triggering and protocol decode and search capabilities with bandwidths from 500 MHz to 3.5 gigahertz (GHz) and a fast digital phosphor display. Other features include rapid capture of signal anomalies, serial triggering and analysis options, and software analysis packages.

Specialized Test Instruments

There are hundreds of specialized test instruments used today for many different applications. Some of the more commonly used instruments are as follows:

1. Transistor tester
2. Capacitor tester
3. Frequency counter
4. Signal generator
5. Insulation meter
6. Voltage tester
7. Clamp-on ammeter
8. Neon voltage tester
9. Test lamp
10. Digital logic pulser
11. Digital logic probe
12. Growler
13. Optical time-domain reflectometer (OTDR)
14. Field strength meter
15. Network analyzer
16. Logic troubleshooting kit

Transistor testers are fairly accurate checkers of diodes and transistors. They also have the capability of checking the performance of these components while in or out of circuit. They are able to measure transistor leakage and beta, and can automatically identify emitter, base, and collector leads (Fig. 2.5). Transistor checkers are often multipurpose instruments with audible and visual test

FIGURE 2.5   A typical transistor checker. (*Courtesy Leader Instruments Corp.*)

indications. Leakage currents can be made with transistors when they are out of the circuit. The probes of these meters contain flexible, spring-loaded pointed tips that allow for convenient and quick measurements. Connections to transistors mounted as printed-circuit boards can also be easily checked.

Capacitor testers not only check the quality of the capacitor but also determine the value of unknown capacitors. Many capacitor testers have the ability to check most capacitors in or out of the circuit, which can speed up the servicing time. Also, the capacitor tester can identify power factor values, leakage, and open circuits, all problems common to capacitors. Keep in mind that the actual capacitance value of the capacitor can only be accurately measured when the capacitor is out of the circuit. Capacitor checkers are very sensitive and can detect even small amounts of leakage. Also, when you check electrolytic capacitors, unlike other types, it is important to measure their power factor. Remember, never touch the terminals of the capacitor tester when the voltage is turned up! Severe shocks can result.

Frequency counters are used to measure the frequency, in hertz (Hz), of an electronic product, and are often used to adjust the frequency of radio receivers and transmitters. They are also very valuable instruments used in research and experimentation. They generally offer automatic triggering, high-stability time base, multiple measuring functions, input voltage protection, and portability. Most instruments have a frequency measurement range between 10 Hz and 120 MHz with optimal ranges up to 1.3 GHz. Some counters include a variety of accessories such as compensated crystal oscillators for stability and memory options (Fig. 2.6).

There are various types of signal generators. The audio-frequency (af) generator provides signals in the audio range, whereas the rf generator provides signals in the rf range. Both meters provide sine wave or square wave outputs, contain built-in attenuators, and provide low-distortion outputs. Typical signal generators generate a range of signals with typical specifications of 20 MHz bandwidth, 14 bit resolution, and 250 MS/s sample rate. They have the ability to create complex waveforms with stable high fidelity signals.

A noise generator is a small handheld signal probe that is handy in signal tracing of radio receivers. This small generator sends out a broadband signal from af to rf. This typical frequency range is from 1 kilohertz (kHz) to 30 MHz.

FIGURE 2.6 A typical frequency counter. (*Courtesy Leader Instruments Corp.*)

The marker generator provides an unmodulated frequency and is used to identify frequencies on a response curve used in television alignment.

The sweep generator is also used in television alignment. It provides a frequency modulation (FM) signal at a desired range of frequencies. The sweep generator and marker generator are usually built in together. Television generators have higher sweep widths than FM stereos. These instruments often contain a color generator that produces test patterns for converging color television. They provide such test patterns as purity, white raster, dots, crosshatch, gated rainbow, and horizontal and vertical lines.

The TV/stereo signal generator is a special multipurpose signal generator used for aligning and troubleshooting television and stereo receiving equipment. The generator can also provide modulation requirements for multiple channels (Fig. 2.7).

When you use signal generators, there are several precautions to observe. Ensure proper grounding of the generator, as done with other test equipment, such as an oscilloscope or multimeter. Inaccurate readings may result from poor grounding. Move the ground position when tracing video, rf, and pulse signals every time the signal probe is moved. Also, remember to match the output impedance of the generator with the test circuit. Reduction in gain can occur, for example, if the input impedance of the generator is slanted across the circuit. Matching impedance pads and multi-impedance probes are available. Also, other helpful practices can be found in the manufacturer's manual regarding output amplitude consistency, calibration, linearity, and distortion factors. Finally, remember to avoid premature realignments prior to thoroughly troubleshooting the product. Correcting the defective circuit may resolve the alignment problem, or it may require only fine-tuning.

Another popular and valuable test instrument is the insulation resistance meter. It is used to check the electrical resistance of an insulator by indicating the resistance on a scale as the meter supplies a voltage. This instrument provides the voltage source from a self-contained hand-operated generator, from a battery, or from a power supply. A variety of insulation meters are also available, such as pocket and midget hand-cranked brushless generator types. The quality, measured in resistance, of an insulator is determined by its ability to withstand the voltage without leakage. Insulation breakdown and degeneration can be caused by such factors as temperature variations, corrosion, water, vibration, contaminants, and

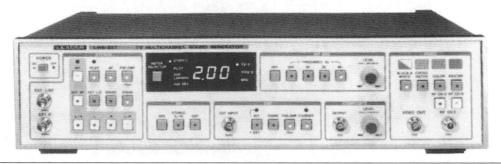

FIGURE 2.7   A typical TV/stereo signal generator. (*Courtesy Leader Instruments Corp.*)

wear. The insulation resistance meter can be used for such items as motors, generators, transformers, cables, wires, switches, circuit assemblies, and terminals. Other uses include power utilities, telecommunications, street lighting, panelboard testing, railway testing, avionics ground testing, military hardware, and cable testing. These meters are also available in either analog or digital types, or both. The test voltages generally range from 50 to 5000 V dc and from 100 to 500,000 megohm (MW) for insulation tests. Some insulation testers provide continuity testing at 200 milliampere (mA) or 20 mA down to 0.01. Other features include measurement of capacitance and distance by capacitance, CAT IV 600 volt rating, analog arc and dual-digital display, etc. Figure 2.8 illustrates a popular Megger® 400 series instrument that is used for insulation testing in power utility, industry, and telecommunication applications.

There is no industrial troubleshooter who could work without a voltage tester. Voltage testers are handy, rugged meters that are commonly used to test ac voltages between 110 and 600 V (Fig. 2.9). They serve as a quick and convenient method for checking line and terminal voltages. Accuracy is not the concern for the industrial troubleshooter using this portable voltmeter. The troubleshooter is generally concerned with whether the circuit is live and the approximate amount of voltage, such as 120, 240, 480, or 600 V. A special type of voltage tester, referred to as a *high-voltage probe*, can measure voltages up to 40,000 V dc. Those probes are generally used in servicing X-ray machines and television and computer terminal CRTs.

High-voltage transmission meters are specialized voltmeters used for checking high-voltage transformers and power lines. These meters, often referred to as "hot sticks" because of their long handles, contain high-voltage resistors encapsulated in epoxy, thereby limiting the current at full voltage. These meters have the capability

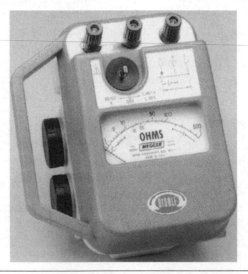

FIGURE 2.8   An insulation resistance meter. (*Courtesy of Megger®.*)

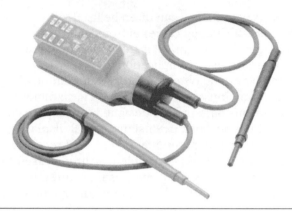

FIGURE 2.9   A voltage tester. (*Courtesy ETCON Corp.*)

of measuring voltages in excess of 145,000 V. Accessories include various isolated and insulated bushing probes, proof testers, extension hot sticks, chart recorders, and phasing equipment.

The clamp-on ammeter is similar to the voltage tester, except that it is used to measure ac or dc amperage by the industrial troubleshooter. This meter contains transformer jaws that clamp around a conductor without interrupting the circuit. These meters are portable and rugged. Figure 2.10 illustrates a typical clamp-on

FIGURE 2.10   A clamp-on ammeter. (*Courtesy Simpson Electric Co.*)

ammeter. This type of meter can often be used with analog VOMs and DMMs. They generally handle alternating currents from 100 mA to 500 amperes (A). Like the voltage tester, its main advantage is the ability to make quick and accurate current measurements in industrial applications.

Specialized clamp-on type ammeters that include a large hook and jaw are used for high-voltage transmission applications. These meters can measure single or multiple lines or bus bars and are designed to measure currents in excess of 2000 A. Accessories include hot-stick extensions, digital displays, variac controls, distribution and substation capacitor, meter attachment, and recorders.

The neon voltage tester is a simple instrument used to check for the presence of voltage in a circuit. This instrument is most often used in troubleshooting house wiring. The tester is very simple to operate, inexpensive, and easy to carry, and it uses a neon bulb to indicate the presence of voltage rather than a meter. Figure 2.11 shows a typical example of a neon voltage tester.

A test lamp is a simple test device used to check continuity. This device is sometimes preferred to the ohmmeter because the luminance of the bulb allows the troubleshooter to watch the test lamp leads and device under service rather than on the scale of the ohmmeter. Typical applications include electric motor and generator repair and small-appliance applications. Figure 2.12 shows an example of the construction of a typical test lamp.

The logic pulser (Fig. 2.13) provides a one-shot positive and negative pulse by depressing the pulse button. A continuous pulse train can be injected by keeping the pulse button depressed. Other types of logic pulsers have additional features such as automatic polarity selection and pulse variations.

The logic probe is used to diagnose the logic states (e.g., high or low) by comparing the voltage to reference thresholds for the desired logic family. Logic probes offer speedy testing and eliminate the need for expensive, bulky oscilloscopes or voltmeters. There are many different types of logic probes offering such applications as memory, overload protection, and very high-speed testing.

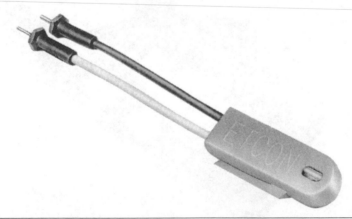

FIGURE 2.11   A neon voltage tester. (*Courtesy ETCON Corp.*)

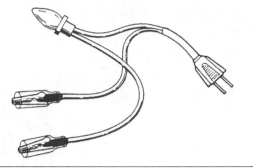

FIGURE 2.12   A simple test lamp.

FIGURE 2.13   A handheld digital logic pulser. (*Courtesy Global Specialties Corp.*)

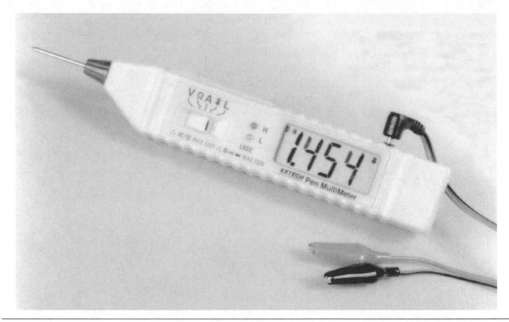

FIGURE 2.14 A handheld multimeter with logic test. (*Courtesy Extech Instruments.*)

Figure 2.14 illustrates a combination multimeter and logic test instrument. This instrument features an eight-function DMM with a large-digit lowest common denominator (LCD) display with full function indication and automatic or manual range selection. Logic monitors are similar to logic probes but offer the ability to clip on to the IC pins, test points, or bus line when multipoint testing is desired. Two common types of monitors are the 16-channel and 40-channel. These instruments have a wide variety of applications in troubleshooting microprocessors and process control systems.

Growlers consist of two kinds—internal and external—and are used to test armatures and stators of electric motors, generators, and other windings for short circuits. The growler contains a coil and iron-core assembly. When alternating current is applied to its coil, a magnetic flux is created between the growler and the tested device. The magnetic flux produces a growling noise.

A narrow piece of steel (called a feeler), such as a hacksaw blade, is used in combination with the growler to test a device. The feeler is placed in parallel with the tested device and the growler. If the tested device is short-circuited, a strong magnetic field is produced directly over the shorted coil, causing the feeler to aggressively vibrate and pull down. In noisy environments, some troubleshooters use a light bulb in series with the growler, which will brighten when a short circuit is encountered.

Some growlers have additional accessories such as a built-in ammeter or feeler that tests for small defective coils, reverse coils, or power-factor ratings. The growler is a very practical, easy-to-operate, and very durable instrument. A typical internal growler is shown in Fig. 2.15.

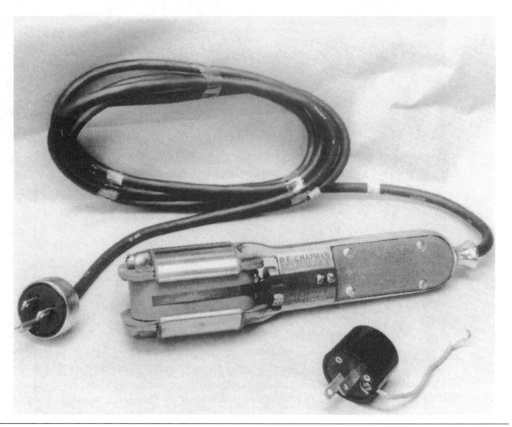

FIGURE 2.15 A typical internal growler. (*Courtesy P. E. Chapman Electrical Works, Inc.*)

An instrument that is used in troubleshooting and installing TV distribution systems is the signal level meter (i.e., field strength meter). This meter provides accurate very high-frequency (VHF) and ultra high-frequency (UHF) measurements and features a built-in loudspeaker. This meter is a tuned-rf voltmeter with a microvolt or decibel scale. This meter has many uses, its most common being to position antennas and measure signal levels. Other uses include measuring insertion loss, amplifier gain, automatic gain control settings, subchannel signals, return loss (i.e., standing wave ratio), and noise levels.

The *optical time-domain reflectometer* (OTDR) is a very specialized instrument used in fiber-optic network testing (Fig. 2.16). It can be used to test signal loss and disruptions on multimode fiber and to identify reflective breaks. Some units are able to incorporate two dual-wavelength modules simultaneously, thereby allowing dual-wavelength multimode optical testing. For example, the TFS 3031 Ranger 2, by Tektronix, Inc., features a 3.5-meter (m) multimode event dead zone, 15- to 60-decibel (dB) selectable reflectance threshold, internal memory for 100 waveforms, and the flexibility of locating a fiber end at over 175 kilometers (km).

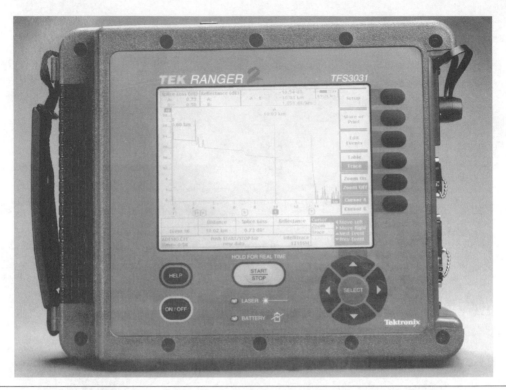

FIGURE 2.16 An optical time-domain reflectometer (OTDR). (*Courtesy Tektronix, Inc.*)

The network, spectrum, and waveform analyzer is a sophisticated module used in a number of applications such as medical speech research, hydroacoustics and sonar, rotating machinery analysis, and structural analysis. The unit features waveform digitizing up to 3.5 million samples and spectrum analysis up to 90-dB dynamic range (Fig. 2.17).

Logic troubleshooting kits provide dynamic, static, and multipin testing of ICs. The kits can be very convenient and effective in pinpointing circuit faults. Figure 2.18 illustrates a digital logic troubleshooting kit.

Another type of logic analyzer, illustrated in Fig. 2.19, is a super high-speed instrument for debugging complex system problems. It has the capabilities of allowing all incoming data to be oversampled at a 2-GHz rate, processed in real time for timing and state acquisition, and triggering without missing critical timing data. These instruments are compatible with standard plug-and-play accessories, are fully interchangeable between mainframes, have a broad range of microprocessor support, and provide simultaneous state and timing analysis through the same probes.

There are several types of calibrating instruments used for troubleshooting and maintenance applications. Figure 2.20 shows an example of a multifunction calibrator. This multifunctional instrument measures temperature, pressure, voltage, current, and resistance. It features a two-line backlighted alphanumeric LCD display of output

FIGURE 2.17   A network, spectrum, and waveform analyzer. (*Courtesy Tektronix, Inc.*)

FIGURE 2.18   A digital logic troubleshooting kit. (*Courtesy Hewlett-Packard.*)

source and input measured value, memory locations for simulation program, digital output, and optional bidirectional RS232 interface/software.

There are a number of other specialized electronic test products available for the troubleshooter. Some of these provide testing for base-band video test

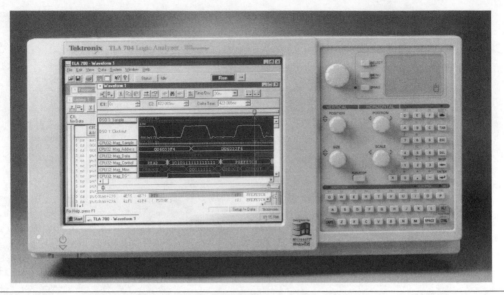

FIGURE 2.19 A logic analyzer. (*Courtesy Tektronix, Inc.*)

FIGURE 2.20 A multifunction calibrator. (*Courtesy Extech Instruments.*)

solutions, imbedded instrumentation, optical modulation canalization, and MPEG video test solutions. In addition, there are several thermal imagers that have the capabilities of using infrared images with visible light images on the same display. These instruments are able to detect heat patterns and temperature differences in electrical devices and objects, which allows the troubleshooter to monitor and maintain electrical equipment and devices.

Using Test Probes

All test instruments come with some type of probe or test lead. However, it is very important that the troubleshooter select the proper probe for the instrument and circuit to be tested. Cheap probes may consist of only a wire lead or a simple *RC* network. Active voltage probes contain more elaborate internal elements, such as FET circuitry for measuring circuits requiring reduced loading, special logic voltage levels, and high-speed analog and digital circuitry. Passive voltage probes do not contain the active internal circuitry and are used for general-purpose analog and digital applications. Oscilloscopes utilize a variety of test probes. The most common is the conventional 10:1 attenuating probe. Other probes include various bandwidths, automatic compensating CRT readouts, demodulator and low-capacitance, and high-voltage/impedance probes.

Every troubleshooter should have an assortment of clip leads, clamps, and alligator clips for making temporary connections and bridging suspected defective components. Other troubleshooting accessories include miniature diagonal cutters, large-nose pliers, spline key and hex key wrenches, magnifiers, and tweezers.

Self-Examination

Select the best answer:

1. Which meter can interface with a personal computer and is commonly used for very accurate bench testing?
   A. Transistor voltmeter (TVM)
   B. Volt-ohm-milliammeter (VOM)
   C. Vacuum tube voltmeter (VTVM)
   D. Field effect transistor voltmeter (FETVM)
   E. Digital multimeter (DMM)
2. The oscilloscope is largely used to measure
   A. Root mean square (rms) voltage
   B. Average voltage
   C. Effective voltage
   D. Peak-to-peak voltage
   E. None of the above

3. The control that adjusts the frequency of the horizontal sweep oscillator is the
   A. Horizontal gain
   B. Z axis
   C. Attenuator
   D. Sync selector
   E. Sweep control
4. The necessary probe commonly used on the oscilloscope when measuring high-frequency or high-impedance circuits is the
   A. Demodulator probe
   B. Low-capacitance probe
   C. Voltage divider
   D. Noise probe
   E. None of the above
5. The control on the oscilloscope that adjusts the height of the waveform is the
   A. Sweep control
   B. Z axis
   C. Sync selector
   D. Horizontal gain
   E. Vertical gain
6. The meter often used in testing an armature for short circuits is the
   A. VOM
   B. VTVM
   C. Megohmmeter
   D. Growler
   E. None of the above
7. A meter device that tests insulation resistance is the
   A. Transistor checker
   B. Voltage tester
   C. Gauss meter
   D. Megohmmeter
   E. None of the above
8. A small handheld probe generator used in signal-tracing receivers is the
   A. Audio-frequency (af) generator
   B. Radio-frequency (rf) generator
   C. Sweep generator
   D. Marker generator
   E. Noise generator
9. The accessory probe used on the oscilloscope to detect the signal is the
   A. Voltage divider
   B. Low-capacitance probe
   C. AF probe
   D. Direct probe
   E. Demodulator probe

10. The control on the oscilloscope that permits use of an internal or external sync is the
    A. Sweep control
    B. *Z* axis
    C. Focus
    D. Sync selector
    E. Intensity

Question and Problems

1. Explain the difference between a VOM and the FET meter.
2. What is a DMM?
3. Explain the procedure of setting up an oscilloscope for operation.
4. Explain how to calibrate an oscilloscope.
5. What is a low-capacitance probe?
6. What is a voltage-divider probe?
7. What is a demodulator or rf probe?
8. Explain what an insulation resistance meter is used for.
9. Where is the marker generator used?
10. What is a neon voltage tester?
11. Explain the purpose of an optical time-domain reflectometer (OTDR).
12. What is the difference between a digital logic probe and a digital logic pulser?
13. Explain the purpose for a network, spectrum, and waveform analyzer.
14. Explain the functions of a TV/stereo signal generator.
15. The high-voltage probe is typically used to test what types of equipment?
16. Explain the operation of a growler.
17. Explain characteristics of the high-voltage transmission ammeter.
18. What special requirements does the biomedical diagnostic EMG unit require for the oscilloscope?
19. Explain the applications and types of high-voltage meters.
20. Explain the typical applications of the test lamp.

Chapter 3

Troubleshooting Electric Motors and Generators

Electric motors are among the most widely used electrical devices in domestic, business, and industrial applications. Understanding the basics of troubleshooting electric motors can provide the carryover knowledge to service other industrial equipment. With the advent of high technology, the need for electric motors becomes even greater. High technology has demanded new designs and capabilities of the motor.

Electric generators, like electric motors, are also used in hundreds of applications from industrial applications to the automotive industry. Though the generator is similar to the electric motor, both have their unique differences and purposes. In this chapter, a review of the basic principles of electric motors and generators is presented, as well as methods of troubleshooting and repairing them.

Fundamentals

The construction and theory of the electric motor are very similar to those of the generator. An electric motor is a device that changes electric energy to mechanical energy. The generator does the complete opposite. A simple direct-current (dc) generator can be converted to an electric motor by connecting a battery to the terminals of the brushes. Figure 3.1 illustrates a simple electric motor.

The current supplied to the armature from the battery makes it an electromagnet. The armature has a north pole and a south pole adjacent to its respective field poles, as illustrated. Therefore, the armature rotates, like poles repel each other, as illustrated in Fig. 3.2.

The armature continues to rotate, as the commutator continuously changes the polarity of the armature poles. This type of motor is called a *repulsion-type* motor. To increase the efficiency of the electric motor, many coils of wire are used in the armature and field poles. Adding these coils of wire makes the motor more powerful and makes it run more smoothly.

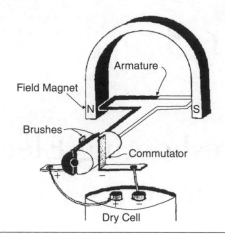

FIGURE 3.1   The construction of a simplified electric motor.

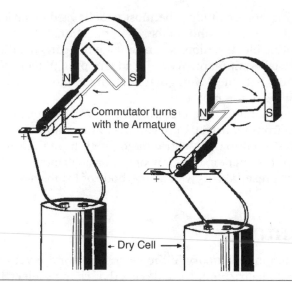

FIGURE 3.2   The rotating action of a simplified electric motor.

The basic parts common to most electric motors include the armature, field windings, end plates, bearings, frame, brushes, starting switch, and base. Figure 3.3 shows some of the common parts of a motor. Whereas electric motors vary in design and operation, most contain a stator (the shell of the electric windings of the motor), a rotor, and end bells (or end plates).

The stator is usually cast or rolled, constructed with multiple steel laminations. These series of identical laminations are compressed and welded together with an oxide coating inside the stator shell, which reduces eddy currents and the core heat during operation. The laminations are insulated and oxide-coated reducing eddy currents and heat.

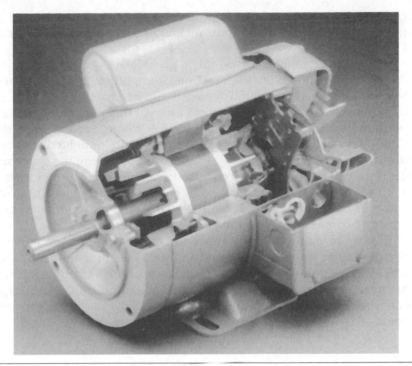

FIGURE 3.3 Typical parts of an electric motor. (*Courtesy Marathon Electric.*)

The windings consist of multiple turns of insulated wire. It is important that each coil of wire be carefully assembled; or ultimately the entire winding will "ground out," and the motor will overheat and cease operation.

Rotors are basically dc wound or ac induction types. The dc wound rotor contains a commutator and is also used for ac universal motors. Like the stator, this rotor contains metal laminations with coils of wire all insulated by varnish. The ac induction rotor contains no wires or commutator. The ac rotor consists of several metal laminations with aluminum, copper, and/or steel bars and dowels. This construction provides low-heat induction. Cooling fans are sometimes constructed on the rotor shaft for heat reduction. The rotors are assembled by "skewing" the rotor slots at an angle to provide smoother operation. The rotor is also balanced with weights to the fan blades or shaft.

Types of Motors

Basically, all electric motors operate on the principle of repulsion or induction. The repulsion-type motor, as you already know, works on the principle that two like magnetic fields oppose each other. The magnetic field of the armature opposes the magnetic field of the stationary field windings and causes the armature to rotate. The commutator continually reverses the polarity of the armature, and the armature continues to rotate. All dc motors and a few ac motors operate on the principle of repulsion and need an armature, commutator, and set of brushes.

The induction motor, as you might guess, works on the principle of induction. Almost all induction motors operate on alternating current (ac). The induction motor has a rotor that generally consists of a laminated iron cylinder containing copper bars buried in slots. The appearance of this rotor resembles a squirrel cage, and therefore it is called a *squirrel cage rotor*. When ac is applied to the stator field windings, a current is induced in the rotor because of induction. This induced current produces a rotor field polarity that is opposite to the stator field windings. The rotor will not begin rotating by itself; therefore, most single-phase motors require a starting winding and switch. Three-phase motors do not require a starting switch, as each phase is displaced 120°. Also, the induction motor does not need an armature, commutator, or set of brushes to operate.

There are many different types and classes of electric motors, each having its own particular characteristics and abilities. Recent technological developments have increased the supply of electric motors with varying capabilities. Some of the most common types of electric motors are the split-phase, capacitor, shaded-pole, repulsion, dc, synchronous, universal, polyphase, gear motor, and stepper motor.

Split-Phase

A split-phase motor is an ac, single-phase induction motor that is generally operated on 115 V. This motor uses a squirrel cage rotor and operates on the principle of induction. It is used on many devices, such as washing machines, water pumps, refrigerators, and fans. Size generally ranges from 1/20 to ½ horsepower (hp). Figure 3.4 illustrates a split-phase motor.

The split-phase motor has two field windings—a running winding and a starting winding. It is called a *split-phase motor* because the starting winding coils are displaced 90° from the main running windings. Figure 3.5 illustrates the two windings of a split-phase motor.

The starting winding, which is connected parallel to the running winding, is composed of fine, insulated copper wire. The starting windings, or auxiliary windings, are responsible for starting the motor and are usually left in the circuit for only a split second. After the motor has reached approximately 75 percent of its speed, a centrifugal switch disconnects the starting windings from the circuit. The main running windings take over. Figure 3.6 shows a picture of the centrifugal mechanism (governor) of a centrifugal switch. Also, this figure shows a rotor with the centrifugal mechanism in place.

The stationary part of the centrifugal switch contains two contacts where the starting windings are connected in the circuit or disconnected from the circuit. Figure 3.7 shows a picture of the stationary part of the centrifugal switch.

FIGURE 3.4   A typical split-phase motor.

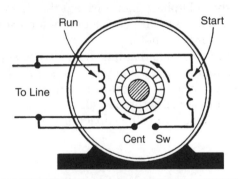

FIGURE 3.5   The run and start windings of a split-phase motor.

FIGURE 3.6   Typical centrifugal mechanism assembly of a centrifugal switch. (*Courtesy Franklin Electric Co.*)

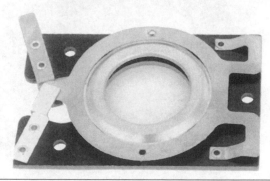

FIGURE 3.7   The stationary part of a centrifugal switch. (*Courtesy Franklin Electric Co.*)

Capacitator

The capacitor motor is an ac, single-phase induction motor. It is almost identical in construction to the split-phase motor except that it contains one or more capacitors and generally ranges in size from a fraction of a horsepower to 20 hp. Figure 3.8 illustrates a capacitor-type motor.

The capacitor is a device that has the ability to store electricity; it is rated in farads (F), microfarads (μF), and picofarads (pF). The most common types of capacitors used in electric motors are the paper capacitor and the electrolyte capacitor. There are basically three kinds of capacitor motors: capacitor start, capacitor start/run, and two-value capacitor run motors.

FIGURE 3.8   A typical capacitor motor. (*Courtesy Marathon Electric.*)

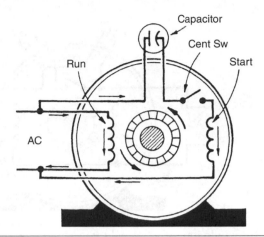

FIGURE 3.9   The internal circuit of a capacitor-start motor.

The capacitor-start motor uses a capacitor that is placed in series with the starting winding. When the motor is turned on, the capacitor causes the current in the starting winding to lead the current in the running winding. This effect induces a current in the rotor, and it begins to rotate. Figure 3.9 shows a schematic diagram of a capacitor-start motor.

In the capacitor-start/run motor, the capacitor and starting winding is left in the circuit at all times. This type of motor is a very quiet and smooth-running motor. It is often used in fans, refrigerators, and air conditioners, where the noise level needs to be at a minimum.

The two-value capacitor-type motor is also a very quiet motor. It is very similar to the capacitor-start/run motor, except that it usually has two capacitors of different values. The high-value capacitor is used to start the motor, and then the low-value capacitor replaces the high-value capacitor once the motor has started running. Also, the two-value capacitor motors are often used on compressors where high starting torque is needed. More than one speed usually can be obtained from this type of motor. Figure 3.10 shows a schematic diagram of a two-value capacitor motor.

Shaded-Pole

The shaded-pole motor is perhaps the least expensive motor and is generally found in sizes ranging from approximately 0.004 to 0.25 hp. Figure 3.11 shows a shaded-pole motor. The shaded-pole motor has very low starting torque. It is used in such devices as fans and blowers where low cost and low maintenance are the most important factors.

The shaded-pole motor is a simple ac, single-phase induction motor that requires very low maintenance. The rotor of the shaded-pole motor is considered a

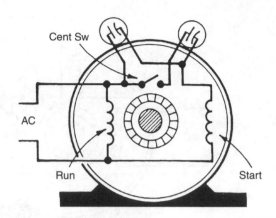

FIGURE 3.10   Internal view of a typical two-value capacitor motor.

FIGURE 3.11   A typical shaded-pole motor. (*Courtesy Franklin Electric Co.*)

typical squirrel cage type. Its poles tend to project past the laminated iron cylinder and are often referred to as *salient* poles. The shaded-pole motor does not have a starting winding that is similar to that of most single-phase induction motors; rather, it has heavy solid copper loops that act as the starting winding (Fig. 3.12).

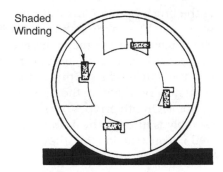

FIGURE 3.12   The shading winding of a shaded-pole motor.

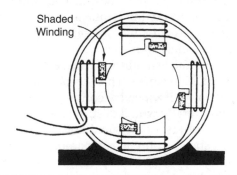

FIGURE 3.13   The running and starting coils of a shaded-pole motor.

When current is applied to the shaded-pole motor, the copper loops or shading coils produce a magnetic field that is out of phase with the main field windings. This magnetic field induces a current in the rotor, and the rotor begins to rotate. After sufficient speed is reached, the main field windings take over and the rotor continues to rotate. Figure 3.13 shows the two windings of the shaded-pole motor.

Repulsion

Basically, repulsion motors can be narrowed down to two kinds: (1) repulsion motors and (2) repulsion start, induction run motors.

As you may recall, the repulsion motor requires an armature, a commutator, and a set of brushes, and operates on the principle of repulsion of like poles. It is very similar to a dc series motor, and its size varies from 0.5 to 10 hp. The repulsion motor has excellent starting torque and variable speed control characteristics. It is commonly used on such devices as compressors, air conditioners, and pumps where high starting torque is required. The speed of the repulsion motor can often be changed by shifting the brush holder. This will cause the rotor poles to be induced closer or farther away, thereby changing the speed of the motor.

The repulsion start, induction run motor starts on the principle of repulsion; then once it has started rotating, it continues to run by induction. The brushes and commutator are only used during the starting of the motor. Once this type of motor has started running, manually lifting up the brushes would not affect the running performance of the motor. On other types of repulsion start, induction run motors, the entire brush assembly is lifted up by a centrifugal switch assembly. These motors are more complicated in construction, but they eliminate unnecessary brush wear.

Direct Current

Direct-current (dc) motors range in size from motors with a fraction of a horsepower to motors of several thousand horsepower. They are used quite extensively in elevators where excellent starting torque and speed regulation are required.

There are three types of dc motors: series, parallel, and compound. The basic difference in the types of dc motors is the type of circuit connection made between the field windings and the armature. In the series motor, the armature and field windings are connected in series. This series motor has the ability to start under heavy loads, and it varies its speed according to its load. Series motors without a load would run wild and literally would destroy themselves. Series motors are generally used in automobile starters, cranes, and hoists where high torque is needed at low speeds. Figure 3.14 shows a simple series motor circuit.

In a shunt, or parallel, motor, the armature and field windings are connected in parallel. This motor maintains a constant speed under varying loads but does not have as good a starting torque as a series motor does. Figure 3.15 shows an example

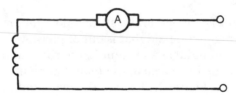

FIGURE 3.14   A simple series motor circuit.

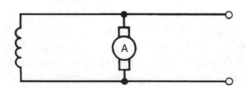

FIGURE 3.15   A simple parallel motor circuit.

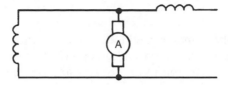

FIGURE 3.16   A simple compound motor circuit.

of a shunt motor circuit. Shunt motors are generally used in pumps and elevators where constant speeds under varying loads are needed.

The compound motor, or series-parallel motor, has its armature and field windings connected in a compound circuit. Both the field windings and armature are connected in series and in parallel. Figure 3.16 shows an example of a compound motor circuit.

As you might expect, compound motors combine the features of the series and shunt motors. They offer fairly good starting torque and fairly good speed regulation. Compound motors are generally used in factories to drive large power equipment where good starting and stalling torques are needed.

Universal

Universal motors can operate on either dc or ac. They generally come in a small fraction of a horsepower. The universal motor is a series-wound motor. It has excellent starting torque and variable-speed characteristics. Universal motors are generally used in vacuum cleaners, sewing machines, food mixers, fans, hair dryers, and other appliances in the home. Figure 3.17 illustrates a simple series circuit of the universal motor.

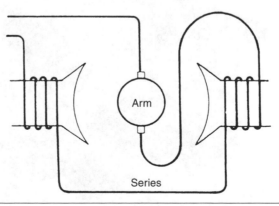

FIGURE 3.17   A simple series circuit of a universal motor.

Polyphase

The basic polyphase motor used today is an ac, three-phase induction motor. These motors range from a fraction of a horsepower to several thousand horsepower. Most three-phase motors are found in industry and range from 10 to 100 hp. Figure 3.18 shows a polyphase motor.

Three-phase motors require little maintenance and repair and are very rugged in construction. The three-phase motor contains a number of coils divided into separate windings, called phases. Each phase has an equal number of coils in its group. The three groups of coils, or phases, are arranged in either a star or a delta connection. Figure 3.19 displays a diagram of a star and a delta connection.

When a three-phase current is applied to the stator windings, a rotating magnetic field is induced within the metal bars of the squirrel cage rotor. An induced magnetic field will cause the rotor to rotate. The continuous flow of three-phase current, which is 120° apart, maintains the rotor revolving because of induction. Three-phase motors have varying degrees of torque, speed, size, and enclosures; therefore, they are quite versatile. They are generally used in driving industrial equipment.

Synchronous

Synchronous motors are induction motors that operate at a constant synchronous speed. This synchronous speed is determined by the frequency of the power source and the number of motor poles. Synchronous motors have a wide variety of shapes,

FIGURE 3.18   A typical three-phase motor. (*Courtesy Marathon Electric Co.*)

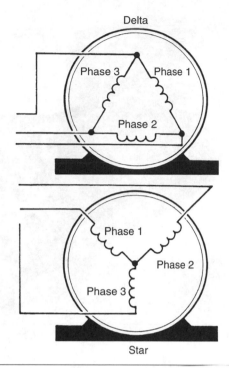

FIGURE 3.19 A simplified diagram of a star and delta motor connection.

sizes, and applications. They are made as small as a fraction of a horsepower for a small electric clock or as large as 3000 hp to drive a steel mill.

Synchronous motors are only suitable for operating on ac. The speed is constant and will not change with varying loads. The basic principle of the synchronous motor is that the rotor, which contains salient poles, rotates in step with the rotating magnetic field. The rotor tends to "lock in" with the rotating magnetic field and remains in constant rotation, uninterrupted by varying loads. Some rotors of synchronous motors need to be excited by dc; others do not. The excitation of the rotor forms definite poles in the rotor, which lock in step with the rotating magnetic field. Often, this type of motor will have a small dc generator attached to the shaft of the motor where it supplies dc to the rotor.

Gearmotor

The gearmotor is a special motor used for reduced-speed and reduced-power applications (Fig. 3.20). Gearmotors eliminate the need for external chains and belts and offer precise speed reduction ratios and efficiency. The gearmotor can be either an induction or a repulsion type. The selection of a gearmotor is generally determined by the speed and torque output, rather than by the horsepower rating.

FIGURE 3.20 Cross section of a gearmotor. (*Courtesy Bodine Electric Co.*)

Many other factors can determine the selection of a gearmotor, such as the mounting, load, and braking requirements.

Three special kinds of gears used in the gearmotor are the spur, helical, and worm gears. The spur gear offers high-power application but is noisier than the other two types. The helical gear is less noisy than the spur gear and transmits a nearly constant motion. The worm gear has minimal noise, and high ratios can be obtained, although it is less efficient than the other two types. Gears are manufactured from both metallic and nonmetallic material. Nonmetallic gears are quieter but offer less strength than metallic gears.

Stepper

The stepper motor is used in digital motion control applications (Fig. 3.21). Some of these applications include computer printer, medical X-ray equipment, photo typesetting, business machines, and industrial process controls. Stepper motors offer fixed and precise movement, rather than continuous running motion, as do conventional motors. The operation of the stepper motor is based on the theory of induction. Figure 3.22 illustrates the basic operation of a four-step sequence stepper motor. The motor's shaft moves one step each time a switch is activated. One full revolution is achieved after all steps have been taken.

The drive system used for the stepper motor contains a pulse source and a translator. The pulse source is generally a computer or microprocessor.

FIGURE 3.21　A dc stepper motor. (*Courtesy Bodine Electric Co.*)

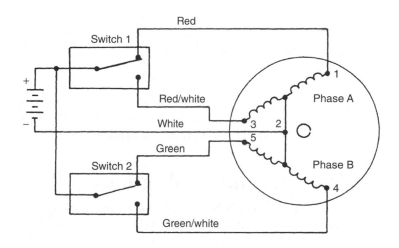

SWITCHING SEQUENCE*

STEP	SWITCH #1	SWITCH #2
1	1	5
2	1	4
3	3	4
4	3	5
1	1	5

*To reverse direction, read chart up from bottom.

FIGURE 3.22　Four-step switching sequence. (*Courtesy Superior Electric Co.*)

FIGURE 3.23   Translator used to convert pulses to switching sequence for the stepper motor. (*Courtesy Bodine Electric Co.*)

The translator, energized by a dc power supply, converts the digital pulses to the proper switching sequence for the stepper motor (Fig. 3.23). The stepper motor, in turn, converts the electrical information to mechanical motion to operate the load.

Special Motors and Applications

Special electric motors include brushless, linear motion, high-peak force, vertical and horizontal motors, and advanced electric motors for special applications. For example, NetGain Technologies, LLC, uses an advanced, high-power electric motor in an electric race car. This dc electric motor operates at 336 volt (V) and 2000 amperes (A) which propels the race car to estimated speeds of 150 miles per hour (mph) and with a torque of over 2000 foot-pounds (ft lb). Also, one of the rapidly growing areas in electric motor technology is motor drives and their power electronic converters.

These systems allow for more complex and efficient motor control functions as used in vehicle, manufacturing, domestic, and industrial applications. For example, the Power Electronics and Motor Drives Program at the Illinois Institute of Technology is developing research advances in power electronics, electric motor drives, switched reluctance motor drives, adjustable speed drives, and brushless dc, which can be used in numerous applications such as robotics, electric vehicles, computer technologies, telecommunications, and modern industrial automation systems.

Types of Generators

The construction of generators is very similar to that of electric motors. A motor converts electric energy to mechanical energy, and the generator converts mechanical energy to electricity. Generators have a wide range of applications and can be found in airports, hospitals, transportation vehicles, computers and telecommunications, construction sites, and industrial operations. The basic two types of generators are ac and dc. The ac generators are often referred to as *alternators*, because they produce ac. And dc generators, as expected, produce a dc. Figure 3.24 illustrates a cutaway of an electric generator.

Most generators are constructed utilizing permanent magnets with single-piece four-pole rotor lamination, a digital voltage regulator, surge and overload protectors, exciter stator and armature, rectifier assembly, bearings, and housing. They are generally specified by their frame size, kilowatt (kW) output, and other National Electrical Manufacturers Association (NEMA) insulation classes.

Generators also serve a critical function as backup standby power sources for lighting, computer control rooms, public areas, elevators, temperature controls, and health-support systems. When the main power source terminates, a control system activates the standby generator.

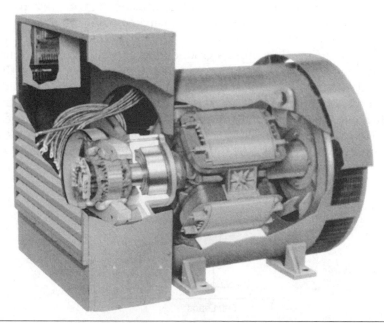

FIGURE 3.24   A cutaway of an electric generator. (*Courtesy Marathon Electric.*)

Motor Repair

When you diagnose motor troubles, it is important to follow a logical, systematic procedure, in order to eliminate wasteful time, tests, and replacement of parts. Most of the common motor troubles can be easily checked by using simple test equipment. It is also important that the troubleshooter have a complete understanding of the use of this equipment to analyze and repair the motor.

The ideal procedure in analyzing motor troubles should begin with an auditory and visual inspection. First, inspect the motor for obvious troubles, such as broken end bells, frames, tight or frozen shaft, or burned lead wires. Any one of these can allow the troubleshooter to quickly isolate a problem. A noisy motor or frozen shaft can be an obvious sign of defective bearings. Check the motor for defective bearings by turning the shaft and trying to move the shaft up and down. A shaft that does not rotate, feels sloppy, or has considerable play when moved up and down probably indicates bad bearings.

The basic tools used in troubleshooting electric motors include the following:

1. Test lamp
2. Amperage measurements
3. Growler
4. Insulation resistance meter (megohmmeter)

Before the troubleshooter tries to run the motor, she or he should first test the motor for such faulty circuits as grounds, shorts, and opens. As you recall, a ground results from the windings making electrical contact with any iron part of the motor. Common grounds result from poorly insulated wire contacting the stator and end bells. A motor that has a grounded winding may blow fuses, run hot, or lack power. Shocks can be obtained from a grounded motor; therefore, extreme care should always be taken when you are testing a grounded motor. To test the motor for a ground, connect one lead of the test lamp to one of the motor leads. Connect the other test lamp lead to the stator or to the frame of the motor. If the lamp lights, this indicates that the motor is grounded. Figure 3.25 illustrates this test procedure.

Here again, as you recall, an open circuit results from a break in the motor circuit, which prevents current from flowing in a complete path. A motor will not

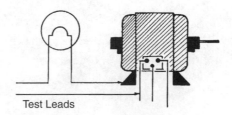

Test Leads

FIGURE 3.25 Using a test lamp to check a motor for grounds.

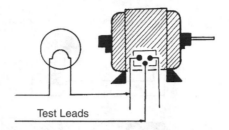

FIGURE 3.26   Using a test lamp to check a motor for opens.

run with an open circuit. Many times, if one of the three phases of a motor is open, the motor has no action; it simply hums. To determine if the motor has an open circuit, connect the test lamp leads to the lead wires of the motor. If the lamp fails to light, the motor has an open circuit. If the lamp lights, the circuit is complete. Figure 3.26 illustrates this test procedure.

A short circuit in a motor is caused by a defect in the motor circuit where two wires of the circuit connect and cause a bypassing of the normal current flow. An ammeter (use a clamp-on type) can often detect a short-circuited motor. If the amperage reading exceeds the rated amperage value stated on the nameplate of the motor, the motor may be short-circuited. Keep in mind that other factors, such as low line voltage, bad bearings, or an overloaded motor, can cause the motor to draw excessive current.

A hot, smoky motor that blows fuses is probably short-circuited. Also, a motor with a short circuit may heat rapidly, run hot, fail to start, or run slowly. A growling noise often accompanies a shorted motor. If power is applied to a single-phase motor and the motor just hums, spin the shaft with your hand. If the motor starts running, the problem lies in the starting circuit. However, if the motor starts, but runs unevenly, slows down, then starts again, the problem lies in the running circuit.

Besides using a test lamp, grounds and opens can best be checked by using a megohmmeter (Fig. 3.27). To test a motor for a ground, connect one lead of the megohmmeter to the motor frame and the other end to one of the motor terminals. A grounded motor will have a zero or near-zero reading. To test for an open circuit, connect the megohmmeter to each pair of phases of the motor. An open motor will show a high reading on the megohmmeter. An ohmmeter can also be used for testing a motor for grounds and open circuits.

Another way to check field windings for a short circuit is to dismantle the motor and apply a small voltage to the stator winding. Each coil now becomes an electromagnet. Place a screwdriver shank by each coil and slowly draw it away, noting the magnetic pull. Each coil should have the same amount of pull. The coil with the least magnetic pull probably is short-circuited. Also, if you touch each coil and find that one is hotter than the rest, the hottest coil is probably short-circuited.

Before you dismantle the motor, mark the two end bells and frame in reference to one another. Generally, two punch marks indicate the front end of the motor,

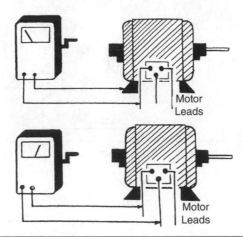

FIGURE 3.27   Using a megohmmeter to test a motor for grounds or opens.

and one punch mark indicates the rear of the motor. Marking the motor enables the troubleshooter to correctly reassemble it. The front-end shaft should also be marked. This can be done with a punch mark or by scratching an X onto the end of the shaft. The base should also be marked in reference to the front end bell of the motor. Many troubleshooters scratch a mark on the rotor shaft with a knife or small file, indicating the correct position of the rotor. This scratch or mark is usually placed on the front-end shaft closest to the front end bell.

To correct a ground in a motor, it is usually necessary to dismantle the motor and trace the windings to locate the part of the circuit that is making contact with the metal of the motor. After you locate and correct the problem, it is generally common practice to clean the windings if they appear dirty or charred. They should be cleaned with a solvent. Reinsulate the windings by spraying or brushing them with a coat of epoxy or other air-drying insulating enamel. If it appears that the grounded motor was caused by moisture, dry out the motor in a warm oven or with a fan.

Common causes of open circuits are a defective or poorly aligned centrifugal switch, a defective capacitor, or a broken wire in the motor circuit. In locating the open circuit of a motor that has a capacitor, it is generally a good idea to check the capacitor first. There are several ways to check the condition of a capacitor. One way is to substitute for it a new capacitor having the same rating. If the open circuit no longer exists, the capacitor was at fault. Another method of checking a capacitor is by the spark test. Connect the capacitor across the terminals of a 115-V line for just a second. After you remove the 115-V line, use a screwdriver blade to short-circuit the two terminals of the capacitor (Fig. 3.28). A good capacitor will show a spark. Absence of a spark indicates a defective capacitor.

To test the capacitor for a ground, a simple test lamp can be used. Connect one of the leads of the test lamp to one of the terminals of the capacitor. Connect the other test lamp lead to the metal case of the capacitor. If the lamp lights, the

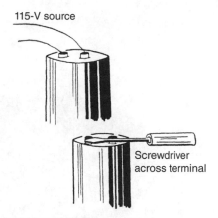

115-V source

Screwdriver
across terminal

FIGURE 3.28   A spark test is performed by using the blade of a screwdriver to short-circuit the terminals of a charged capacitor.

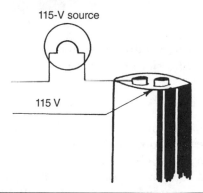

115-V source

115 V

FIGURE 3.29   Using a test lamp to check a capacitor for a ground.

capacitor is grounded and should be discarded (Fig. 3.29). Other methods used to test a capacitor include the use of an ohmmeter, a capacitor tester, or a combination ammeter-voltmeter.

The centrifugal switch often causes a single-phase motor to be open. The switch should be checked to see whether the contacts are closing. If the contacts are not closing, washers may need to be added to the rotor shaft to correct the problem. Also, check the condition of the centrifugal switch, as it may be in bad condition and may need to be replaced.

The motor windings should also be inspected for possible breaks. One or more broken wires may cause the open circuit. If the windings are badly burned or broken and beyond repair, it may be necessary to replace the windings of the motor.

A short circuit in the stator windings can be checked with an internal growler. Place the growler on the laminations of the stator at one end of a coil. Together, the

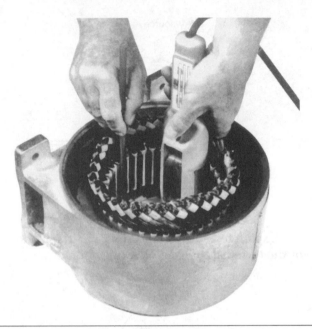

FIGURE 3.30   Checking a stator for a shorted coil by using an internal growler. (*Courtesy Crown Industrial Products Co.*)

growler and the coils of the stator act as a transformer. The coils of the growler act as the primary circuit, and the coils of the stator act as the secondary circuit. The growler, which may have a built-in feeler blade, vibrates excessively when placed on the short-circuited coil (Fig. 3.30). When a motor has tested out as short-circuited, either it should be discarded or the windings should be replaced.

A motor that has a short-circuited armature may jerk, severely vibrate, hum, growl, fail to run, or blow a fuse. The shorted coil of an armature can often be identified by its discoloration and insulation breakdown.

The armature of a motor can be checked for short circuits by using an internal growler. Place the armature on the growler with a strip of metal placed on top of the armature. Rotate the armature. If the metal strip rapidly vibrates, the armature is shorted. Figure 3.31 illustrates this principle, using an internal growler and a hacksaw blade.

The armature can be tested for a ground by the use of a test lamp. Place one lead of the test lamp on the commutator and the other lead on the armature shaft. If the lamp lights, the armature is grounded (Fig. 3.32).

Whereas the actual rewinding of stators and armatures is beyond the scope of this book, the decision to rewind the product is an economic one. Generally, it is not as cost-effective to rewind smaller products as the larger ones. Advances in stator and armature winding equipment have produced affordable, high-quality windings and a large variety of configurations to offset the needs of motor applications.

FIGURE 3.31   Checking an armature for a short by using an internal growler. (*Courtesy Crown Industrial Products Co.*)

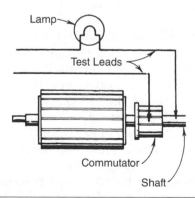

FIGURE 3.32   Using a test lamp to test an armature for a grounded circuit.

FIGURE 3.33   Winding an armature using a computerized electronic/pneumatic armature winder. (*Courtesy Globe Products, Inc.*)

Modern-day winding machines use computerized touch commands and display screens that allow easy, flexible, and high-production-rate operations. Figure 3.33 shows an example of a programmable electronic pneumatic armature winder.

Bad bearings in a motor can cause the motor to run noisily, hot, or not at all. Generally, bad bearings cannot be saved by cleaning or reconditioning. They are usually replaced. If a ball bearing fails to rotate smoothly, it should be replaced. To remove a ball bearing, a gear puller or a special bearing-removing tool that knocks off the bearing is commonly used. To install a new ball bearing, a drift punch or a piece of round stock is often used to press the bearing on the shaft. The sleeve bearing is generally removed; the new one is installed by using a piece of round stock and a press. Sometimes the sleeve bearing is removed from the end bell by using the special tool illustrated in Fig. 3.34. Often, the inner diameter of a new bearing is made slightly smaller. It is then necessary to ream the bearing to the correct size, using a tool called a reamer.

If a repulsion motor fails to start, severely sparks at the brushes, or runs intermittently or with low power, the causes may be a dirty or worn commutator, worn brushes, misaligned brushes and brush holder, or defective brush spring. To correct any of these problems, check the condition of the brushes, brush holder and spring, and commutator.

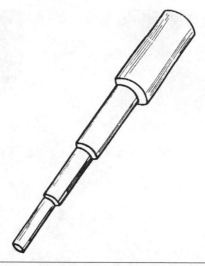

FIGURE 3.34 A special tool used to remove sleeve bearings from an end bell.

If the brushes are badly worn, they will need to be replaced. When replacing brushes, always use the correct replacement. The exact curvature and size of the brush are extremely important for satisfactory operation of the motor.

Before you insert the brushes, make sure the brush holder is clean, allowing the brushes to move freely. Also, the spring tension should be sufficient to keep a constant pressure, allowing the brush to make good contact with the commutator. Once the brushes have been secured in place, the last step is to match the curvature of the brushes with the commutator. This is called *seating* and is done using a brush seater stone.

To seat the brushes, operate the motor at normal speed. Place the brush seater stone directly against the rotating commutator. Make sure the brushes are riding firmly against the commutator. Hold the brush seater stone against the commutator for only a few seconds. The stone will give off a powder of granules that ride under the brush, shaping the brush to best match the curvature of the commutator. Never overseat brushes! This will cause excessive wear to both the commutator and the brushes. Last, it is a good idea to blow off or vacuum the brush area to remove any powder granules from the commutator. The method used to seat brushes is illustrated in Fig. 3.35.

If the spring tension or position of the brush holder is wrong, the motor may not operate properly. Check the spring tension. If the spring is not pressing the brushes firmly against the commutator, the brushes should be replaced. Make sure the brush holder is in a position to allow the brushes to make a firm, even ride on the commutator.

If the commutator appears to be out of round or dirty or contains high mica, it will need to be resurfaced and undercut. A commutator can be resurfaced with a hand stone or by turning it on a lathe. Depending on the condition of the

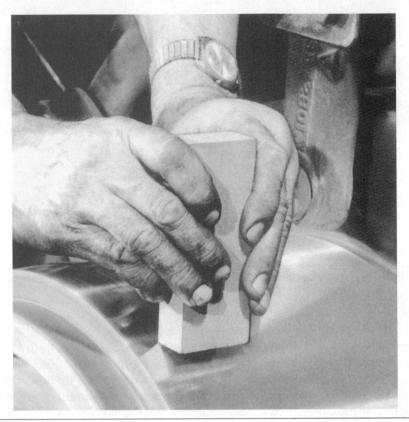

FIGURE 3.35   The method used to seat a commutator by hand with a stone. (*Courtesy Ideal Industries, Inc.*)

commutator, hand stone grinding tends to be the more effective and quicker method (see Fig. 3.35). This method allows the commutator to be resurfaced with the motor operating at its normal speed. Lathe turning, on the other hand, requires the motor to be taken apart at the armature and placed on a lathe where it is "turned down." Never resurface a commutator more than is needed for a clean, concentric surface. Overgrinding or resurfacing will completely wear out the copper bars of the commutator.

If a commutator has been resurfaced, the mica should also be reconditioned. This process is called *hand slotting* or *undercutting*. Undercutting can be done by hand, using a hacksaw blade or a direct motor-powered mica undercutter. Hand slotting with a hacksaw blade is not done much today because it is more time-consuming and less efficient. Undercutting is performed to remove the mica between the commutator bars to a depth approximately equal to the width between the bars. The mica is removed to allow the brushes to ride smoothly on a commutator that is uniform, concentric, and free from ridges and burrs. Figure 3.36 shows an illustration of a direct-drive universal mica undercutter.

FIGURE 3.36   Removing mica from the commutator slots with a direct-drive mica undercutter. (*Courtesy Ideal Industries, Inc.*)

Often, a single-phase motor needs its shaft rotation reversed. The direction of rotation of most single-phase motors can be changed by reversing the starting or running wires of the motor (Fig. 3.37).

To change the direction of rotation in a shaded-pole motor, it is usually necessary to dismantle the motor and reverse the stator end for end. This is done because the direction of rotation is dependent on the shading coil effect from the main pole to the shaded pole (Fig. 3.38).

To change the direction of rotation of a dc motor, simply reverse the polarity of either the field poles or the brushes.

The rotation direction of a three-phase induction motor can easily be changed by switching any two of the three motor leads. Usually the outer two leads are interchanged (Fig. 3.39).

When you reassemble a motor, it is important that the wires not be in contact with the iron part of the motor. It is extremely important not to pinch any wires between the end bells and frame. This will result in a ground or short circuit.

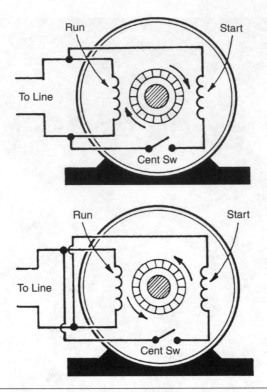

FIGURE 3.37   Reversing the rotation of a single-phase motor.

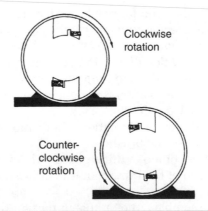

FIGURE 3.38   Reversing the rotation of a shaded-pole motor.

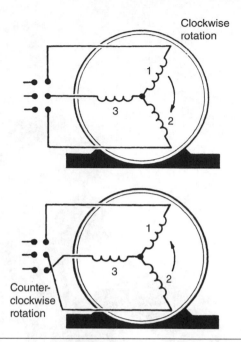

FIGURE 3.39   Reversing the rotation of a three-phase motor by switching the outer two leads.

When you reassemble the motor, it is important to align all the marks punched before the motor was dismantled. It is common practice to tap the end bells with a rubber mallet or rawhide hammer to prevent misalignment.

Problems unique to gearmotors relate to improper lubrication, bad seals, gaskets, and gears. Oil is used for lubricating most gearmotors. Oil provides constant lubricating film to the gear teeth. However, smaller gearmotors use grease instead of oil because of sealing problems. Sufficient, clean lubricant must be maintained; otherwise, damage will result to the seals and gears. Also, excessive and abnormal operation (e.g., too low or high an ambient temperature) can reduce the life of gears.

Some of the problems associated with stepper motors include bad bearings, short-circuited windings, defective transmission lines, and drive system malfunctions. It is important to first isolate the problem by determining whether it is the stepper motor, transmission line, or drive system. The stepper motor can be checked for shorts, opens, and grounds similar to a conventional motor. Substituting a new stepper motor for the suspected defective one is also a good way to check its performance.

Another servicing operation for motors is *core loss testing*. Core loss testing provides information about the efficiency of stators, rotors, and armatures. Core loss testing records data in watts per pound (W/lb). The degree of hysteresis and eddy current are factors determining the losses. These testers can also help locate hot spots and defects that can be repaired. Figure 3.40 illustrates a setup procedure for

FIGURE 3.40 A setup procedure for measuring core losses using a core loss tester. (*Courtesy Lexseco, Inc.*)

measuring core losses using a core loss tester. It is important that transformers or other looped-type windings not be tested, because the coils can produce dangerously high voltages. Always follow the manufacturer's instruction manual.

Generator Repair

Troubleshooting generators is similar to that of electric motors. Begin by discussing the malfunction with any operators using the generator. Typical problems of generators are a blown regulator fuse, inoperative regulator, low or high voltage output, or fluctuating voltage. Remember that high residual voltages may be present within the regulator. Make a thorough inspection of the generator, looking for broken connections, pinched or broken wires, corroded terminals, foreign objects, or burned or frayed components. A common occurrence is foreign particles such as dirt and industrial debris entering through the cooling screen and clogging or disrupting the generator.

With the power off, an ohmmeter can be used to check the regulator fuse. The input and output voltages can be measured to help determine the operation of the regulator. If the regulator is found to be inoperative, it may need to be adjusted or replaced. A typical problem arises when a load is applied to the generator and the output voltage is too low or fluctuates. Assuming the meters are accurate and there are no defective or loose connections, it may be necessary to dismantle the generator and test its components.

When you dismantle the generator, remove the power and mark and identify all wires and parts for reinstallation. Use appropriate hoist, straps, lifting fixtures, and other equipment to prevent damage to the parts, especially for heavy, large generators. Figure 3.41 shows a hoist and strap being used to remove an exciter stator. Always check the stator for loose, frayed, or burned windings, and measure the resistance between the leads and compare with the manufacturer's values. For example, 20 V of resistance may be expected. Zero ohms would indicate a short circuit, and infinite resistance would indicate an open winding. Also, check for grounds between the windings and frame, using a megohmmeter.

When removing heavy generator rotors, you should use a combination of a hoist and a special lifting fixture (Fig. 3.42). Remove the rotor carefully by guiding it out without touching the associated parts, to prevent damage to the rotor or generator windings.

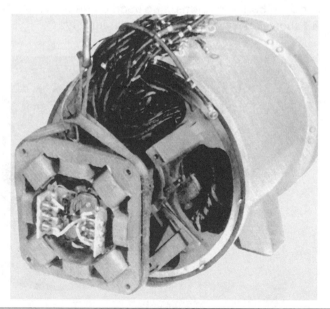

FIGURE 3.41   Removing the exciter using a hoist and strap. (*Courtesy Marathon Electric.*)

FIGURE 3.42 Using a hoist and special lifting fixture to remove the main generator rotor. (*Courtesy Marathon Electric.*)

A common problem is defective diodes. As discussed before, test the diode by measuring its resistance. The diode should show a high reading in one direction and a low reading when the meter leads are reversed. A short-circuited diode reads low resistance in both directions. The open diode reads high in both directions. Figure 3.43 shows an exciter and rectifier assembly.

FIGURE 3.43 Typical generator exciter armature and rectifier assembly. (*Courtesy Marathon Electric.*)

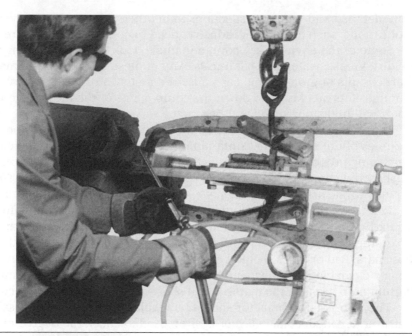

FIGURE 3.44   Removing a bearing with a bearing puller and heat. (*Courtesy Marathon Electric.*)

Remember that the resistance meter may not have enough internal voltage to conduct the diode, and a false diagnosis may result. Never use a megohmmeter to check diodes or the regulator. If the diode or rectifier needs replacement, observe the polarity markings. Never apply too much torque to the terminal nuts.

If bearings need to be replaced, use an appropriate bearing puller (Fig. 3.44). Sometimes bearing caps stick and need to be heated. Also, always replace the old grease in the bearing cap cavity, often to half its capacity. Do not overfill the cavity.

Preventive Maintenance

The life of an electric motor or generator is often determined by the type of maintenance it receives. A poorly maintained motor or generator can generally be detected by its dirty, corroded appearance. A thorough maintenance program includes the periodic inspection, recording, and servicing of these devices. A minor adjustment, replacement of the bearings, or a simple cleaning of the motor and generator can prevent expensive and time-consuming repairs later on.

The frequency of breakdowns and abuse of electric motors can be reduced through periodic inspections. All inspections should be accompanied by a logbook, where the condition and services administered can be recorded. Some motors need more detailed and regular inspections than others. Motors used in very dirty or wet applications, such as those in industry, need to be inspected more regularly.

Some of the obvious, key contaminants to look for when you inspect motors are dirt, grease, water, and chemicals. Any of these key contaminants can cause short circuits or grounds and can cause the motor to run excessively hot. Also, these contaminants quickly wear down the components of the motor. Eventually, the motor will break down and will need to be repaired.

Dirt and grease and other contaminants have a tendency to block the ventilating openings of the motor. The contaminants often accumulate on the commutator, preventing proper electrical conduction. They also add to the wearing down of the brushes and commutator. Excessive water accumulation can cause short circuits or grounds in the armature or the stator windings, and the motor may break down.

Repulsion-type motors should have their brushes and commutator inspected regularly. Check the brush tension and the alignment of the brush and holder. Gently tap on the commutator bars with a small rawhide mallet to test for the presence of loose bars. Bad brushes should be replaced. A dirty commutator should be cleaned by using a clean cloth, sandpaper, or cleaning stone. *Note:* Never use emery to clean a commutator, because the material will short-circuit the bars on the commutator.

The rotor shaft of a motor should be regularly inspected for misalignment. An instrument called a *dial indicator* is often used to check the degree of rotor shaft misalignment.

Switch contacts should be given regular attention by cleaning and reshaping the contacts with sandpaper, a file, or a cleaning stone whenever they are dirty or disfigured. Most electrical and electronic distributors sell chemical contact cleaner.

All bolts and nuts should be kept tight, and all motor lead wires and windings should be inspected for dirt, breaks, and worn insulation; this prevents major breakdowns later on. It is common practice to clean and reinsulate motor windings by using an air-drying epoxy or some other insulating enamel.

Regular inspection for bearing wear is essential in preventing motor breakdowns. Sleeve bearings should be regularly oiled but should not be overlubricated. Ball bearings can be lubricated with either oil or grease, depending on the manufacturer's specifications. Grease, which is a combination of oil and soap, is generally the lubricant for ball bearings. Do not forget that overlubrication of ball bearings can cause overheating, resulting in premature ball bearing failure. Squeaky or tight bearings should be replaced with new bearings.

Incorrect motor end play should be given immediate attention. The motor end play can be tested by pulling and pushing the motor shaft back and forth. Maximum end play is usually about 1/64 inch (in). Incorrect end play can be corrected by adding or removing washers, replacing bearings, lubricating bearings, or tightening nuts or bolts.

Any decrease or increase in motor temperature, increase in noise, discoloration, or disfiguration in appearance is a common danger sign of motor troubles. These conditions should be given immediate attention and the causes identified.

If it is determined that the motor cannot be repaired or that it is the wrong motor for the operation, a replacement motor is necessary. To select the correct motor, several factors need to be considered. Match the operating and load requirements with the motor design. The National Electrical Manufacturers Association (NEMA) provides standard guidelines for motors, such as torque, revolutions per minute (rpm), horsepower, enclosure, and mounting dimensions. For example, if the operation requires the motor to run for long periods, a continuous-duty motor should be selected. An intermittent-duty motor can be used for short-running operations. In other cases where temperature or excessive turning force are factors, the proper lubricating and torque features will be important.

Energy efficiency is another factor in motor selection. For example, the difference in annual savings between a 50-hp motor and a 75-hp motor can be substantial in both electricity savings and purchase cost if the 50-hp motor is sufficient. Likewise, an overloaded motor will draw excessive current and eventually burn out if underrated for the operation. Always consult the manufacturer's motor selection and energy-efficiency charts before obtaining a motor.

Self-Examination

Select the best answer:

1. Basically all electric motors operate on the principle of repulsion or
   A. Magnetism
   B. Capacitance
   C. Resistance
   D. Induction
   E. Semiconduction
2. The basic tools of troubleshooting electric motors consist of amperage measurement, test lamp, growler, and
   A. Substitution
   B. Heating and freezing
   C. Bridging
   D. Megohmmeter
   E. Oscilloscope
3. A hot, smoky motor is a good indication of
   A. A ground
   B. An open circuit
   C. A short circuit
   D. (A) and (B)
   E. (B) and (C)
4. The best way to test a capacitor when used in a 115-V electric motor is by
   A. Bridging
   B. Spark test
   C. Voltmeter
   D. Ammeter
   E. All the above

5. The direction of a three-phase motor can best be changed by
   A. Switching two of the three leads
   B. Dismantling the motor and switching two leads
   C. Changing the voltage
   D. Reversing the stator
   E. None of the above

6. When you are cleaning a commutator, never use
   A. Sandpaper
   B. Clean cloth
   C. Both (A) and (B)
   D. Emery
   E. None of the above

7. An instrument often used to check the degree of motor shaft misalignment is the
   A. Voltmeter
   B. Clamp-on ammeter
   C. Growler
   D. Megohmmeter
   E. Dial indicator

8. The bearing that should be oiled regularly is the
   A. Sleeve bearing
   B. Ball bearing
   C. Roller bearing
   D. Both (A) and (B)
   E. None of the above

9. Grease is basically a combination of
   A. Water and soap
   B. Vaseline and soap
   C. Gas and water
   D. Oil and soap
   E. None of the above

10. Maximum end play is usually about
    A. 1/64 in
    B. 1/8 in
    C. 1/4 in
    D. 1/2 in
    E. 1 in

11. The gear that provides the least amount of noise is the
    A. Spur gear
    B. Helical gear
    C. Worm gear

12. The gear that offers very high-power application is the
    A. Spur
    B. Helical
    C. Worm
    D. All the above

13. The lubrication used for most gearmotors is
    A. Oil
    B. Grease
    C. Both (A) and (B)
    D. None of the above
14. The selection of a gearmotor is generally based on
    A. Speed
    B. Torque
    C. Both (A) and (B)
    D. None of the above
15. A drive system used for the stepper motor is the
    A. Microprocessor
    B. Translator
    C. Computer
    D. All the above
16. The device that converts mechanical energy to electricity is
    A. Motor
    B. Core loss tester
    C. Generator
    D. Stepper
17. The ac generator is called
    A. An ac motor
    B. An ac stepper
    C. A regulator
    D. An alternator
18. The degree of hysteresis and eddy currents contributes to
    A. Core loss
    B. Regulation differential
    C. Surge suppression
    D. Armature excitation
19. High residual voltages may be present within the
    A. Regulator
    B. Stator
    C. Droop circuit
    D. Rectifier
20. When you are checking the diode, two high resistance readings in both
    directions indicate
    A. An open circuit
    B. A short circuit
    C. A ground
    D. None of the above

Questions and Problems

1. Explain the basic difference between an electric motor and a generator.
2. Explain the basic principle of the electric motor.
3. Name the fundamental parts of an electric motor.
4. Explain the difference between a repulsion and an induction motor.
5. Name several types of electric motors.
6. What is an open circuit?
7. What is a short circuit?
8. Explain how to test a motor for an open circuit, ground, or short circuit.
9. Explain the best way to dismantle a motor.
10. Explain the different ways to check a capacitor.
11. Explain the difference between an internal and external growler.
12. Name two types of bearings.
13. Explain how to clean a dirty commutator.
14. List some common problems and causes of electric motor breakdowns.
15. What are the different types of gears used in the gearmotor?
16. Why is grease rather than oil used to lubricate small gearmotors?
17. Explain the operation of a stepper motor.
18. Describe the drive system for the stepper motor.
19. List the applications of the stepper motor.
20. Explain the purpose of a core loss tester.
21. Explain how to test a diode of a generator.
22. List a precaution to consider when testing a regulator.
23. Describe several problem symptoms of a generator.
24. List several components of a generator.

Chapter 4

Troubleshooting Industrial Controls

The need to improve industrial control systems is an ever-changing process. Automated devices, such as electronic, hydraulic, pneumatic, and advanced computer-controlled systems, are being developed at rapid rates. Modern-day control systems offer reduced personnel and production costs and increased safety, manufacturing efficiency, and quality control. Industrial power controls are used to regulate such devices as electric motors, lights, robotic operations, audible signals, heaters, conveyors, machine tools, pumps, medical diagnostic and therapeutic equipment, and production operations.

For example, every electric motor must have a control system, whether it is a simple switch to start and stop it or a complex microprocessor for stepping operations. Motor controls provide a wide variety of functions for the motor, such as starting and stopping, reversing, accelerating, decelerating, braking, and performing time-controlled operations. The motor control is as important to the motor as the power that drives it.

In Chap. 3, a review of the most common types of electric motors and their repair was presented. The purpose of this chapter is to examine how electric motors and other power devices can be controlled. The basic theory of industrial power controls is presented, along with types of controllers, testing procedures, and preventive maintenance.

Fundamentals

The basic functions of a motor control are to provide operations such as starting and stopping, protecting, sequencing operations, reversing, and speed variations. The simplest form of a motor control device is a single-pole single-throw switch. This switch, as illustrated in Fig. 4.1, has the responsibility of controlling current to start and stop this alternating-current (ac) squirrel cage induction motor.

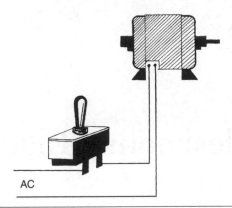

FIGURE 4.1   A single-pole single-throw switch is a simple motor control device.

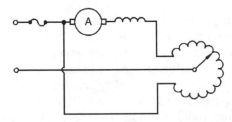

FIGURE 4.2   A fuse and a variable transformer act as a control device.

Besides adding protection to the motor, the motor control helps protect the operator. Figure 4.2 shows how a simple fuse and variable transformer act as a control device. The fuse acts to protect the motor and operator. The variable transformer acts as a speed control for this series-wound direct-current (dc) motor.

Control systems are commonly of two types: open-loop and closed-loop systems. For example, for an outdoor bonfire, the amount of wood thrown on the fire controls the heat—or it is an open-loop system. If the wood-burning stove is controlled with a damper, it is a form of closed-loop system. This feedback action of increasing or decreasing the damper provides better regulation than an open-loop system. Sophisticated commercial building systems work on a closed-loop principle and make use of thermostats, electric motors and fans, regulators, and programmable operations to control the heating.

Many motor controls operate on the principle of electromagnetism. If insulated wire is wrapped around an iron bar and the two ends of the wire are connected to an electrical source, we have an *electromagnet* (Fig. 4.3). Reversing the current flow in an electromagnet reverses the polarity of the iron bar.

A coil of wire connected to a battery has a magnetic flux that surrounds the coil much like that in a permanent magnet. Figure 4.4 illustrates this magnetic flux. This magnetic flux is the key to motor control operations. It creates the mechanical movements that allow for on-and-off operations.

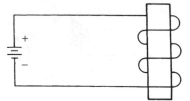

FIGURE 4.3 An iron bar wrapped with wire and supplied with voltage forms an electromagnet.

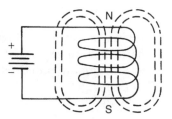

FIGURE 4.4 A magnetic flux surrounds an electromagnet.

The relay is a typical motor control device. It is a magnetic device that is used to open or close circuits. Figure 4.5 illustrates a simple relay schematic diagram. The solenoid is a mechanical relay that uses this magnetic principle to activate a metal plunger, creating linear mechanical motion that provides on-and-off operations. There are hundreds of applications for relays and solenoids. Also, there are hundreds of configurations of relays, but basically they are called either *electromechanical* or *solid-state*.

Circuit breakers are a special type of relay often thought of as a manual switch. They are frequently used in homes, businesses, and industries to protect electric

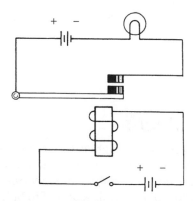

FIGURE 4.5 A simple electric relay.

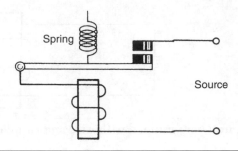

FIGURE 4.6   A simple circuit breaker.

circuits from excessive currents or overloads. Figure 4.6 illustrates the magnetic portion of a simple schematic diagram of a circuit breaker. A large amount of current will cause the electromagnet to pull the lever down, thus separating the contacts and breaking the circuit.

Types of Controllers

There are a wide variety of motor and industrial controllers and power control devices. Each has its own particular characteristics. Some of the more popular types of controllers include the following:

1. Overload protectors
2. Manual starters
3. Magnetic starters
4. Reversing magnetic starters
5. Lighting contactors
6. Pushbutton stations
7. Limit switches
8. Drum switches
9. Timers
10. Electronic drives
11. Programmable controllers
12. Sensors

Overload Protectors

Most of the motor controllers, such as the manual starter, magnetic starter, and reversing starter, have some type of overload protection. Typical overload protectors are melting alloy thermal, bimetallic, and electronic. The electronic overload protectors are the most popular, bimetallic are being used less frequently, and the melting alloy relay is used the least.

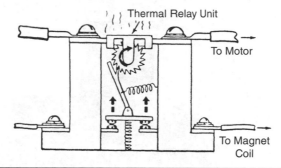

FIGURE 4.7   Melting alloy thermal overload relay. (*Courtesy Square D Co.*)

The overload device called the *melting alloy thermal overload relay* is illustrated in Fig. 4.7. Basically, when too much current is drawn in an overload situation, the *solder pot*, which is made of brass, contains an eutectic alloy that melts. This solder pot, which kept the ratchet from turning, can no longer do so. The melting alloy allows the ratchet to turn. The turning action opens the control contacts, and the motor stops running. When the solder pot cools down, the relay must be manually reset. Repeated tripping and setting of the relay generally do not affect its calibration. The different types of these relays include the slow, standard, and quick-trip. Also, more sophisticated melting alloy overload relays contain isolated alarm contacts, which allow the relay to be used with the starter for computer applications requiring isolated contacts.

The bimetallic thermal overload relay works on the principle of a bimetal strip that expands under heat (Fig. 4.8), opening the contacts. These relays come in automatic or manual reset. They operate, in principle, as a thermostat does. As the temperature cools, the relays reset. It is important that the troubleshooter turn off the power to the automatic reset relay prior to working on the device. If not, the device could turn on and injure the troubleshooter when the relay cools down and resets. Overload relays are available in several different current ratings. Accessories include ambient temperature and noncompensated versions, alarm contacts, and solid-state monitors. This rating should always be matched with that of the operating device. Many motor relays contain self-adjusting setscrews that can be adjusted to match the precise overload protection level.

Electronic (solid-state) overload protectors are very popular. This multipurpose overload and dc-sensing relay is designed for dc motors (Fig. 4.9). When a predetermined current is reached, the relay trips. Both manual and automatic reset operations are available. These overload protectors also are able to sound an alarm and initiate other functions, preventing unnecessary damage to operating systems. Another special type of solid-state relay is the motor logic plus programmable electronic overload relay. This relay provides programmable control functions for motors and other devices.

Another type of overload relay is the magnetic overload relay. It works on the principle of electromagnetism (Fig. 4.10). If excessive current is drawn, the coil pulls

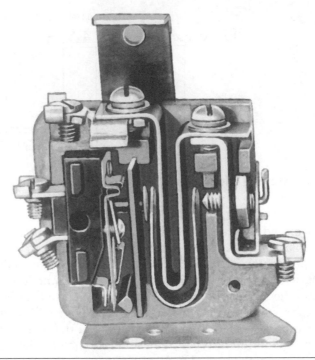

FIGURE 4.8 Bimetallic thermal overload relay. (*Courtesy Square D Co.*)

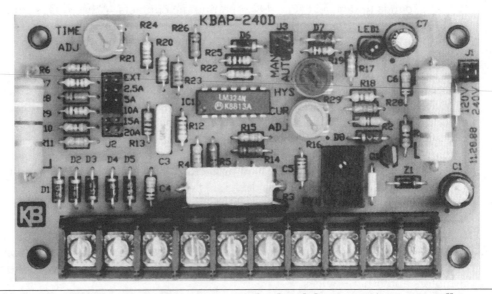

FIGURE 4.9 A multipurpose electronic overload and dc sensor. (*Courtesy Allen-Bradley.*)

FIGURE 4.10 Magnetic overload relays. (*Courtesy Square D Co.*)

in the core that opens the contacts. They are used to control many devices such as electric motors, heaters, lights, and audible signals. Modern relays include transient suppressors, logic-read cartridges, and latching attachments.

Manual Starters

The manual starting switch is a switch manually operated by a toggle handle. It is generally used to turn on or off fractional-horsepower motors of 1 horsepower (hp) or less. Some single-phase manual starting switches provide overload protection through use of a thermal unit that trips when too much current is being drawn. Once the thermal unit has tripped, it must be allowed to cool off before operation can be restored. Thermal units do not provide low-voltage protection. Also, if a power failure resulted, the contacts of the switch would remain closed and operation of the motor resumes when the power returns. The motor could unexpectedly turn on, which might be dangerous to the operator. Other types of manual switches provide a reversing-control option, two-speed operations, and removable-key accessories.

Three-phase manual switches are also available where overload protection is not important. These switches can control up to 10 hp and have a 600-volt (V) rating. They are used to control such devices as small pumps, conveyors, heaters, fans, and electric machines.

Three-phase manual starter switches are sometimes used to turn on or off three-phase motors of less than 10 hp. The switches are generally used on small machines that have separate overload protection. Figure 4.11 illustrates a manual motor starter switch with no overload protection. Some manual starters employ low-voltage protection with automatic start-up features. Continuous-duty solenoids are de-energized upon voltage loss. The solenoid must be manually reset.

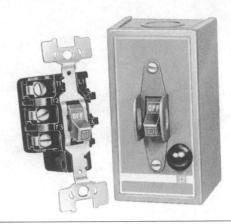

FIGURE 4.11 Manual motor starter switch without overload protection. (*Courtesy Square D Co.*)

Another type of manually operated starter often used to control single-phase motors up to 5 hp and polyphase motors up to 10 hp is a line voltage manual starter (Fig. 4.12). This starter provides overload protection through use of a thermal relay. When tripped, the thermal relay requires resetting after a sufficient cooling-down period.

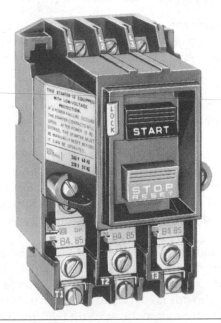

FIGURE 4.12 An integral-horsepower manual starter. (*Courtesy Square D Co.*)

Magnetic Starters

The magnetic starter is generally used for controlling motors, transformers, and heating loads from a distant location (Fig. 4.13). The magnetic starter and magnetic contactor basically are the same thing. Both have the ability to handle high currents. Figure 4.14 shows a magnetic contactor. The main difference between the two is that a magnetic starter contains overload protection whereas the contactor does not. The contactor requires separate overload protection. Magnetic contactors have a number of types of enclosures, such as dust-tight, corrosion-resistant, watertight, and drip-tight. Some also include solid-state monitors used for phase failure and overload protection.

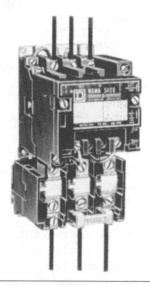

FIGURE 4.13   An ac magnetic contactor. (*Courtesy Square D Co.*)

FIGURE 4.14   A magnetic starter with overload protection. (*Courtesy Square D Co.*)

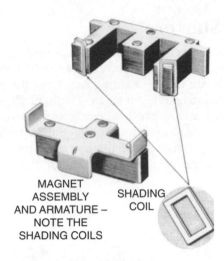

MAGNET
ASSEMBLY
AND ARMATURE –
NOTE THE
SHADING COILS

SHADING
COIL

FIGURE 4.15 **A magnetic armature and shading coils. (*Courtesy Square D Co.*)**

The primary principle of the magnetic starter is that when current is supplied to the magnetic coil, the coil becomes energized and pulls in the armature, which closes the contacts of the starter and starts the motor. To prevent chatter between the magnet and the armature, due to the variation of the magnetic field sinusoidally with time, a shading coil is added. This coil helps seal in the armature by offsetting the phase of the magnetic coil. Also, a small air gap is left in the laminated iron in order to prevent the residual magnetism (magnetism left in the coil after current has been removed) from locking in the armature when the magnetic coil is de-energized (Fig. 4.15).

In addition to overload protection, the magnetic starter contains a holding interlock. This normally open interlock holds in the circuit for the magnetic coil that would otherwise be broken when the pushbutton was released.

Reversing Magnetic Starters

The reversing magnetic starter is used to operate the motor in the forward or reverse direction. Actually, two magnetic contactors are interwired. This starter consists of two magnetic contactors with motor terminals T1, T2, and T3 connected to L1, L2, and L3 on one contactor. Terminals T1 and T3 are reversed on the other contactor. Figure 4.16 presents a picture of the magnetic reversing starter.

Neither the forward nor the reverse contactor can be energized with the other contactor already energized. This is accomplished by mechanical and electrical interlocks that prevent this action. If the forward contacts are closed, the mechanical and electrical interlocks will prevent the energization of the reverse contactor. The forward contactor must first be de-energized in order for the reverse contacts to be closed. Some magnetic starters include solid-state monitors featuring phase failure

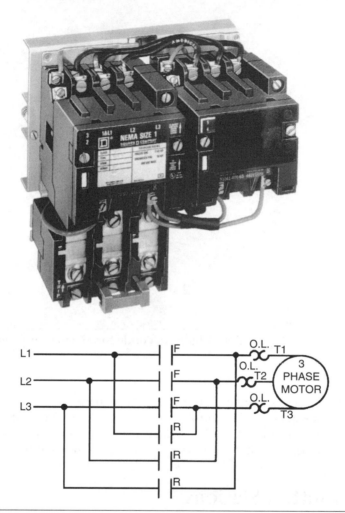

FIGURE 4.16   A magnetic reversing starter with circuit diagram. (*Courtesy Square D Co.*)

protection, ambient temperature overload protection, and underload protection. Also, some reversing contactors consist of two contactors mechanically and electrically interlocked, which are constructed either in horizontal or vertical positions. These contactors are also available as open type, watertight, or hazardous location enclosure type and either for 60- or 50-Hz operations.

Lighting Contactors

There are many different types of lighting contactors. Some common types include the multipole, programmable, and standard ampere-rated contactors. Many use silver–cadmium-oxide contacts, which can hold up to 800 amperes (A). While most control

FIGURE 4.17   A typical 200-A lighting contactor. (*Courtesy Square D Co.*)

tungsten, ballast, and other lighting resistance loads, others are designed for alarm systems, elevators, traffic controls, irrigation systems, and door locks. Most contactors also have available hardware kits containing compression lugs, which can be suitable for both copper and aluminum wire. Figure 4.17 shows a typical lighting contactor.

Pushbutton Stations

There are several types of pushbuttons used in power control operations. Generally, a pushbutton station contains two sets of contacts. One is normally open; the other is normally closed. This means that when one set of contacts closes, the other set must open and vice versa. Pushbuttons are used in conjunction with magnetic controllers. They are not required to be located close to the controller. Pushbuttons help control the functions of starting, stopping, jogging, reversing, and other operations. They also are designed for various operating conditions and may include indicator lights, key operation, or padlocking provisions (Fig. 4.18).

Limit Switches

Limit switches, like pushbutton switches, are commonly used in conjunction with magnetic starters. One of the main differences is that the limit switch is often used to convert motion of machinery to electric control signals (Fig. 4.19).

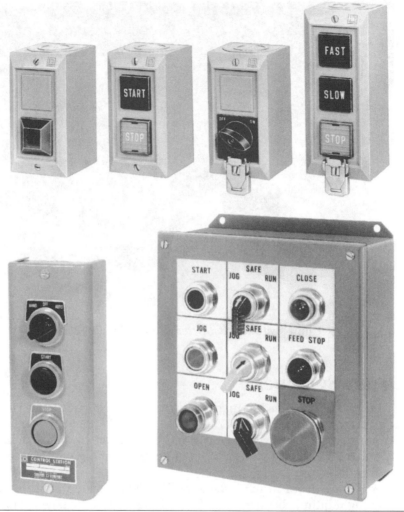

FIGURE 4.18   Pushbutton control stations. (*Courtesy Square D Co.*)

Limit switches are very popular and perform myriad control operations. They are used on production lines to stop, start, and lower and raise speed. The parts of the limit switch include the internal contacts, control assembly, and housing. Limit switches are sometimes repaired instead of replaced. The internal contacts wear down and commonly need replacement. The control assembly, consisting of actuator arm, lever, and roller or plunger, also can wear or break and need to be replaced. Different kinds of limit switches are the push-type, wobble stick, cat whisker, lever, spring return, and plug-in. Many limit switches are epoxy-sealed for contaminant and liquid protection.

FIGURE 4.19 Limit switches. (*Courtesy Allen-Bradley.*)

Drum Switches

Drum switches are used in many industrial control operations. They operate on high current and are typically three-pole manually operated switches used for reversing single- or three-phase motors. Some drum switches contain up to 16 poles and 7 throws. Figure 4.20 shows the internal switching positioning of the switch and the overall picture.

When the drum switch is used for three-phase motor control, the internal switching requires that two of the three leads be interchanged. This is easily accomplished by following the connecting diagram on the switch housing. Typical types of drum switch enclosures include watertight, rainproof, and oil-tight.

Timers

There is a large variety of timers used in motor control operations, such as interval, pulse, percentage, and revolution timers. The pneumatic timer, illustrated in Fig. 4.21, is a timing relay that works on an air-diaphragm principle. The needle valve controls the flow of air back into a partially evacuated chamber. When the diaphragm returns to its initial position, with equal air pressure on both sides, the contacts operate.

Once these contacts operate, the circuit will turn on or off depending on whether the normally open or normally closed contacts are being used. Pneumatic

FIGURE 4.20   Drum switch shows internal switching position and overall view. (*Courtesy Square D Co.*)

FIGURE 4.21   A pneumatic timing relay. (*Courtesy Allen-Bradley.*)

timers range from approximately 1/20 second (s) to 3 minutes (min) in duration. A surge suppressor can be added for protection against high-voltage spikes. Many pneumatic timers have been replaced with solid-state timers, which offer programmable timing and counter options.

Electronic Drives

Electronic drives are industrially rugged control systems designed to provide adjustable speed control for motors (Fig. 4.22). The typical functions of electronic drives include start/stop, forward/reverse, run/jog, auto/manual, and other speed variations. The driver contains a microprocessor with multiple preset speed controls, braking, overload, and torque control features. A digital display is generally provided to illustrate various faults such as overcurrent, grounds, inadequate voltage, overload, live dip, and function or memory loss.

Variable-frequency drives (VFDs), also called adjustable-frequency drives, variable-speed drives, microdrives, inverter drives, etc., are remarkable energy-saving and efficiency devices. These are solid-state systems that comprise topologies

FIGURE 4.22   An electronic driver for motors. (*Courtesy Reliance Electric.*)

of power-factor corrections, harmonic distortion management, filtering, phase-controlled current source inversion, and load regulation. Electric motors, without a doubt, utilize the vast amount of electricity in the United States and these drives help reduce electric power usage and dependency. In fact, the electric motor is probably one of the few devices that actually cost more to operate in a given year than the cost of the device. It is possible that an electric motor may cost more than four times to operate in 1 year than the initial cost of the motor.

Programmable Controllers

The programmable controller, often referred to as a *programmable logic controller* (PLC), is the most sophisticated motor control device (Fig. 4.23). This device is a microprocessor-powered computer. Prior to the PLC, an elaborate array of relays and switches and magnetic controllers were used.

The PLC offers many advantages, such as flexible programming, digital fault display, printout options, key locks for security, and recorder options. The PLC is discussed in greater detail in later chapters.

Sensors

The use of sensors has dramatically increased in the industrial field, especially in manufacturing. There are several types of sensors such as photoelectric, thermocouple, crystal, proximity, vision, advanced digital fiber-optic, digital

FIGURE 4.23   Programmable controller. (*Courtesy Reliance Electric.*)

pressure, bar code readers, laser, and digital video microscopic. Sensors have a wide range of applications such as detecting workpieces in assembly lines, measuring fan belt runout in an automobile, measuring the height of manufacturing products to ensure quality standards, measuring sand mold ripple in a casting, counting connector pins for sockets in semiconductors, measuring eccentricity of a shaft, checking for accurate soldering connections, measuring automobile tire alignment, and even checking for the absence of chewing gum in a wrapper on a production line.

One of the more common sensors is the fiber-optic sensor. It appears that fiber-optic sensors are becoming more popular than other photoelectric sensors. The basic operation of the sensor involves a light beam traveling down a high-refractive-index material called a *core material* bounding off a low-refractive-index material called the *cladding*. The variation of detecting distance of the fiber-optic sensor is based on the fiber length. Proper mounting and alignment of the transmitter and receiver are critical for effective operation. Sensors generally have a sensitivity adjustment for controlling the operation. Adjustment of the sensitivity can compensate for placement variations or extraneous light.

Repair and Testing Procedures

Common techniques used in the troubleshooting of motor control circuits include substitution of controls, internal parts, and amperage, voltage, and resistance measurements.

Basic manual switches can be checked with an ohmmeter. The switch should have continuity in one position and not in the other. Continuity or infinite resistance when the switch is in both positions would indicate a defective switch. Also, if the thermal overload relay fails to reset or continues to trip with normal current operation, the device may need to be replaced. Keep in mind that an incorrect thermal unit selection may have been made in the first place. Also, do not disregard the fact that there may be parts broken inside the overload relay while absolutely nothing is wrong with the thermal unit itself or its selection.

Relays can also be checked by using an ohmmeter. First begin with a visual inspection. Look for burned contacts or charred coils. When the relay is manually pushed in, the ohmmeter should indicate either continuity or no continuity, depending on whether the contacts are normally open or closed. The ohmmeter will show continuity one way and not the other. Also, the total resistance of the winding can be checked. Depending on the coil size, a relay coil should not read 0 V. Zero ohms might indicate a short-circuited coil. On the other hand, infinite coil resistance would indicate an open circuit. Likewise, if power is supplied to the coil, but the coil fails to pick up the armature, the coil is probably open.

Manual and magnetic starters can easily be checked by visually inspecting the contacts (Fig. 4.24). When contacts appear to be dirty, deformed, or pitted, they should be replaced. Most contacts can be cleaned with acceptable cleaning solvents and can be dressed with a file. *Note:* Silver contacts should never be filed!

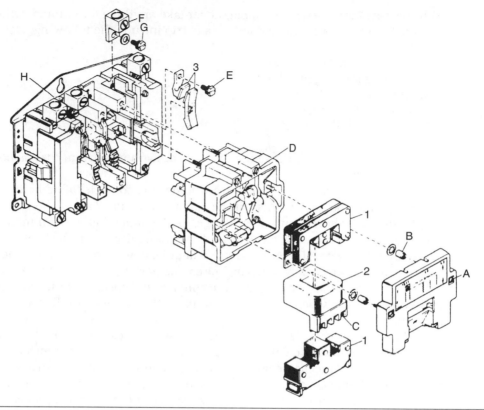

FIGURE 4.24 An exploded view of a magnetic starter. (*Courtesy Square D Co.*)

Dirty, gummy starters, coils, and other devices may need to be enclosed in a more appropriate enclosure. Although hot soapy water sometimes is used to clean control equipment, it is not recommended by the authors. Control equipment should never intentionally be exposed to water. There are many excellent cleaning solvents available from motor control distributors that would be more appropriate to use.

If the thermal overload relays fail to hold under normal current or fail to reset, they should be replaced. Also, make sure that there are no broken parts inside the overload relay that might be causing the problem.

Always inspect the starter for loose connections, and check for any type of physically damaged parts. Replace any broken wires or terminals with the same type of replacements. Also, remember that positioning a contactor or magnetic starter too far from the control device can cause impedance to the control wiring and negatively affect starter performance. Therefore, it is best to minimize the distance between contactors and control devices.

When you are troubleshooting overload relays, check to see if the coil continues to trip. Also check the ambient temperature. It may be necessary to replace the coil with the next larger size. Remember, never take anything for granted. Other factors to consider when an overload continues to trip include the following:

1. Service factor of the motor
2. Jogging for plugging duty
3. Long acceleration time
4. Overloaded motor
5. Slow-speed motors

Any of these factors may cause the motor to draw more current than is normal for a given horsepower.

When you check contacts, observe whether the contacts fail to hold. If they do, the voltage may be too low. Also, if the contacts tend to wear quickly, the problem may be due to a short circuit, low voltage, loose connection, foreign matter, or an abnormal voltage surge. Never disregard the possibility of physical abuse of the starter by its operator! Excessive abuse will quickly wear down the contacts. If testing fails to identify any abnormal circuit or power problems, double-check and tighten all connections and replace the contacts and springs. If the contacts continue to wear prematurely, you may need to replace the entire controller with a higher-rated one.

When you are troubleshooting magnetic starters, common problems include dirty gummy parts, worn contacts, loose connections, defective thermal overloads, damaged or worn mechanical parts, etc. For example, failure of a magnetic coil to pick up could indicate a defective coil, wrong coil, low voltage, incomplete control circuit, a break in wiring continuity, or some type of mechanical problem. On the other hand, if a magnetic coil appears to be overheated, check for incorrect voltage, incorrect amperage, wrong coil, or starter abuse by the operator. If a coil fails to drop out, check the physical position of the coil and inspect the starter for dirt or gum, wrong voltage or amperage, welded contacts, or mechanical damage. A noisy magnetic starter can often be attributed to a broken shading coil, low voltage, or any type of dirt or corrosion.

If a magnetic starter continues to trip the overloads, check for any possible short circuits or grounds in the circuit. Also, the overloads may need to be replaced. If the contacts chatter, check for low voltage or defective contacts. Remember, any type of starter abuse by the operator or dirty conditions can cause premature breakdown of the magnetic starter.

Typical problems associated with pneumatic timers include erratic timing or failure of the contacts to operate. Check the operating mechanisms for foreign matter that might be clogging the timer. Consider adjusting the actuating switch to correct the timing problems. Other remedies might include replacing the timing head, snap switch, or coil.

Figure 4.25 represents a typical three-phase 240-V motor with a 120-V (low-voltage) control circuit. Here, the fused transformer provides low voltage for safer operation. If the motor is completely dead when the start or pushbutton is pushed, first make sure that the overload relay is reset. If the relay is set, begin by checking the transformer

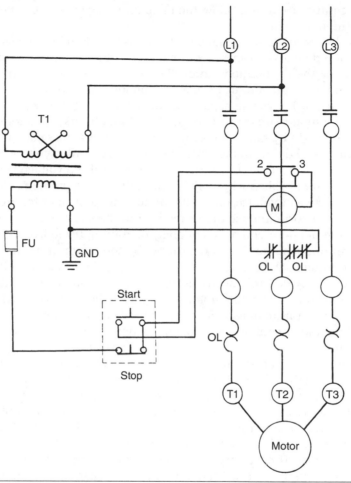

FIGURE 4.25 A schematic diagram of a three-phase motor control circuit.

line fuse for continuity. Check each fuse in the disconnect box, control overloads, and/or main panel box. If a blown fuse is found, it is advisable to determine what caused the fuse to blow. To do this, check the motor for possible grounds, short circuits, bad contacts, or possible foreign material or contaminants. Examine the motor control for possible loose connections and overheated coils.

If the motor fails to run when the start pushbutton is depressed, manually push in the magnetic control armature. If the motor now operates, check the continuity in the pushbutton station, transfer line fuse, magnetic coil, holding contacts, wire to and from transfer line to magnetic control, etc. A defective component or loose connection must be the problem. Isolating and testing each stage will determine the cause. If an open fuse is found, replace the fuse only with one of the same rating.

If necessary, use a fuse below the rated amperage value, but never one with more than the rated value.

If the motor runs, but fails to stop consistently when the stop button is pushed, look for a defective or miswired switch or welded holding circuit interlock. Examine and check the pushbutton station. The stop contacts may be badly worn and corroded, thus preventing proper conduction. Finally, check all wires for possible breaks, by both a visual inspection and a continuity check.

If the motor runs at half speed or "sits and hums," look for one of the phases to be open. Begin by taking a voltage measurement across each phase by the motor terminals and determining which phase is open. After you locate the open phase, trace back in the circuit with the voltmeter until the voltage is restored. A loose connection or broken wire may be the cause of the problem. *Note:* If all line phases have the correct voltage, check the motor with a Megger for grounds or short circuits. A short-circuited motor may act in the same manner as an open-line phase.

If the operator complains of noisy operation, try to isolate the noise by determining whether it is being caused by the motor or by the control. A chattering or humming in the control could indicate a broken shading coil, dirty or worn contacts, dirty control, mechanically worn or misaligned armature, or low voltage. If the noise comes from the motor, this could indicate worn bearings, a bent shaft, or excessive vibration due to loose motor base connections.

If the motor continues to trip the magnetic control overloads, check the amperage at the motor terminals to determine if excessive current is being drawn. The motor may be shorted; the overloads may be faulty, partially shorted or grounded; or wires may be loose. Remember, do not overlook the obvious. Excessive moisture, dirt, heat, corrosion, and other contaminants can all contribute to tripped overloads or blown fuses.

If you discover frozen or welded contacts, there are a number of possible causes, such as abnormal current rush, sustained low voltage, excessive rapid jogging, or possible contaminants in the contacts. Start by inspecting the contacts for contaminants. If necessary, clean the contacts, using the manufacturer's suggested cleaning solvent, such as Freon. Abnormal voltages and currents can be checked for possible grounds, short circuits, or excessive loads. If necessary, you may need to replace the contacts, springs, carrier assembly, or armature; possibly, the entire contactor may need to be replaced.

Other factors may cause increased heat, and motor problems include sustained jogging (repeated starting and stopping) and plugging (rapid stopping of a motor by momentarily connecting the motor in reverse direction). Operator abuse is also a factor that should be considered.

If the motor fails to run and the transformer is suspected, various methods of determining a defective transformer can be used. Many transformers can be visually checked by looking for charred or burned insulation and windings. A shorted transformer will feel hot when touched and will exude a bad odor. The windings can be checked with an ohmmeter; zero resistance probably indicates a short circuit. An open winding will read infinite resistance. A voltmeter can also be used to check the performance of the transformer. If the correct voltage is read at the primary side, but not at the secondary side, the transformer is probably open. Incidentally, check to make sure the transformer lead connections are secure. A loose connection could indicate an open transformer circuit.

Typical problems associated with electronic drives and PLCs involve wiring errors, power line faults, motor and mechanical malfunctions, and controller defects. Step-by-step troubleshooting procedures outlined in the manufacturer's service manual using oscilloscopes, logic pulsers and tracers, digital multimeters (DMMs), etc. can pinpoint circuit faults. Figure 4.26 shows a logic pulser and current tracer being used to test a circuit.

Begin with a close inspection of all wires and connections. Make sure the proper power line voltages, conductors, protection overloads, and isolation transformer requirements are present. Do not overlook the motor. Check for obvious malfunctions, wiring faults, grounds, open circuits, short circuits, and installation problems. Never use a megohmmeter to check the motor while it is connected to the controller, or else damage may result. Always consider obvious mechanical problems, binding, foreign objects, breaks, or other defects.

A common term associated with troubleshooting PLCs is *peripheral failure.* This term is defined as any external devices or hardware connected to the PLC, such as relays, switches, wiring, and pushbuttons. Operators are quick to place blame on the PLC, when the fault may be with the peripheral device. If any of the peripheral devices should fail, the PLC will also fail. Also, any associated disruptive conditions,

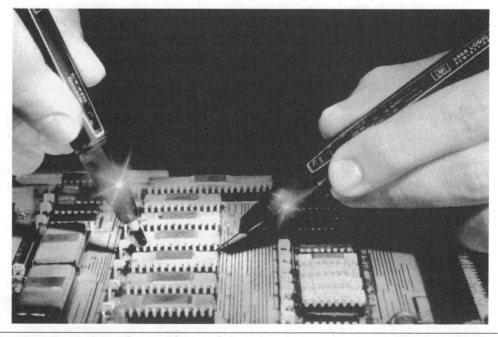

FIGURE 4.26   Using a logic pulser and digital current tracer to troubleshoot a PLC circuit. (*Courtesy Hewlett-Packard.*)

such as vibration, large transients, chatter, and heat, can cause a PLC to falsely trigger and require reprogramming.

For example, if the drive motor does not start, check the disconnect box and fuses and verify the proper line voltages. Check the condition of the brake coils, solenoids, interlocks, diode or rectifier assembly, thermostats, external overloads, and circuit breakers. If a relay or circuit breaker needs to be reset, always check for motor grounds or short circuits prior to resetting. Controller output measurements are often taken by using digital clamp-on ammeters, DMMs, handheld tachometers, or oscilloscopes. The actual troubleshooting of PLCs and related devices will be discussed in later chapters.

Also, when troubleshooting industrial controls, the associated wiring can become defective and impact the control devices. For example, a common problem is insulation breakdown of the wiring and cables. As a general rule when testing insulation high values using an insulation instrument indicate good insulation quality, fair values with constant tendency towards lower values should be examined for faults, and low values probably indicate faults. It is generally necessary to clean, dry out, inspect, and repair any faults before placing equipment back in service.

Preventive Maintenance

It has been generally accepted that the motor control's normal operating life is dependent on the type of maintenance and operating conditions of the control. Dirty, gummy controls should be scrubbed with a stiff brush and cleaning solvent. If control equipment should get wet, it is important to completely dry out the control by heating or drying it prior to reoperation. Finally, check the coils with an ohmmeter or insulation resistance meter for correct insulations resistance.

Controllers should be periodically checked for worn parts, loose connections, or weak spring tension. Dirty, corroded copper or cadmium-plated contacts should be kept clean and refiled only to maintain correct shape. Silver contacts should not be filed. Disfigurements and discoloration generally do not damage or hamper operation.

Lubrication of motors and controls may be needed to avoid breakdown due to excessive friction. However, controllers generally do not require lubrication, because this attracts dirt and other contaminants that wear down the controller. Most motors require periodic cleaning and bearing greasing or replacement.

Most water jackets used in cooling operations should be flushed out periodically and replaced to eliminate rust and corrosion. Also, proper oil level and oil or air pressure should be constantly checked for possible leaks.

Using common sense and a little care can prevent many accidents as well as control breakdowns. Keep a lookout for loose guards, worn belts, worn controls, and cracked, charred, or brittle wire. If a section of wire is brittle, cracked, or frayed, the wire should be replaced. If the entire wire cannot be conveniently replaced, a new section can be spliced in. Use the exact replacement wire, and make sure the connections are both mechanically and electrically secured by soldering or splicing terminals. Last, make sure the connection is well insulated with electrical tape or sleeving.

Electronic drives and PLCs, while rugged, are sensitive to temperature, humidity, chemicals, moisture, and other adverse environmental conditions. Do regular maintenance of each specific device by following the manufacturer's manual.

Self-Examination

Select the best answer:

1. Reversing the flow in a coil
   A. Does not change the magnetic polarity
   B. Destroys the coil
   C. Reverses the magnetic polarity
   D. Does none of the above
2. The simplest form of a motor controller is the
   A. Toggle switch
   B. Magnetic switch
   C. Drum switch
   D. Relay
3. Another name for a magnetic starter is a
   A. Manual switch
   B. Contactor
   C. Manual starter
   D. Magnetic control
4. What type of overload contains a solder pot?
   A. Melting alloy
   B. Bimetallic
   C. Magnetic
   D. Fuse
5. What type of overload operates on the principle of electromagnetism?
   A. Melting alloy
   B. Bimetallic
   C. Magnetic
   D. Fuse
6. A 15-A fuse is found open, but the servicer does not have a replacement and has a choice between a 10- and 30-A fuse. Use
   A. The 10-A fuse
   B. The 30-A fuse
   C. Either (A) or (B)
   D. The 40-A fuse
7. Low voltage often causes the magnetic control to
   A. Gum up
   B. Run more efficiently
   C. Chatter
   D. Do none of the above

8. Rapid stopping of a motor by momentarily connecting the motor in reverse direction is called
   A. Jogging
   B. Inching
   C. Plugging
   D. Sequence operation

9. A hot, smoky transformer indicates
   A. An open circuit
   B. A short circuit
   C. Both (A) and (B)
   D. None of the above

10. A shorted motor will draw
    A. Low current
    B. High current
    C. Both (A) and (B)
    D. None of the above

11. Three-phase manual starter switches are sometimes used to turn on or off motors up to
    A. 5 hp
    B. 10 hp
    C. 15 hp
    D. 20 hp

12. The device used to control a motor from a distant location is the
    A. Manual starter
    B. Drum switch
    C. Magnetic controller
    D. Snap switch

13. Which of the following devices is a type of timer?
    A. Interval
    B. Pulse
    C. Percentage
    D. All the above

14. The device that controls the flow of air back into the chamber of the pneumatic time relay is
    A. The diaphragm
    B. The needle valve
    C. The shading coil
    D. All the above

15. If a motor runs but fails to stop, the problem may be
    A. An open fuse
    B. Welded holding circuit interlock
    C. Jogging or plugging duty overload
    D. None of the above

16. If the motor runs at half speed or "sits and hums," what should you look for in the magnetic controller?
    A. An open phase
    B. Welded holding circuit interlock
    C. Both (A) and (B)
    D. None of the above
17. Repeated starting and stopping is referred to as
    A. Stepping
    B. Jogging
    C. Phasing
    D. Long acceleration time
18. A chattering or humming in a magnetic contractor may be due to
    A. A broken shading coil
    B. Dirty or worn contacts
    C. Both (A) and (B)
    D. None of the above
19. Which of the following contacts should not be filed?
    A. Cadmium-plated
    B. Silver-plated
    C. Both (A) and (B)
    D. None of the above
20. An increase in heat in a motor can be caused by
    A. Sustained jogging
    B. Repeated plugging
    C. Both (A) and (B)
    D. None of the above
21. Control systems commonly are of two types: closed-loop and
    A. Single-loop
    B. Double-loop
    C. Open-loop
    D. Control loop
22. Prior to servicing a motor or industrial device, the troubleshooter should
    A. Shut off the power
    B. Check the wiring
    C. Service the controller
    D. Replace the fuse
23. Limit switches have three parts, which are the
    A. Contacts, lever, and bearing
    B. Driver, switch, and magnet
    C. Driver, drum, and timer
    D. Contacts, control assembly, and housing

24. The drum switch may contain up to
    A. 3 poles
    B. 6 poles
    C. 8 poles
    D. 16 poles
25. PLC stands for
    A. Processing logic circuit
    B. Programmable logic controller
    C. Processing logic controller
    D. Programmable logic circuit
26. One of the most typical problems of PLCs is
    A. Wiring errors
    B. Power line defaults
    C. Motor malfunctions
    D. All the above
27. Never check the motor when it is connected to a PLC with
    A. An insulation resistance meter
    B. An ohmmeter
    C. A tachometer
    D. A thermometer
28. A handy test instrument to check current output of a controller is
    A. A clamp-on ammeter
    B. An insulation resistance meter
    C. A tachometer
    D. An oscilloscope
29. If a PLC overload relay trips, it is a good idea to check the motor for
    A. Grounds
    B. Open circuit
    C. Short circuit
    D. All the above
30. Generally when servicing most PLCs, the manufacturer supplies
    A. A substitute PLC
    B. A test instrument
    C. A service troubleshooting manual
    D. None of the above

Questions and Problems

1. State the basic functions of a motor controller.
2. Explain the basic theory of a circuit breaker.
3. What is a toggle switch?
4. What is an advantage of a magnetic starter compared with that of a manual starter?
5. Describe the characteristics of a magnetic starter.

6. What is a limit switch?
7. Explain the theory of operation of a melting alloy thermal overload relay.
8. Explain the theory of operation of a bimetallic thermal overload relay.
9. Explain the theory of operation of a magnetic overload relay.
10. What is a pneumatic timer?
11. How do you check a relay?
12. List the various methods to test a transformer.
13. What is the proper technique to clean a coil?
14. What might be wrong with a motor that runs at half speed?
15. State different types of noise from a motor and motor control. What are the causes of this noise?
16. What is a shading coil?
17. Define the term *jogging*.
18. Define the term *plugging*.
19. Why is low control voltage often used in motor control operations?
20. Explain the procedure in maintaining or servicing motor control contacts.
21. What is an electronic driver?
22. What is a PLC?
23. List the various functions of the electronic driver.
24. List the various faults illustrated by the digital display of an electronic driver.
25. List the various features of the PLC.
26. Explain the typical problems of PLCs.
27. Why should the troubleshooter never use an insulation resistance meter to check the motor when it is connected to a PLC?
28. List the typical test instruments used to check PLCs.
29. List the typical problems associated with electronic drives and PLCs.
30. What are some devices to check if the drive motor fails to start?

Chapter 5

Troubleshooting Residential, Commercial, and Wireless Communication Systems

Electric wiring provides the medium by which power is transmitted to electric motors, controls, and other electrical and electronic apparatus. In this chapter, the fundamentals of residential and industrial wiring are presented, along with troubleshooting techniques. In addition, testing and repair procedures are presented for television distribution systems and residential and industrial lighting systems.

Fundamentals

Basically, most electric wiring circuits consist of four basic sections:

1. Power source
2. Transmission line
3. Control device
4. Operating device

Power transmission has largely been dominated by alternating current (ac) versus direct current (dc) for the past several decades. Ever since the battle, often called the *War of the Currents*, between Thomas Edison and George Westinghouse in the late 1800s, there has been controversy about which type of current is the best. Many people feel that dc is more efficient and safe whereas others believe that ac is more practical and suitable for large distribution systems. However, even though ac has been the predominant technology in the power market, dc has been gaining more attention especially for high-voltage and computer distribution centers. For example, ac can incur large reactive power losses in long transmissions and underground environments. Also, dc has virtually no reactive power losses as compared to ac and eliminates the need for large-scale rectification. However, this being said, ac is still the predominant power source for most residential and commercial applications.

While most of the electric power is produced by fossil-fuel generating stations, the use of hybrid power is becoming more and more popular with the increased attention being given to green, renewable energy, and emissions-free electricity production. Some examples include geothermal, hydroelectric, solar, and wind turbines (wind farms). In addition, the advent of hybrid solar and wind turbines power installations are becoming more popular in some areas of the country. Many of these hybrid production facilities are being fueled by production tax credits (PTCs). Also, sophisticated software can help regulate solar power production and electric grid operations.

However, in most residential homes, the single-phase ac 120/240-volt (V) power source usually supplies a distribution panel box. It has a total amperage capacity that generally ranges from 60- to 200-ampere (A) service with 200 A generally being the minimum today. In the panel box, each individual circuit is connected to a hot side and a neutral ground side. Figure 5.1 illustrates an example of a 120- and a 240-V transformer supply.

In the panel box, the hot leg that is often called the *power leg* is usually a black wire, or red wire if used for 240 V. The neutral ground wire is a white wire, and the green bonding wire acts as a safety appliance and equipment ground. Fuses or circuit breakers protect the hot side from overloads. The neutral side is almost always grounded to earth by connecting this wire to a ground rod or cold water pipe, depending on the local code. Figure 5.2 shows a typical distribution panel box layout.

An appliance called an *equipment ground* consists of a bonding wire that is connected to the neutral (common ground) terminal block in the panel box. It is then connected to the metal receptacle box and third odd hole of the receptacle. When an appliance, such as a drill, is plugged into this receptacle, it will also have a third prong and wire that is connected to the metal frame of the drill (Fig. 5.3).

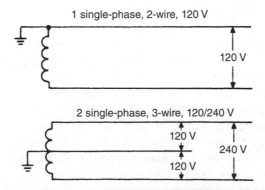

FIGURE 5.1 An example of 120- and 240-V transformer supply.

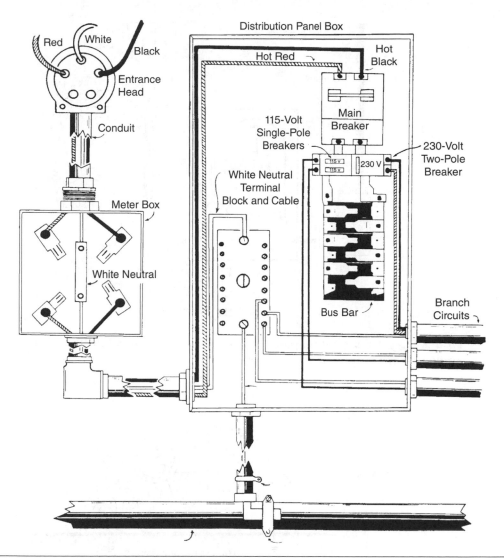

FIGURE 5.2 A typical distribution panel box layout.

If a grounded circuit should be created while a person is operating the drill, the current will not go through the operator. Instead, current will return through this bond wire, back to the source. If there is no protective appliance ground, the current from the hot side of the line may go through the grounded drill, through the operator, and back to the earth ground (Fig. 5.4).

Grounding systems protect operators and also protect equipment. Electronic equipment such as programmable logic controllers (PLCs) and computers are especially sensitive to grounds. Circuits are more prone to grounds when located in high-humidity and moist environments, such as basements and outdoor unenclosed

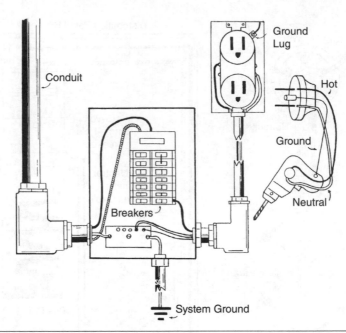

FIGURE 5.3 A distribution panel box, receptacle, and hand drill showing the appliance or equipment ground wire.

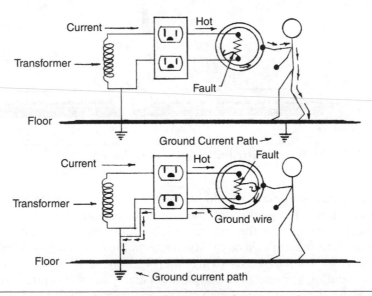

FIGURE 5.4 An example of a grounded circuit and a circuit that is not grounded.

facilities. For example, if a computer-operated system is located in an industrial warehouse or basement and the floor is wet, both the operator and the equipment are more susceptible to grounds.

The troubleshooter should not assume that the receptacle ground third prong on the plug will always protect the equipment or the operator. It is not unusual for the grounding wire to be open somewhere along the entire wiring circuit. Ungrounded plugs are especially common in older industrial facilities. Even if the receptacle ground is complete, it still may not provide adequate protection owing to the damp or wet conditions of the equipment and operator.

To protect industrial equipment such as motors and controls, elaborate programmable protectors have been developed. Figure 5.5 illustrates a programmable motor protector. This protector is designed to protect large and expensive motors. It guards against not only ground faults, but also jams, overloads, phase problems, and other power-load problems.

Less expensive ground-fault protectors are called *ground-fault interrupters* (GFIs). Figure 5.6 illustrates a typical *ground-fault circuit interrupter* (GFCI). The GFCI is similar to a circuit breaker, except that it protects the circuit from ground faults. It triggers whenever small amounts of current flow to ground. For example, some typical GFCI designs include tamper-resistant systems for pediatric areas, children playrooms, patient care areas, and psychiatric and medical facility environments. The interrupters often contain rapidly flashing light-emitting diodes (LEDs) to alert users that a problem exists.

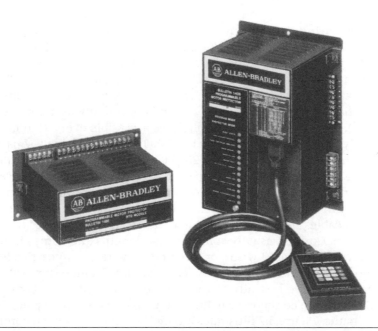

FIGURE 5.5   A programmable motor protector. (*Courtesy Allen-Bradley.*)

FIGURE 5.6   A typical ground-fault interrupter (GFI). (*Courtesy Hubbell, Inc.*)

Most wiring circuits found in the home usually use no. 14 wire for 15 A, no. 12 wire for 20 A, no. 10 wire for 30 A, no. 8 wire for 40 A, and no. 6 wire for 55 A.

The maximum capacity of a 20-A general-purpose appliance circuit is approximately 2400 watt (W), and for a 15-A circuit, about 1800 W. The total number of appliances, if used at one time, should not exceed the rated capacity. In calculating residential wiring circuits, a 15-A circuit is commonly used for every 375 square feet (ft²). A convenience outlet should be accessible every 12 ft of running wall space. Typically, an electric range (8000 to 18,000 W) should have a separate 240-V (40-A) circuit. An electric water heater (2000 to 4000 W) requires a separate 240-V (40-A) circuit. Kitchens usually have two or three separate 20-A appliance circuits. One general-purpose circuit (15 to 20 A) often can supply two bedrooms and a bathroom.

The three basic kinds of cable used are nonmetallic sheathed cable, flexible armored cable, and electrical metallic tubing (EMT) or "pipe" (generally referred to as *conduit*). Conduit usually comes in 10-ft lengths and comes in these sizes: ½ inch (in) carries four no. 14 wires, or three no. 12 wires; ¾ in carries four no. 10 wires or five no. 12 wires, etc. Different types of conduit include thin-wall, rigid, and plastic. Nonmetallic cable basically consists of indoor and outdoor types.

Some of the common switching circuits used are single-pole, three-way, and four-way circuits. Figure 5.7 illustrates a typical single-pole switch controlling a lamp. The current travels through the switch to the lamp and returns to the power source.

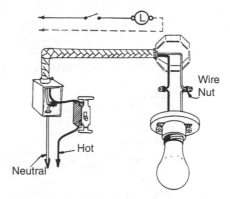

FIGURE 5.7   A typical single-pole switch controlling a lamp.

A three-way switch controls one lamp or more from two locations (Fig. 5.8). In the position shown in Fig. 5.8, the lamp is off; by turning either switch, the lamp will turn on. Three-wire cable is necessary for this circuit between the two switches.

A four-way switch controls one lamp or more from three different locations (Fig. 5.9). Here again, three-wire cable is needed for the circuit between the switches. Also, the middle switch needs to be a four-way switch. If more than three switch control locations are desired, just add additional four-way switches.

A three-phase four-wire system consists of three phases having equal voltage and a fourth grounded wire that is used for protection and reduced voltage. Figure 5.10 illustrates a typical wye and delta system commonly used in schools, businesses, and industry.

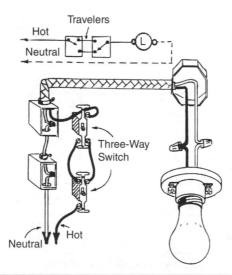

FIGURE 5.8   A typical three-way switch controlling a lamp.

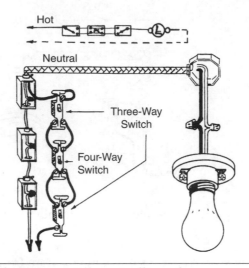

FIGURE 5.9 A typical four-way switch controlling a lamp.

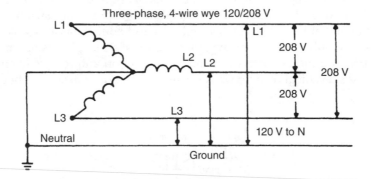

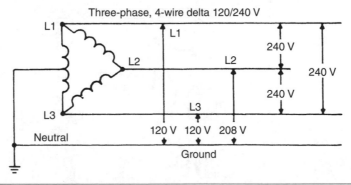

FIGURE 5.10 Common wye and delta circuits.

Electric Wiring Circuit Repair

Common faults that occur in residential and industrial wiring circuits include open circuits, grounds, short circuits, and power-factor problems. While opens, grounds, and shorts have been previously discussed, power-factor problems are unique to industrial wiring and ac power devices. A typical problem found in three-phase industrial systems is the *low power factor.* Any type of ac "inductive reactant" circuits or devices, such as heaters, lamps, motors, and controls, will cause a lagging of current that is expressed in percentage. Power-factor percentages often range somewhere between 30 and 90 percent, with 90 percent being excellent and 30 percent very poor. Power factor actually is the cosine of the phase angle between voltage and current. A low power-factor rating uses excessive current, is wasteful and expensive, and increases heat. A power-factor rating is usually corrected in a circuit by adding capacitance in parallel with the circuit (Fig. 5.11).

Power-factor correction is usually expressed in a unit called *var* (volt-ampere reactive) or kilovars. The correction is accomplished by the addition of capacitors that decrease the phase difference between the voltage and amperage used in the circuit. Therefore, a motor with corrected power factor simply dissipates less power in vars, which costs less money. In Fig. 5.11, note the capacitor box has a fuse in each line for added protection, in case one of the capacitors should short-circuit. Also, bleeder resistors are shunted across the capacitors to decrease shock hazards and increase the life of the capacitor. Notice that actual physical placement of the

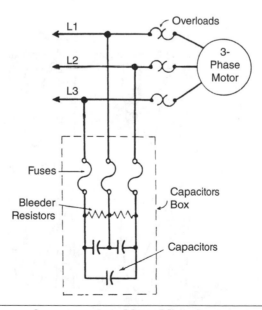

FIGURE 5.11   Low power factor corrected by adding capacitor.

capacitor banks can be located at the power source entrance, across the motor, and/or ahead of the motor loads.

In troubleshooting wiring systems, start by using your senses. If a circuit breaker has just tripped, instead of resetting the breaker, stop and think what might have caused the breaker to trip. Did someone just plug in a device, overloading the circuit? Did you smell any funny odors? Were there unusual noises? Examine the system for cracks, charring, odors, mechanical problems, etc.

If the electric wiring system has a fuse box, check the main fuses by using a voltmeter. Figure 5.12 illustrates the technique of locating a blown fuse in a single-phase system. When the leads of the voltmeter are placed across the power line (top right figure), with meter indication, we know voltage is present. When the leads are placed across the terminals past the fuses (top left figure) and no indication is obtained, we know one or both fuses are defective. When one lead is placed on the power line and the other past the fuse (bottom right figure) and no indication is present, we know this fuse is blown. The last fuse should be checked in this same manner.

In a three-phase system, fuses can be checked with a voltmeter or ohmmeter in the same manner as they are checked in a single-phase system. Figure 5.13 shows the procedure for checking fuses in a three-phase system.

Another common device used to check the final installation of a house wiring circuit is a receptacle tester. It indicates typical circuit faults such as open hot, open neutral, hot and neutral reversed, and reversed polarity. A typical receptacle tester is shown in Fig. 5.14.

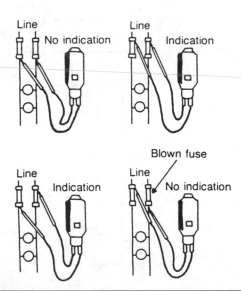

FIGURE 5.12   Checking for blown fuse in a single-phase system. (*Courtesy Square D Co.*)

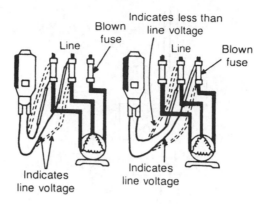

FIGURE 5.13   Checking for a blown fuse in a three-phase wiring system. (*Courtesy Square D Co.*)

FIGURE 5.14   A typical receptacle tester. (*Courtesy ETCON Corp.*)

Shorted wires, open wires, and wires grounded to conduit are common problems in industry. Using Fig. 5.15 as a typical industrial circuit with a fault, we explore a step-by-step procedure for troubleshooting the circuit.

1. Check for the presence of power from the distribution line coming from the main panel box.
2. Check for dirty or loose connections or blown fuses. Inspect the quality of the disconnect box, looking for possible problems, such as worn or corroded blades, defective fuse connectors, and faulty drive mechanisms.
3. Check the motor control and wires leading to and from the control. Check for loose connections, dirt, oil, shorted windings, bad contacts, etc. Using an ohmmeter, check for continuity from the disconnect box to the motor control. Also, the wires can be disconnected at the motor controller and power restored at the disconnect box. If voltage is present at the motor controller

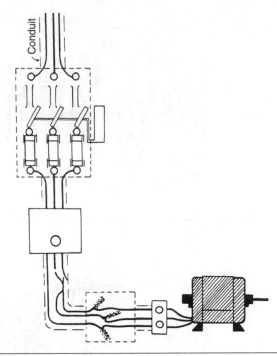

FIGURE 5.15   A typical industrial wiring circuit.

and it operates correctly with power restored, we know the defect must exist somewhere between the motor controller and the motor.

4. Now check for open, shorted, or grounded wires in the conduit between the motor controller and terminal box. The wires at the pigtail connections can be disconnected, and the wires checked from the motor controller and terminal box. In this case, we find that the wires in the conduit are open and shorted to the conduit itself.

5. Also, while you examine the junction box, look for poor pigtail connections, wire nuts, wires pinched inside the box, etc. Check for the presence of moisture, dirt, oil, etc.

6. After you repair the line, make sure to finish checking the line between the terminal box and electric motor. Also, test the operation of the pushbutton switch and motor. This point-by-point troubleshooting is an easy and logical approach to identify faults.

Keep in mind that if a cable blows fuses or trips a breaker without a load attached to it and all control devices have been checked, you can assume that the cable is shorted. Here is where conduit has an advantage over nonmetallic cable. The old shorted wires can be easily pulled out of conduit and new ones fished through, using a steel flexible coil of wire called *fish tape*. If nonmetallic cable needs to be replaced, it may be difficult to do so without cutting into the walls. If at all possible, when you pull out the old nonmetallic cable, attach the new cable to it and pull it through as you pull the old cable out. Conduit is also mechanically stronger and protects the copper wire better. With this in mind, all exposed wiring should be installed using conduit.

If a GFI is used in a circuit, it will be necessary to trace the entire branch circuit, looking for moisture and/or shorted or grounded wires. The main advantage of the GFI is that it can save lives and prevent injury by automatically shutting off the power. One of the disadvantages of the GFI is that any type of moisture can trip it. For example, many times on a hot, rainy day or during a rainstorm, the GFI will trip, although there is no immediate threat to anyone. Another disadvantage is that even though the GFI trips, a surge of voltage (even a microsecond in duration) can pass through the GFI and shock a person, causing him or her to fall and be injured. This could be especially dangerous to a person working in high places. Other protective devices when used with an isolation transformer could eliminate this disadvantage of the GFI.

A resistance meter can be used to check the insulation resistance in three-phase distribution cables. Test one conductor at a time by connecting the meter to one line and tying the other two lines together and grounding them (Fig. 5.16). Compare the ohmmeter reading with that of the specified standard of the cable, which depends on its size, length, and operating conditions.

Besides the actual line cable, broken or charred insulators may need to be replaced or rebuilt. Often, a troubleshooter may need to design and build new insulators. This situation most often arises when a particular insulator is no longer available.

Another method of troubleshooting wiring systems is through the process of elimination. Using an insulation resistance meter or voltmeter, trace a system from one point to another, disconnecting one apparatus at a time, until the short, open, or ground is found (Fig. 5.17). For instance, to locate a ground, first disconnect the motor at point *C* and check the line voltages, using insulation meter. Second, disconnect the motor control and check the voltages. Third, proceed to the panel box.

Every industrial troubleshooter will encounter a variety of distribution panels, disconnect boxes, and other electric wiring and control enclosures. Problems associated with panel boxes are generally limited to poor installation

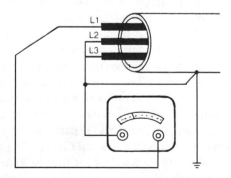

FIGURE 5.16   Checking insulation resistance with an insulation meter.

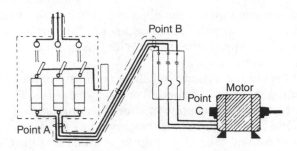

FIGURE 5.17   Troubleshooting procedure in a typical industrial wiring circuit.

connections, broken wires, corroded terminal pads, and defective fuses or circuit breakers. Visual inspections will identify most mechanical hardware problems. Look for any signs of contamination or broken wires or connectors. Test equipment is available to check circuit breakers, although if a circuit breaker fails to reset, routinely trips, or is mechanically sloppy, it should be replaced.

Large power distribution systems sometimes use a combination circuit breaker with visible isolation blades, called a *circuit interrupter*. This device uses a spring-operated mechanism and vacuum interrupter to provide overall protection against phase overcurrents and phase losses and shorts.

Disconnect boxes use a flange-mounted switch with operating handle to disconnect power from motors, pumps, controls, and other power devices. These disconnect switches are generally very dependable. However, if they become inoperable, look for a defective drive mechanism, broken or detached spring or connecting linkage, or worn or corroded pressure plate and fuse connector.

Enclosures, like panel and disconnect boxes, are subject to mechanical damage, atmospheric corrosion, chemicals, oil, moisture, dirt, and other foreign matter. Repair any loose wiring connections, shorted wires, poor pigtail connections, physical damage, and corrosion.

Figure 5.18 shows a typical industrial control enclosure. Notice the neat and well-positioned wires and connections. Always make sure the wire lengths are not too tight or loose, which will prevent possible wiring faults, such as wires shorting together or loosening apart. Also, when you make cable connections in electrical boxes, allow 6 in of wire for the pigtail connections. These pigtail connections are usually made by using plastic wire connectors.

Enclosures and disconnect boxes come in a variety of sizes and designs depending on the application. These remarkable devices allow the troubleshooter to isolate and shut down the power for one piece of equipment instead of shutting down an entire circuit. By adding a disconnect switch the troubleshooter can end power at desired points in a circuit and repair can be performed. Using the correct switch that conforms to *National Electrical Code* (NEC) requirements for industrial motor applications also supports safety for the troubleshooter in performing testing and maintenance procedures. Figure 5.19 shows a typical circuit lock 30-A fused disconnect switch.

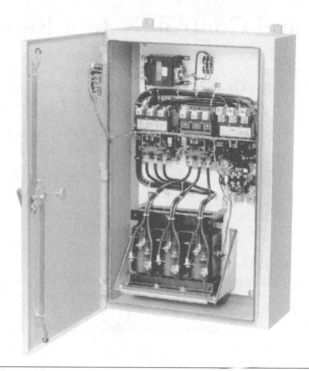

FIGURE 5.18   A three-phase industrial wiring and control panel enclosure. (*Courtesy Allen-Bradley.*)

FIGURE 5.19   A typical 30-A fused disconnect switch. (*Courtesy Hubbell, Inc.*)

Lighting and Control System Repair

Residential and industrial lighting systems often need servicing. Some of the common lighting systems include incandescent, fluorescent, mercury vapor, high-pressure sodium, LED, induction lamps, and high-intensity discharge types. The incandescent lamp (i.e., light bulb) that is becoming obsolete is generally used in homes and uses a tungsten filament. Typical problems with this lamp are cracked lamp sockets, low voltage, and vibration. Low voltage greatly reduces the efficiency (i.e., brightness) of the lamp. Vibration is a major problem of this lamp, because it causes the tungsten filament to break. Measures should always be taken to safeguard against vibration.

There are several types of fluorescent lighting systems. Some include the preheat, rapid start, and instant start. Preheat lamps require an auxiliary starting switch. The starting switch, when activated, allows the ballast to generate current for preheating the electrodes of the fluorescent lamp. After a few seconds, the starter opens and the lamp ignites.

Some common problems of preheat lamps include flickering and failure of the lamp to ignite. These problems may be due to a broken socket, improper lamp seating, low voltage, dirty lamp, improper wiring, or a defective lamp, ballast, or starter switch. The most common problems are low voltage, dirty lamp, or a defective starter switch.

Rapid-start and instant lamps differ from the preheat lamp, because they do not require a starter switch. The ballast incorporates a built-in winding, which provides the necessary current to ignite the lamp. A typical problem with these lamps is that when the lamp becomes dirty, especially under high humidity, the lamp does not ignite but rather flickers. When this flickering occurs, the lamp needs to be cleaned or replaced. Other problems include defective sockets, low voltage, improper wiring, or defective ballast.

Mercury vapor and high-pressure sodium lamps have traditionally been used for industrial applications. These lamps offer high brilliance but require a constant and reliable voltage source. If the voltage varies, the lamp will not operate. Common problems include low voltage, power interruptions, and defective transformers. If the voltage varies, the lamp will shut down and will require 5 to 10 minutes (min) to cool before it will return to full brilliance.

The high-intensity lamp operates similarly to the mercury vapor lamp. These lamps also require a constant, reliable voltage source and sufficient cooling time for restarting. Common problems include inadequate voltage, power interruptions, or defective lamps or ballasts.

Other lamps include metal halide, high-pressure sodium lamps, LED, induction lamps, and lunar resonant streetlights. There are advantages and disadvantages to different lights. For example, the halogen incandescent is about 25 percent more efficient than the traditional incandescent bulb; the compact fluorescent (CFL) can have a life span of nearly 10,000 hours (h) and is about 75 percent more efficient than traditional bulbs but has trace amounts of mercury and lacks some of the dimming capabilities. The LED lamps can last over 25,000 h more than traditional bulbs, are dimmable, and are available in several colors, but they tend to be more expensive than others.

The repair and maintenance of lighting often involves the retrofitting. The replacement lighting often comes in a retrofit kit that allows the troubleshooter to quickly and conveniently install the new lighting fixture. Many retrofits are done to save cost and future maintenance of the lighting. Some of the specifications of retrofit kits include lumen efficiency ratings, lumens per watt, output protection, autorecovery output short-circuit protection, temperature protection, color characteristics, location rating (e.g., dry and wet environments), estimated lamp life, construction qualities, warranty, and specific application. When installing lighting, ensure that the replacement lighting does not encounter physical stress; the power supply is suitable for the lamp; adhere to proper electrostatic discharge (ESD) precautions; power is disconnected; abide by all NEC and applicable federal, state, and local regulations and Underwriters Laboratories (UL) safety standards.

TV Distribution System Repair

The field of television and radio communications continues to develop with new technology such as home and commercial satellite systems, medical telemetry products, digital broadcasting, and advanced antenna distribution systems. The basic field of radio and television reception systems can be mastered by understanding the basic operation and troubleshooting of the television antenna distribution system. There are many types of antenna systems depending on how old the system is and location of receivers. Some of these include parabolic antennas (satellite dish), mobile satellite systems, traditional TV antennas, multisatellite dishes, very small aperture terminals (VSAT), and cable TV coaxial cable transmission and high-speed Internet systems. Depending on the type of application, the antenna is the first primary component of the system that is responsible for receiving the signal. The preamplifier is sometimes mounted in the housing of the antenna and is responsible for increasing the signal level at the point of reception. The distribution amplifier is used to boost the signal to provide a signal sufficient for good reception. For example, Fig. 5.20 illustrates a single-stage distribution amplifier for use in medium to large CATV and MATV (cable and master antenna) systems. It features a variable-gain control, a selectable FM trap, and maximum signal stability. The splitter divides or couples signals to transmission line. The most common types of splitters are the two-way and four-way splitters. Figure 5.21 shows a two-way splitter.

Taps are used to route a portion of the signal from the trunk line to the receiver. The two main types of taps are the line tapoff and the drop tap. Line tapoffs are mounted in the wall near the TV set. The line tapoff provides isolation, so that receivers are unaffected by each other, preventing such problems as smears and ghosts. The drop tap is used when there are several outlets (receivers) a long distance from the main trunk. Drop taps provide lead signals directly to TV sets, which simplifies installation. Figure 5.22 shows an example of a single-outlet directional line drop tap.

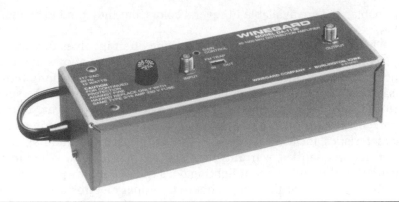

FIGURE 5.20 A solid-state distribution amplifier designed to amplify very high-frequency (VHF) and ultra high-frequency (UHF) signals. (*Courtesy Winegard.*)

FIGURE 5.21 A two-way hybrid splitter for coupling or dividing signals on 75-V coaxial line. (*Courtesy Winegard.*)

FIGURE 5.22 A single-outlet directional line drop tap. (*Courtesy Winegard.*)

Couplers are designed to match the signal(s) from the antenna(s) to the TV set(s). There are several different types of couplers. For example, if separate VHF and UHF antennas are used, a coupler designed to combine these separate signals into a single 75- or 300-V output eliminates the need to run two separate lines to the TV set. Couplers are often mounted on the antenna or in the attic.

Traps are used to attenuate specific signals (i.e., frequencies), eliminating interference. The citizens' band (CB) and FM traps are common ones used to eliminate citizens' band and FM radio interference. Most FM traps are designed to attenuate FM signals from 88 to 108 megahertz (MHz). Attenuators are also used to filter out undesirable frequencies.

An antenna is a conductor that picks up electromagnetic waves and induces current. The length of the antenna and its resonance effect are directly related to the frequency and wavelength. The length of the antenna is inversely proportional to the frequency; for example, the lower the frequency, the longer the wavelength, and the lower the frequency, the longer (larger) the antenna. The decibel is used to measure the gain, loss, or isolation in antenna and cable distribution. The decibel is a ratio of two voltages (i.e., input and output voltages in an antenna/cable system). One thousand microvolts (µV) is equivalent to 0 decibels (dB), when the reference voltage is 1 millivolt (mV).

The two basic kinds of antenna wire commonly used are 75-V coaxial cable and 300-V twin-lead cable. Coaxial cable is generally used for commercial wiring and is the more popular wire. Coaxial cable is round, smaller, and more expensive than twin-lead cable. It is ideal for running through pipes, is weatherproof, and has excellent noise rejection and high durability. It also is a shielded wire that protects against noise and interference.

There are basically two types of twin-lead wire—the standard flat twin-lead and the foam-filled tubular twin-lead wire. The standard flat twin-lead wire is cheaper than, but does not stand up as well in poor weather conditions as, the foam-filled twin-lead type. Neither twin-lead wire gives significant shielding against noise and interference.

Figure 5.23 shows an 82-channel, 96-outlet distribution system. This system illustrates the use of splitters, line taps, amplifier, power supply, preamplifier, and antenna. Terminating resistors are also used to prevent signals from feeding back up the transmission line.

More advanced antenna systems include digital satellites. Satellites provide very clear video receptions and CD-quality sound with capabilities of accessing more than 175 channels. For example, the Winegard mobile digital satellite contains an 18-in galvannealed steel antenna with ultraviolet and weather protection. Various types of satellites include mobile crank-up and automatic digital systems, and portable home digital systems with optional installation kits.

There are many factors that contribute to faulty TV distribution systems. The antenna must be of good quality and free from rust and corrosion. It must be mounted securely to maintain proper antenna orientation. If the antenna is not pointing in the proper directions, weak reception or ghosts may result. Ghosts are duplicate images seen on a television receiver. The ghosts are caused by the

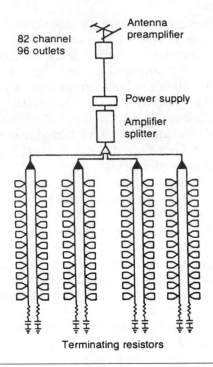

82 channel
96 outlets

Antenna
preamplifier

Power supply

Amplifier
splitter

Terminating resistors

FIGURE 5.23   Master antenna TV system for an apartment complex. (*Courtesy Winegard.*)

television's picking up multiple signals from the transmitting station, which are caused by the wave's being reflected by an object such as a large building. The reflected signal is actually being delayed and arrives shortly after the direct signal and is displaced slightly to the right on the television screen. To remedy this problem, reposition the antenna or install a better antenna system. Another method of reducing ghosts might be to install a smaller antenna that has high front-to-back ratio, but with medium gain.

The antenna must be mounted away from trees, metal, or large objects. The major cause of antenna faults is corrosion from weather conditions. When the antenna is corroded and signal loss is evident, the antenna should be replaced.

The antenna can be mounted in the attic of a home, but aluminum, insulation backing, and other metals may dampen the signal strength.

The major cause of distribution line faults is insulation breakdown. This is most often caused by prolonged exposure to the elements of the weather, such as rain, ice, and snow. The transmission line can be checked by an ohmmeter. Also, by using a small television or monitor, the antenna system can be traced step by step. In this way, the transmission line can be diagnosed and, if necessary, replaced.

The distribution line can also be checked by substituting a known good cable. Keep in mind that a short circuit in coaxial cable does not always create a short circuit. The shorted cable often acts as a trap, apparently because of the dc resistance and self-inductance developed by the short circuit. Unusual effects can be produced on the television receiver, such as color missing on one channel but not on the others or only some receivers having weak signals. An open circuit in coaxial cable can also produce strange results, such as smearing, ghosts, poor reception, or absence of signal. Remember, preamplifiers can only boost the signal that is received by the antenna. It cannot extend the receiving range or increase the signal to the antenna.

Sloppy installation of the distribution system can cause poor reception, ghosts, or smears. Some of the problems include failing to install a terminating resistor at the end of a run, pinching the cable with staples, making sharp bends, creating impedance mismatches, and making loose or poor connections.

Actually, whenever a good television has a snowy picture, the reason must be due to a weak signal. A signal of 0 dBmV, which is equivalent to 1000 mV, is usually needed for good, clear, sharp reception. A common problem of many antenna systems is that too many television receivers are connected to the antenna without proper signal amplification. Each television connected to the antenna will decrease the signal strength. Also, if the distribution cable is too long, the signal will be decreased. For example, 100 ft of common coaxial cable will attenuate a loss of approximately 6 dB at channel 14. This is a decrease of approximately one-half of the 1000 mV needed for a good picture.

When you are servicing or installing antenna systems, it is essential to calculate the signal strength and use the most suitable amplifier. A strong signal could overdrive one amplifier but not another. Also, it is important to design a system in which all decibel losses have been taken into account, in order to have adequate signal strength. Also, when troubleshooting high-speed Internet cable TV systems, remember to check all ancillary equipment that can impact performance such as routers, wireless routers, cable modems, and Ethernet switches (covered in Chap. 11).

In areas where undesired noise or reception occurs, it may be necessary to use a filter to attenuate this undesired frequency. In areas of exceptionally strong FM signals, an FM trap may need to be inserted. If necessary, the troubleshooter may need to connect several FM traps together in cascade.

Lightning is a major threat to antennas. Lightning often strikes an antenna, causing any number of problems, such as open transmission line, open preamp, open amplifier, and open splitter or other component. Usually the preamp that is located in the antenna housing has a protective diode that decreases the chance of more serious problems when struck by lightning. Also, lightning arresters and grounding rods can reduce the chance of more serious problems arising upon being struck by lightning.

Fiber-Optic Communications Repair

There are many types of transmission methods available for data communications. Communications data can be transmitted through telephone lines, coaxial cable, fiber-optic cable, microwave, and satellite. Besides the traditional coaxial cable transmission, the use of fiber-optic and wireless transmission has become common. The transmission of pulses of light that forms and electromagnetic carrier wave is a basis of fiber-optic communication. Fiber optics has been found to be useful in a number of communications such as telephone, Internet, and cable television. There are many advantages of fiber-optic versus traditional wiring systems in that they generally have significantly lower attenuation and less signal interference. Fiber-optic communication systems have particularly been useful for long-distance applications due to the associated higher costs of traditional communication wiring systems. The fiber-optic communication systems commonly use a light-emitting diode (LED) transmitter. These devices have the ability to produce an incoherent light as compared to laser dialed that produces coherent light. The use of vertical cavity surface-emitting laser (VCSEL) devices provides significantly higher power and communication properties than LEDs.

There are various types of fiber-optic receivers. The photo detector is the main apparatus of the optical receiver. It converts the light into electricity from the transmission line. The purpose of the receiver is to extract the communications data from the fiber-optic cables. Once data is received, the optical receiver converts the light signal to an electrical form through the use of photodiode or avalanche photodiode (APD) and other optical detectors. These devices are used in conjunction with transimpedance and limiting amplifiers to produce desired digital signals which can be controlled. Fiber-optic cable has revolutionized the long-distance communications. The cable generally consists of hair-sized class core embedded within cladding and buffer coating insulating material. Fiber-optic cable offers many benefits over traditional copper cable such as higher bandwidth capacity, signal transmission distance, speed, and less sensitivity to electromagnetic interference. However, fiber-optic cable initially had been more expensive and more time-consuming to install. Various types of construction of fiber-optic cable have afforded hybrid and composite materials that offer greater indoor and outdoor applications. For example, specialized cable includes armored cable, which is sensitive to rodent intrusion and ground and aerial applications. In addition, low-smoke zero halogen cables can provide less toxic and quasi fire protection capabilities. Also, depending on the type of cable, the use of fiber connectors and work station housing can impact the installation time.

There are a number of types of connectors that can be used such as non-epoxy and polished, and hybrid connectors. The connectorizing process varies depending on the type of connector. The installation time and skill level of working with these connectors vary. For example, epoxy is a common connectorization method, as well as cleaves and crimp that uses a preloaded fiber stud and ferrule ultraviolet adhesives and epoxyless connectors.

There are many factors that impact the troubleshooting and proper operation of fiber-optic communication. Some of the basic testing equipment includes a power meter, optical time domain reflectometer (OTDR), and fault tracers. For example, OTDR is used to transmit fiber-optic pulses through the fiber cable that scatters the light signal back. The instrument measures the rate of the pulse return, which provides detection of the fiber-optic cable length, attenuation, and potential breakpoints.

There are many features of these instruments such as the ability to test up to three wavelengths in one module, high-resolution and short dead zone for distribution of fiber qualification, automatic traffic detection, automatic macrobend detection, and connectivity to computer software that can be used for post-processing test results. In using the OTDR you must ensure that all connectors, cable, jumpers and adaptors are operational and properly installed.

Described here are some of the typical troubleshooting and installation problems associated with fiber-optic cable.

- Excessive unstripping of cable near the terminating blocks
- Untwisting of cable at the terminating blocks
- Cinching of cables too tightly through use of tie and cable wraps
- Squishing cables when installing with staples
- Overstressing cables by over-tightening multiple cable ties
- Sharp bending of cables versus a sweeping uniform bend
- Kinks, knots or snags in the cable
- Deformation of the pair-twist, altering cable performance
- Improper grounding and bonding that does not conform to NEC requirements

There are many NEC requirements and practices that help ensure proper installation of fiber-optic cables. One good source of cable installation can be found in the manufacturer's instructions. One of the basic methods of repair is to conduct a visual inspection to ensure the proper wire color matching, good connections, obvious breaks or defects in the cable, or obvious smell of burnt equipment or cabling.

Look for obvious damage to the cable due to improper installation typically done by excessive pulling methods, tension damage, kinks, and slicing problems. Close visual inspection can often reveal broken or cracked cable jackets or improper slicing connections. There are several other test instruments other than the OTDR such as the optical loss test set (OLTS) light source and power meter and fault test instruments. If cables are found to have very high loss, they should all be tested for defects and damage. The OTDR (visual fault locator) can be used for this purpose or the OTDR is especially useful for long distances. Always look for the obvious damage to cable such as scratches, dirt, and debris. Never take for granted that manufactured products are good even if new. For example, patchcords often contain heat-tiered epoxy connectors on each end. Potential problems with patchcords include damage from excessive handling or microscopic fiber breakage, material stressors, or manufacturing defects.

Like fiber-optic cables, patchcords can be tested using an optical power meter. Remember to test in both directions. Besides the obvious cable issues, remember to thoroughly test all the communication equipment. It is useful to start by checking the receiver to ensure proper operation. Using an optical power meter, you can disconnect the receiver and conduct power measurements as compared to manufacturer's specifications. Local problems might include low- or high-power measurements due to cable and connector patchcord defects.

There are several faults that can occur with power cable insulation. The objective of insulation is to protect the wire in a power cable. Good insulation

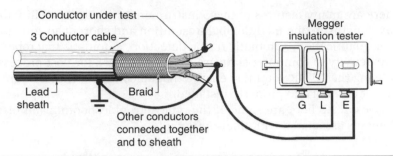

Conductor under test

3 Conductor cable

Megger insulation tester

Lead sheath

Braid

Other conductors connected together and to sheath

G L E

FIGURE 5.24 Testing the insulation resistance of a power cable. (*Courtesy Megger®.*)

basically equates to high resistance, and poor insulation equates to low resistance. Insulation can breakdown due to several factors—moisture and water, excessive cold or heat that causes cracks or deterioration, dirt and petroleum products, and physical damages. The insulation tester is an instrument that is used to measure the quality of the insulation. The Megger® insulation tester is a popular instrument that can be used for testing a wide range of electrical devices and applications such as power utilities, cables, motors, panelboards, street lighting, avionics, and military hardware. The insulation tester develops a high dc voltage and can be produced by hand cranking or line generation. For example, Fig. 5.24 illustrates a typical setup for testing a power cable. When testing the cable itself, both ends of the cable are generally disconnected to avoid potential leakage from switchboards or panelboards.

When testing insulation, some general guidelines include that high values indicate good insulation quality, fair values with constant tendency toward lower values should be examined for faults, and low values probably indicate faults. In this case, one of the first steps is to clean, dry out, inspect, and correct any faults before placing equipment back in service. Also, when obtaining fair or high values, previously well maintained but now showing a sudden lowering in values, conduct tests at frequent intervals, correct the fault, and secure values that are steady and safe for operation. Figure 5.25 shows an example of a troubleshooter using a Megger® insulation resistance and continuity tester for panelboard testing.

Preventive Maintenance

An ideal preventive maintenance program should consist of periodic and routine inspections of all distribution systems, panel boxes, and operating devices. Logbooks should be kept to record all breakdowns.

Chattering, noisy controls should be given immediate attention and corrected. Dirty, gummy wet lines, motor controls, and panel boxes should be periodically cleaned.

Line voltage should be periodically checked for abnormally low or high readings. Line voltages should never vary more than 10 percent of the specified voltage. During hot summer days when low line voltage tends to be a problem, the maintenance

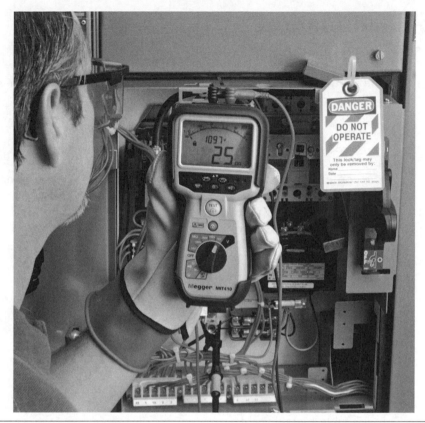

FIGURE 5.25   Panelboard testing using the Megger® insulation resistance tester. (*Courtesy Megger®.*)

department should try to shut down any nonessential operating lights, motors, and other equipment. Low line voltages cause motors to draw more current, which in turn produces more heat. This eventually ruins cables, motors, and other equipment.

The ambient temperature (the temperature close to the area of the operating device) should be routinely checked for abnormal temperatures. Abnormally high temperatures cause breakdowns and can indirectly be a fire hazard, as high heat increases current and causes insulation breakdown. Also, extremely cold temperatures can cause cracking.

Semiconductors are very sensitive to temperature change. Any change in temperature can destroy diodes, transistors, silicon-controlled rectifiers (SCRs), and other semiconductors. Make sure that solid-state devices have adequate cooling and are not subjected to harsh temperature conditions.

Any unnecessary physical or mechanical abuse should be avoided, as it poses a safety hazard to the operator and also premature equipment breakdown. Look for methods to reduce or eliminate any unnecessary vibration, mechanical motion, foreign materials, and contaminants.

Self-Examination

Select the best answer:

1. The minimum amperage capacity panel box used in most homes today is
   A. 30 A
   B. 60 A
   C. 100 A
   D. 150 A
   E. 200 A

2. The common neutral wire is usually _____ in color and connected to a _____.
   A. Black, cold water pipe
   B. White, circuit breaker or fuse
   C. Black, circuit breaker or fuse
   D. White, cold water pipe
   E. Red, circuit breaker or fuse

3. Number 8 wire is generally used for
   A. 15 A
   B. 20 A
   C. 30 A
   D. 40 A
   E. 60 A

4. Another name for electrical metallic tubing (EMT) is
   A. Nonmetallic cable
   B. Flexible metallic cable
   C. Conduit
   D. Armored clad cable
   E. None of the above

5. Another name for flexible steel cable is
   A. Conduit
   B. EMT
   C. Synthetic cable
   D. Nonmetallic cable
   E. None of the above

6. A four-way switch controls a lamp from _____ different location(s).
   A. One
   B. Two
   C. Three
   D. Four
   E. Five

7. The power-factor rating of an inductive reactive circuit can be increased by adding
   A. Capacitors
   B. Inductors
   C. Coils
   D. Fuses
   E. None of the above

8. The device used to pull wire through conduit is called
   A. Wire tongs
   B. Straps
   C. Connectors
   D. Fish tape
   E. None of the above

9. The letters OTDR stand for:
   A. Optional testing digital recorder
   B. Optical time digital resistance meter
   C. Optical time domain resistance meter
   D. Optical time domain reflectometer
   E. All the above

10. To decrease signal strength in strong-signal areas, use
   A. A splitter
   B. An amplifier
   C. A coupler
   D. A new antenna
   E. An attenuator

11. Which of the following lamps uses a starting switch?
   A. Preheat
   B. Rapid-start
   C. Instant-start
   D. All the above

12. Which of the following lamps requires a cooling period prior to restarting?
   A. Incandescent
   B. Fluorescent
   C. Mercury vapor
   D. None of the above

13. The device used to attenuate specific signals is the
   A. Drop tap
   B. Line tapoff
   C. Splitter
   D. Trap

14. The wire that is shielded is
   A. Flat twin-lead wire
   B. Foam-filled twin-lead wire
   C. Coaxial
   D. All the above

15. Which of the following wires has 75-V impedance?
    A. Flat twin-lead wire
    B. Foam-filled twin-lead wire
    C. Coaxial
    D. None of the above
16. The device that is used to prevent signals from feeding back up the transmission line is the
    A. Trap
    B. Line tap
    C. Preamplifier
    D. Terminating resistor
17. A cause of ghosts and smears is
    A. Open transmission line
    B. Improperly positioned antenna
    C. Multipath reception
    D. All the above
18. The preamplifier is generally mounted in the
    A. TV receiver
    B. Antenna
    C. Attic or wall
    D. None of the above
19. The device used to boost a signal is the
    A. Line drop tap
    B. Amplifier
    C. Attenuator
    D. Trap
20. The maximum capacity for a 20-A general-purpose appliance circuit is approximately
    A. 1500 W
    B. 1800 W
    C. 2400 W
    D. 2800 W

Questions and Problems

1. Explain the different types of cable commonly used in residential and industrial wiring.
2. Explain the operation of a three-way and a four-way switch.
3. Explain the appliance and equipment ground system.
4. What are the differences in voltages between a delta and wye transformer power supply system?
5. Explain the functions of a programmable motor protector.
6. What are some disadvantages of a ground-fault interrupter (GFI)?
7. Explain some techniques used in the troubleshooting of wiring systems.

8. State some reasons why a television may have a snowy picture.
9. Explain the effect of ambient temperature on an operating device.
10. State some preventive maintenance procedures one should follow in industry.
11. Explain how ghosts are developed in a television receiver.
12. What can be done to attenuate strong FM signals?
13. State some common problems in antenna and cable systems.
14. What is the difference between coaxial cable and twin-lead?
15. How many microvolts are generally required for a good, clear television picture?
16. Describe what may happen when coaxial cable develops a short circuit.
17. What are some possible results of an open circuit in coaxial cable?
18. Explain the difference between a line tapoff and the drop tap.
19. What is an attenuator?
20. What is the purpose of a splitter?
21. Describe the benefits of coaxial cable.
22. Describe the difference between the standard flat twin-lead wire and the foam-filled tubular twin-lead type.
23. What is a decibel?
24. What is the difference between an amplifier and a preamplifier?
25. Why are terminating resistors used?
26. Explain what an OTDR is.
27. Describe typical insulation problems associated with fiber-optic cable.
28. Explain how to test the insulation of a power cable.
29. What are some typical safety guidelines in testing power cables?

Chapter 6

Troubleshooting Radio and Television

A radio receiver is an electronic device that picks up electromagnetic signals from the air, amplifies them, extracts the intelligence, and reproduces the sound. The receiver must match the type of transmitter. The type of receiver is specified by the type of modulation the receiver is designed to handle: amplitude modulation (AM), frequency modulation (FM), frequency multiplexing (FM MUX), stereo, etc.

The field of television is a very special branch of receivers and uses sophisticated technology. The transmission and reception of pictures are a tremendous feat in black and white and a marvelous development in color. In this chapter we examine some principles and the troubleshooting of radios and television receivers. We first cover the radio and then move to the more complicated television receiver.

AM Fundamentals

A radio wave is basically an electromagnetic energy vibration. This wave travels at a speed of 300 million meters per second (m/s) or 186,000 miles per second (mi/s), with each wave having a certain length. The lower the frequency, the longer the wave; the higher the frequency, the shorter the wave. An audio wave is approximately between the frequency of 20 hertz (Hz) and 20 kilohertz (kHz), or 20,000 Hz, and can be heard by most people. Waves above this frequency, which are called *radio-frequency* (rf) waves, cannot be heard by people.

Because the physical size of an antenna is proportional to the length of the wave, using an antenna to transmit and receive audio waves would be impractical. Therefore, a high-frequency continuous wave (cw) produced by an oscillator is mixed with the low-frequency audio wave. This produces a modulated wave used to transmit the intelligence from the transmitter to the receiver. Figure 6.1 illustrates the amplitude modulation principle in which the intelligence, or low

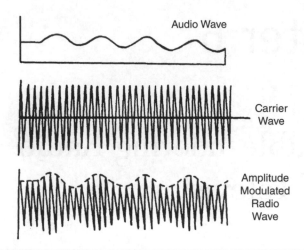

FIGURE 6.1 The two waves that form a modulated rf wave.

audio frequencies, varies the amplitude of the carrier. The AM wave or signal is a composite of the carrier and the upper and lower sidebands.

Most AM receivers are called *superheterodyne* receivers because of the mixer stage. Figure 6.2 shows a block diagram of an AM receiver with the signals that are processed at each stage. The antenna receives many radio frequencies within its frequency band. The tuner, which consists of a variable capacitor and a coil, selects a desired band of frequencies and passes them on to the mixer stage.

At the mixer stage, the incoming rf signal is combined with a constant-amplitude continuous wave that oscillates at a frequency of (generally) 455 kHz [the usual intermediate frequency (i.f.)] above the incoming rf signal. The beating of the signals together produces many frequencies composed of the sums and differences of these signals. The output of the mixer passes through a tank circuit tuned to 455 kHz.

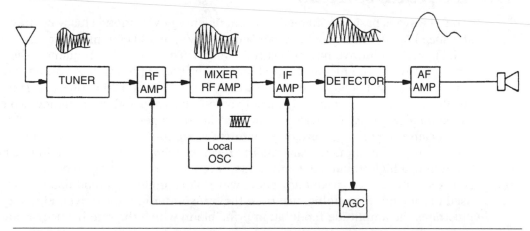

FIGURE 6.2 An AM receiver block diagram. (*Courtesy Howard W. Sams & Co., Inc.*)

Suppose that the rf tuning circuit is set to receive a signal at 1000 kHz and the i.f. circuit is set at 455 kHz. Then the local oscillator will be set at 1455 kHz above the incoming signal so that the new intermediate difference frequency of 455 kHz will be created. The 455-kHz i.f. is the difference between the incoming frequency of 1000 and 1455 kHz. Note that 455 kHz is the most common i.f.; many receivers are designed to operate at other intermediate frequencies.

The next stage of signal processing is a single stage or multiple stages (up to three) of i.f. amplification. Each i.f. stage is tuned to the receiver's specific i.f. for improving selectivity.

The function of the detector stage is to separate the audio intelligence from the i.f. carrier. It does this in two steps. First, the composite signal is rectified, which leaves the upper envelope of the composite AM signal; then the i.f. carrier component is filtered through a capacitor to ground, and only the low-frequency audio signal is passed. The *automatic gain control* (AGC) circuit feeds back a portion of the signal to provide control for constant volume.

The audio signal from the detector is passed through an *audio-frequency* (af) amplifier stage. Here, enough power is given to the audio signal to drive the speaker.

FM Fundamentals

The frequency modulation (FM) technique of transmission begins with an rf carrier wave and an audio frequency called the *modulating signal.* When the rf carrier is modulated with the audio signal, the frequency of the rf carrier varies with the amplitude of the modulating signal. Figure 6.3 illustrates the principles of frequency modulation.

A block diagram of an FM receiver is shown in Fig. 6.4. The antenna receives FM signals within its band, and the tuner stage selects a specific band of frequencies. The rf amplifier strengthens the FM signal. The local oscillator generates a constant-amplitude rf signal that is mixed with the FM signal, producing an i.f. The i.f. stages are generally tuned to 455 kHz. The i.f. stage or stages pass as well as amplify the i.f. carrier.

The FM detector stage is different from the AM detector stage. The FM detector must convert the frequency variations to its audio representation. Then the detected audio signal is fed through a deemphasis network. The deemphasis network restores the relative amplitudes of the signal's frequency components. At the transmitter, the high frequencies are further amplified—called *preemphasis*— to improve the signal-to-noise ratio for transmission. Therefore, at the receiver, a reverse process must be done.

After deemphasis processing, the audio signal is amplified to drive the speaker at the audio amplifier stage. Notice that there is an *automatic frequency control* (AFC) that keeps the receiver oscillator properly tuned.

Some of the AM circuit can be used by an FM receiver. Figure 6.5 illustrates a block diagram of a combination AM-and-FM receiver. When you switch from AM to FM reception, the unique circuits are all switched in simultaneously.

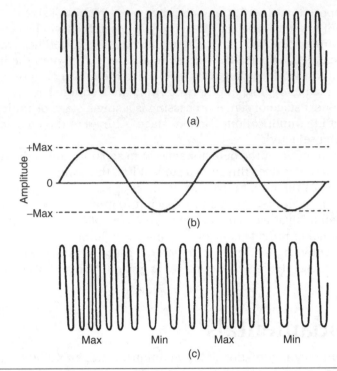

FIGURE 6.3 Frequency modulation principle—an audio wave mixed with a carrier wave, producing frequency variations.

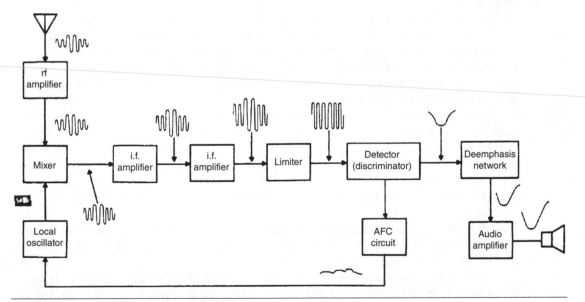

FIGURE 6.4 FM receiver block diagram.

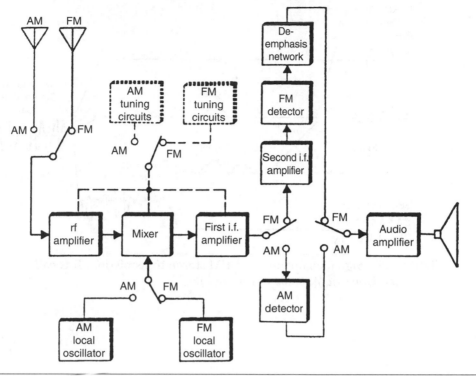

FIGURE 6.5   AM/FM receiver block diagram.

FM Multiplex Fundamentals

When you listen to an FM stereo station, it is very apparent that there are two separate audio channels coming from the two speakers. Figure 6.6 shows the components of an FM stereo multiplexed transmitted signal. Back at the radio station, the FM stereo multiplex signals begin with two separate microphones picking up the audio signals. The signals are designated L (left) and R (right), based on the relative positions of the microphones. The L and R signals are sent to a circuit, called a *matrix network*, which produces two new outputs. One output is the sum of L and R, called the *L1R signal*, and the other is L minus.

The L2R output is amplitude-modulated by a 38-kHz subcarrier, which produces sidebands above and below 38 kHz. The 38-kHz subcarrier is then suppressed after the modulation, leaving only the sidebands. The L2R sidebands are sent to an FM transmitter. The FM transmitter is frequency-modulated by the L1R output and the L2R output, combined with a 19-kHz pilot signal carrier. Figure 6.7 shows a block diagram of a stereo FM transmitter. Notice the matrix network.

A frequency spectrum analysis of a modulated FM stereo multiplex carrier is shown in Fig. 6.8. Notice the low end of the frequency spectrum contains the

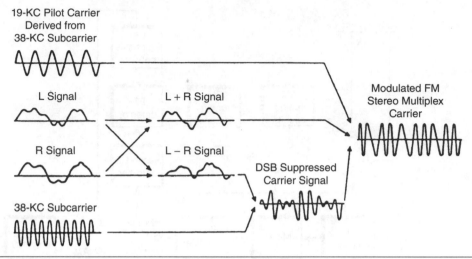

FIGURE 6.6   Signal components of FM stereo transmission. R is called L2R signal. (*Courtesy Howard W. Sams & Co., Inc.*)

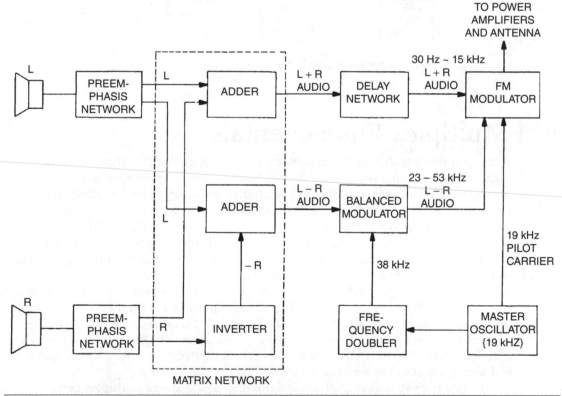

FIGURE 6.7   FM stereo transmitter block diagram.

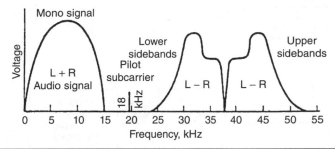

FIGURE 6.8   FM stereo frequency spectrum.

L1R signal for monophonic receivers (30 Hz to 15 kHz). The L2R sidebands with suppressed carrier (23 to 53 kHz) occupy the high end. A 19-kHz pilot carrier is also transmitted as part of the FM multiplexed carrier to be used by the receiver for synchronization and re-creating 38 kHz for demodulation. The purpose of the suppressed subcarrier is to reduce the energy content of the composite stereo signal for an optimum signal-to-noise ratio.

A receiver works "backward" from the transmitter in demodulating the composite stereo signal. When the composite stereo signal is received, it consists of the L1R signal, the sidebands of the L2R signal, and the 19-kHz pilot carrier. If a receiver is not equipped for FM stereo, it responds only to the L1R signal and processes this as a monophonic signal.

However, when the receiver has FM stereo, the L2R signal is recovered by mixing the L2R modulation sidebands with a 38-kHz carrier and then extracting the original L2R signal. Remember that the 38-kHz carrier is generated in the receiver and uses the 19-kHz pilot carrier for synchronization. Then the L2R signal and the L1R signal are processed in a matrix network circuit similar to the one used at the transmitter.

In the matrix network, the L1R and the L2R signals are added, which produces the original L signal. Also, in the same matrix the L1R and L2R signals are subtracted, which produces only the original R signal. Then the original L and R signals are amplified and sent to their respective speakers.

The bandpass and matrix method of demodulation is shown in Fig. 6.9. After the discriminator, the three signal components are separated by filter circuits. The L1R signal is obtained through a low-pass filter (30 Hz to 15 kHz) and then is passed through a delay circuit so that it reaches the matrix network at the same time as the L2R signal. A high-bandpass filter (23 to 53 kHz) "selects" the L2R double sideband signal. A narrow-bandpass filter of 19 kHz "selects" the pilot carrier and feeds it to an amplifier and then to a doubler circuit, which yields a 38-kHz output. The 38-kHz output is the precise carrier frequency of the double sideband suppressed carrier.

Passing the L2R signal and the 38-kHz signal through the nonlinear circuit of an AM demodulator produces the sums and differences of the signals, of which one is the 30-Hz to 15-kHz L2R signal that is selected by a low-pass filter. The L2R audio signal and the L1R audio signal are sent to the matrix network and processed as shown in Fig. 6.10. Notice that one channel adds L1R and L2R signals, which

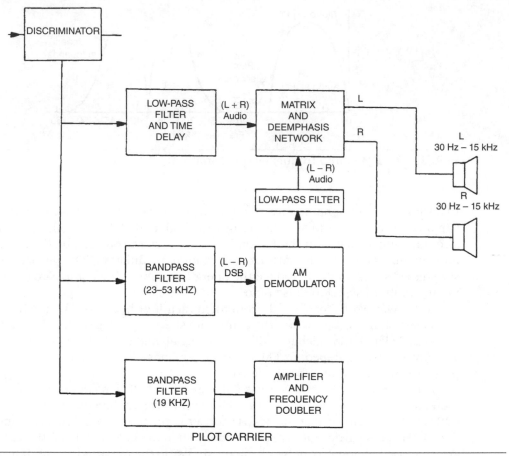

FIGURE 6.9   Bandpass and matrix demodulation block diagram.

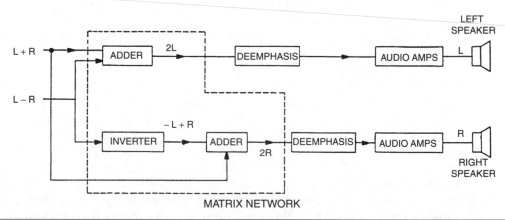

FIGURE 6.10   Matrix network block diagram.

produces only the L signal. The L2R signal is passed through a phase inverter that changes the signs of L2R to 2L1R. The 2L1R signal and the L1R signal are then added, which produce only the R signal. The L and R signals are sent to respective audio amplifiers and then to their speakers.

Another method of demodulation used is electronic switching. A block diagram is shown in Fig. 6.11. The signal coming from the discriminator is passed through a subcarrier regeneration circuit and is then fed into an electronic switch. An illustration of the signal that is fed into the electronic switch is also shown. The electronic switch samples the positive and negative halves of the composite audio stereo waveform and, in effect, demodulates the stereo signal. Figure 6.12 shows a schematic of an electronic switching multiplex unit. The composite audio signal comprises the L1R signal, L2R sidebands, and the 19-kHz pilot carrier.

The 19-kHz pilot carrier is amplified and sent to a frequency doubler (CR1 and CR2), which regenerates the 38-kHz subcarrier. The subcarrier signal is then mixed with the audio signal to reconstruct the composite stereo wave. A switching bridge circuit consisting of CR3, CR4, CR5, and CR6 is turned on by the instantaneous polarity of the 38-kHz subcarrier voltage; and the two sides of the bridge alternately conduct. Thus, the positive and negative envelopes of the composite waveform are

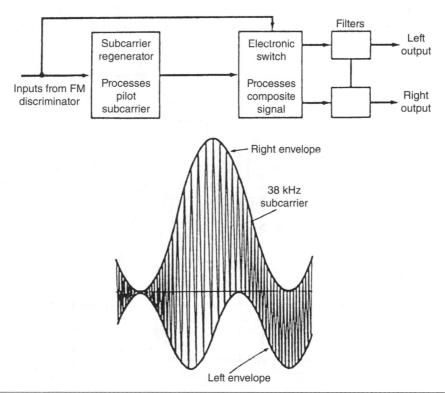

FIGURE 6.11   Electronic switching demodulation block diagram and FM signal.

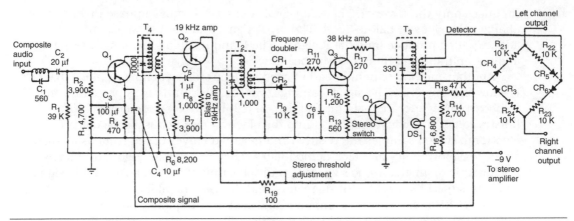

FIGURE 6.12   Schematic of electronic switching multiplex unit. (*Courtesy Howard W. Sams & Co., Inc.*)

sampled at a 38-kHz rate, and the bridge output yields the demodulated L and R signals.

Tape players, compact disc(CD) players, stereos, and other sound equipment require one or more amplifiers to increase the signal so that it can be heard in the speaker. Figure 6.13 illustrates a typical push-pull amplifier common to many types

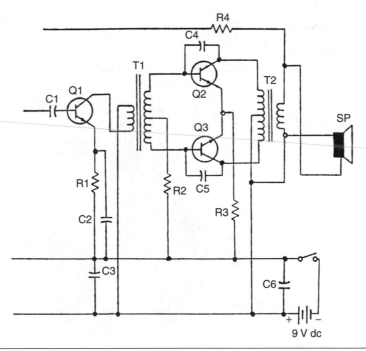

FIGURE 6.13   Typical push-pull amplifier.

of sound equipment. Transistor Q1 is the driver stage for the push-pull amplifier. Capacitor C1 is a coupling capacitor that passes the signal to the driver from the preamp stage. Capacitor C2 and resistor R1 are the *RC* emitter network that properly biases the transistor for amplification. Transformer T1 splits the signal phase by 180° for Q1 and Q3. These two balanced transistors receive the signal alternately for each half-cycle, adding the total power together at the output. Capacitors C4 and C5 couple part of the collector signal back into the base of the transistor. This negative feedback prevents the transistors from oscillating and reduces distortion. Transformer T2 matches the impedance of Q2 and Q3 with the speaker. Capacitors C3 and C6 are filter capacitors used to isolate each section from stray, unwanted signals in the line.

Another type of push-pull amplifier that eliminates the need for the transformer is the quasi-complementary type. It is the most popular one used. A schematic is shown in Fig. 6.14. The upper pair of *npn* transistors, Q2 and Q4, conduct when the signal at the output of Q1 is positive. The *pnp* transistors, Q3 and Q5, conduct during the negative portion of the signal's cycle. Both portions of the cycle are reconstructed as a complete cycle across the load resistor *RL*, thereby implementing the push-pull.

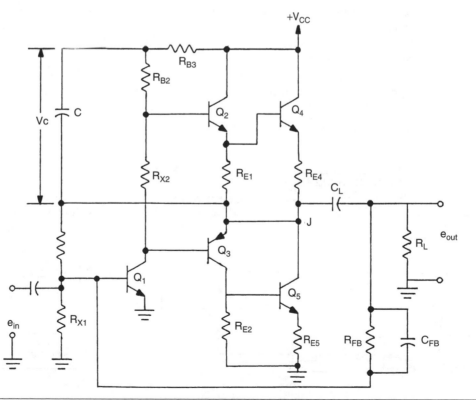

FIGURE 6.14   Quasi-complementary amplifier circuit.

Older equipment will contain circuits made up of discrete transistors, as shown in these figures. In more modern equipment, the audio amplifier is often contained in a single integrated circuit (IC) or sealed module. Audio modules with high-power audio output capabilities are commonly used in receiver and stereo amplifier combinations. For these modules, the troubleshooting consists of verifying that power is applied to the module and that the correct signal is applied to the inputs. If there is no output, the module must be replaced. There are still some current designs that use discrete transistor amplifier circuits for the high-current (high-power) output stages.

TV Transmitter and Receiver Fundamentals

The TV analog are only used in special applications today and have been largely replaced by digital broadcasts. Chapters 8 through 11 cover the fundamentals and troubleshooting of digital devices and networks. However, the basic analog transmitter is actually two separate transmitters. The video or picture signal is amplitude-modulated onto a carrier, while the aural or sound transmitter is actually an FM system very similar to a broadcast FM station. Therefore, the composite transmitted signal is a combination of both AM and FM principles. A simplified block diagram of a TV transmission and reception system is shown in Fig. 6.15. The TV camera acts as a transducer that converts light energy to electric energy, whereas the picture tube [cathode-ray tube (CRT)] is a transducer that converts the electric energy back to light. The microphone and speaker are related transducers for the sound system.

The TV camera has a thin electron beam that is moved horizontally across a light-sensitive surface, producing a voltage proportional to the light. This creates a video line. The electron beam is made to retrace and scan another line. This is done 525 times per second. The TV receiver must have some means of synchronizing the

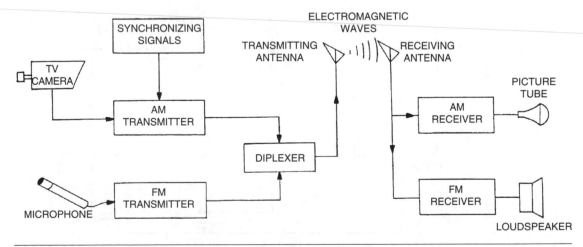

FIGURE 6.15 Simplified TV transmission system.

traces made at the camera. Therefore, the transmitted signal carries synchronization pulses.

The analog television receiver actually reproduces a series of dots. These dots travel at such high speed that the viewer sees the total overall sequence as a picture on the screen. The degree of speed and the intensity of the dots vary. The electron gun produces a stream of electrons that magnetically scan left to right and from top to bottom. Specific phosphors on the screen illuminate when struck by electrons.

The scanning system used in television is the interlace system that starts at the top left, scanning the odd lines from left to right and completing 262½ lines. The interlaced scanning is illustrated in Fig. 6.16. The scanning returns from the bottom of the screen back to the top center and completes the scanning of the even lines. Each odd or even set of scanning lines represents a field, and both an odd set and an even set represent a frame. Therefore, there are a total of 525 lines per frame, and the frequency is 30 frames per second.

Each time the scanning beam moves from the left side to the right, it must quickly return. This is called *horizontal retrace.* When the scanning beam gets to the bottom of the screen, it also must quickly return to the top of the screen. This is called *vertical retrace.* During retrace, the picture is black. Only during the trace scanning period is a picture visible. The vertical oscillator running at 60 Hz deflects the scanning beam upward. The horizontal oscillator runs at 15,750 Hz and higher, and deflects the scanning beam from left to right across the screen.

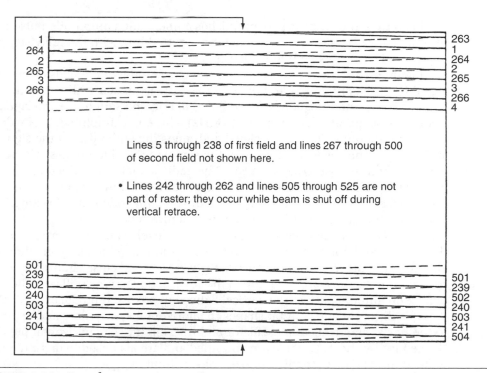

Lines 5 through 238 of first field and lines 267 through 500 of second field not shown here.

• Lines 242 through 262 and lines 505 through 525 are not part of raster; they occur while beam is shut off during vertical retrace.

FIGURE 6.16   Interlace raster scanning pattern.

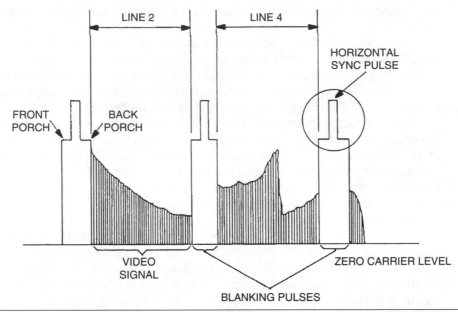

FIGURE 6.17 Simplified video signal showing lines of video, blanking pulses, and horizontal sync pulses.

Each time the scanning beam finishes a line, a high-amplitude pulse is introduced. The pulse "syncs in" each transmitted line with the television receiver. Figure 6.17 shows a simplified video signal.

Figure 6.18 shows a block diagram of a black-and-white television receiver. The signal from the antenna is amplified in the rf stage, mixed with a continuous wave of a predetermined frequency from the oscillator, and sent to the intermediate-frequency stages at 45.75 megahertz (MHz) where it is amplified. The video detector then demodulates the signal and sends the audio part of the signal to the audio stages and the video portion of the signal to the video stages. The sound detected from the video signal is FM. The audio signal is amplified in the audio i.f. amp, demodulated in the FM detector, again amplified in the audio or af amp, and reproduced as sound by the speaker. Meanwhile, the video signal is amplified by the video amplifier and sent to the grid of the picture tube.

The AGC maintains the signal at a constant level. The sync separator removes the vertical and horizontal pulses, and applies them to integrating and differentiating circuits. The integrating circuit shapes the vertical sync pulses into a series of triangular pulses and applies them to the vertical deflection oscillator. The vertical deflection amplifier drives the vertical deflection yoke and produces vertical scanning.

The differentiating circuit shapes the horizontal sync pulses and applies them to the horizontal deflection oscillator and AFC circuit. The horizontal pulses are "locked in" by the AFC. They are amplified and are also used to drive the horizontal deflection yoke that produces horizontal scanning.

The high voltage needed to light the picture tube is derived from the horizontal amplifier. This voltage from the horizontal amplifier is stepped up to approximately

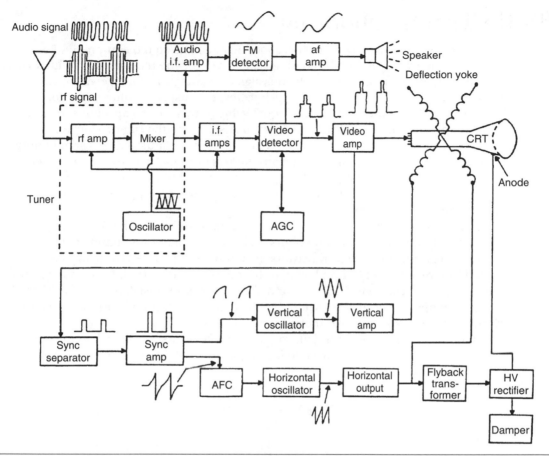

FIGURE 6.18   Block diagram of a black-and-white receiver.

30,000 volts (V) or more by an autotransformer (often called a *flyback transformer*). The voltage is rectified by the high-voltage rectifier and finally sent to the anode of the picture tube. The damper is a diode placed in the path of the kickback pulse (ringing) from the yoke. The function of the damper is to prevent the kickback action from happening more than once for each applied "kick."

The circuit descriptions of modern receivers vary according to the manufacturer. Most use a microprocessor that interfaces with the operator controls and video processor. Many video processors contain a number of functions such as audio and video i.f., source switch, sync separator, and mixers and drivers for color units. In most receivers, the horizontal and vertical deflection circuits are driven by a pulse from the video processor. Other sections using ICs include the power supply, video output, audio system, high voltage, and tuner. In addition to conventional televisions, digital high-definition televisions (HDTV) contain sophisticated system control functionality that provides higher picture quality. Plasma display technology has also given superb picture quality for either conventional or high-definition video sources.

Radio Troubleshooting

Several techniques are used in troubleshooting radio and sound equipment. For example, when you are troubleshooting a superheterodyne receiver, start with a visual and hearing inspection. Look for obvious signs of damage. If the receiver has a hum, most likely it has a defective filter capacitor. Test the suspected filter capacitor by bridging it with a known good one of equal value or by using a capacitor substitution box, as shown in Fig. 6.19. If the hum disappears, replace the filter capacitor.

In a dead set, check the switch with an ohmmeter. Naturally, this should only be done with the unit disconnected from the power line. Also, use the ohmmeter to check the fuse, power supply diode, thermistor, and filter coil. Any one of these components could cause an open circuit. A simple power supply showing the above components is seen in Fig. 6.20.

If a radio or stereo plays for a while, then stops and comes back on later, check for thermally intermittent components. Using a combination hot-and-cool blower, carefully heat the suspected transistors or ICs. When the defective component is heated, the receiver will stop playing or become badly distorted. Now cool off the transistor, using the cool-air blower or a chemical freeze mist. The receiver should once again operate normally. When the thermally intermittent transistor has been found, replace it with a recommended model.

If a receiver has power but fails to produce sound, first try to isolate the defective stage. Signal injection can be used to locate the faulty stage, as shown in Fig. 6.21.

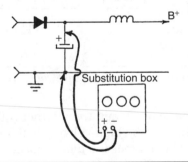

FIGURE 6.19   Using a substitution box to bridge a capacitor.

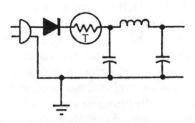

FIGURE 6.20   Simple power supply.

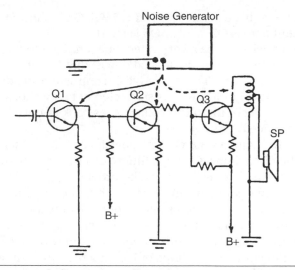

FIGURE 6.21   Isolating a defective stage by using a noise generator.

First, using an af signal generator or a noise generator, connect the ground lead to the chassis ground and the other lead to each stage of the receiver. Start by injecting an af signal around 400 Hz into the speaker; this should produce a tone. If you hear a tone, the speaker is operating. Proceed to the amplifier stage and inject a signal into the base of this transistor. If a tone is heard, proceed to the next stage toward the detector. When you inject a signal into the detector stage or back to the antenna, use an rf signal. If no tone is heard when the rf signal is injected into the detector stage, you can assume this is the faulty stage. At this time, proceed to make resistance checks of each component until the faulty component is located.

Should the faulty stage be the quasi-complementary amplifier circuit, the problem usually exists in the output and/or driver transistors, Q2 through Q5. To troubleshoot the circuit, check the resistance between ground and point J and between $1V_{CC}$ and point J without power applied. Refer to Fig. 6.14 for these reference points. Connect the negative lead to ground and the positive probe lead of the voltmeter to point J. Note the measured value. Then reverse the voltmeter leads. The second resistance value should be considerably less than the first. A dead short indicates that transistor Q3 and/or Q5 is shorted.

Repeat this procedure, but this time check the resistance between $1V_{CC}$ and point J. You will get a lower reading when the positive lead is at $1V_{CC}$. A short indicates that Q2 and/or Q4 are defective. Now short RX2. Turn on the amplifier. A voltage measured at point J should be $V_{CC}/2$. If the measurement is within 25 percent of this condition, try different values of RB1 or RX1 to achieve a more reasonable tolerance. Sometimes RX2 is substituted with diodes to improve bias stability with varying temperatures. Also, if your audio equipment has diodes, check them for shorts or opens.

Check the voltage across Q3. If it is $1V_{CC}$, Q4 is shorted. Similarly, if the voltage measurement across Q2 is $1V_{CC}$, Q5 is shorted.

Voltage checks can be a very effective technique to locate a problem. For example, if a power supply is "loaded down" due to a shorted component, it will draw excessive current. This excessive current will bring down the voltage. If the Zener diode shown in Fig. 6.22 is suspected of being short-circuited, you lift one of its leads from the circuit. If the voltage goes back to normal, the diode is shorted and should be replaced.

If a receiver squeals, howls, or "motorboats," the signal feedback is appearing in one of the circuit stages. Check filter capacitors and battery connections; most likely, a defective filter capacitor is the problem.

Intermittent operation can be caused by other components besides transistors. For example, loose connections and soldered connections can cause intermittent operation. To locate the intermittent problem, wiggle the wires and tap on the printed-circuit side of the board. Sometimes it may be necessary to resolder several soldered connections in the hope of locating the problem. Remember, always turn off the power before you resolder. Otherwise, the heat from the soldering iron will cause a transistor to draw more current and destroy itself. Also, always use a heat sink when resoldering diodes and transistor connections.

If a receiver continues to blow fuses, turn off the power and start taking resistance measurements. Keep in mind that if a shorted capacitor is suspected, do not bridge it with another. Instead, substitute a different capacitor in its place. Check for shorted rectifiers by using the ohmmeter and the manufacturer's resistance specifications. Check each specification with the reading obtained with the ohmmeter until the short circuit has been identified. A short circuit has a near zero or zero reading, as shown in Fig. 6.23.

An oscilloscope can also be effective in locating a defective stage. The oscilloscope allows the troubleshooter to actually see the signal waveform.

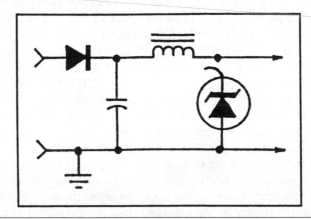

FIGURE 6.22   Lifting one lead of a Zener diode to determine its effect on the circuit.

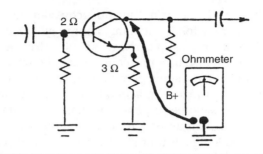

FIGURE 6.23   Measuring the collector resistance of a transistor by using an ohmmeter.

Frequency response, gain, phase shift, noise, and hum can all be identified by the oscilloscope.

When you are repairing tape players, a common problem that causes low volume, lack of treble control, and distortion is dirty tape heads. Use isopropyl alcohol or tape head cleaner to clean the head of the tape recorder. Also, use a cotton swab and the cleaner to clean off the capstan shaft and roller.

Laser Disc and DVD Players

Laser disc players (LDs) are now the standard for music and are becoming more popular again for video recordings. The major advantage of the laser disc is that the information is digitally recorded. This means that the voltage of the signal coming from a microphone or video camera is sampled at regular intervals, and the value of the signal at that instant in time is stored as a binary (digital) number. These digital numbers are stored by burning pits in the surface of a disc, using a laser beam. The circuitry necessary to reproduce sound or video images from digital information is very sophisticated. A semiconductor laser reflects its beam off the surface of the disc as it rotates. As the beam strikes the pits, the light does not reflect, and this change is sensed by the circuitry. Combinations of these reflecting and nonreflecting points are used to represent numbers, and a sequence of these numbers represents the original analog signal (audio or video). Digital signal processing in the CD player's circuitry is able to minimize noise and distortion while reproducing a signal very similar to the original analog input.

The actual circuitry that makes up the spindle speed control, the pickup carriage control, and the audio signal processing on a CD player is not generally repairable. It must be replaced if it is found to be defective. The input and output actuators such as switches, sensors, and motors should be checked.

Laser disc players, in many respects, have similar mechanical components to those of CD and tape players. Typical problems involve the mechanical parts, such as the loading mechanism, spindle motor elevator mechanism, lever assembly, tilt sensor, tilt motor and cam gear, drive gear, and other limit switches, levers, and brackets.

For example, if the disc is not loading, consider the loading mechanism or whether the trigger switch is engaging. Poor playback may be a result of misaligned or defective tilt sensor, pickup carriage, spindle, drive assembly, or circuit problem. Foreign material can also be a problem. Check the pickup lens for dust and dirt. Clean the lens and surrounding components, using a cotton swab and isopropyl alcohol. Gently wipe the lens in a spiral motion from the center of the lens to the outer edges. Never use other types of alcohol, or else damage may result.

The basic test for a multiplex unit is to measure the decibel (dB) separation between the L and R signals that it produces. The setup for a stereo multiplex separation test is illustrated in Fig. 6.24. A stereo multiplex signal generator supplies the L and R signals, which are the same but different in phase relationship. Therefore, when the R signal is applied to the multiplex unit, the R output channel should be maximum, while the L channel should theoretically be zero.

On the other hand, when the L signal is applied to the multiplex unit, the L output channel should be at maximum and the R channel should theoretically be zero. In practice, about a 20- to 30-dB difference should be observed between meter readings for proper operation of the multiplex unit. If a separation of 10 dB is obtained, the multiplex unit is borderline and is probably defective.

Inadequate separation is usually caused by a component failure in the multiplex unit such as unmatched diodes or diodes with poor forward- and reverse-bias ratios in the switching bridge, a collector leakage in a transistor, or a shorted or open capacitor. Also, misalignment can cause poor separation, especially in a matrix network unit. Consult an appropriate service manual for multiplex unit alignment procedures.

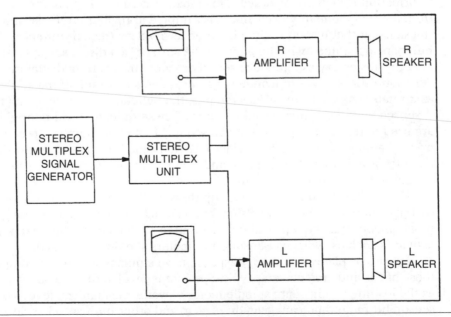

FIGURE 6.24   Stereo multiplex separator test setup.

An off-value resistor in a matrix network circuit can reduce separation. If you find that the stereo multiplex unit has appropriate separation when tested with a composite audio signal from a generator but does not when it is operated normally, the problem may be poor alignment of the high-frequency signal channel in the receiver.

Distortion of the output signal can be caused by a defective diode or an off-value resistor in the switching bridge. A faulty electrolytic capacitor in the subcarrier regeneration circuit can cause a distorted output signal (i.e., humming).

If the left channel is distorted and the right channel is good, compare the voltage, resistance, and waveform measurements and try to isolate the defective component. For example, in Fig. 6.25 if Q15 has a collector voltage of 7.38 V, this transistor is not operating and could be defective. Also, if the voltage at this same point is down to 4 V, a shorted transistor is suspected. If the output is low from the driver transistor Q15, a faulty emitter bypass capacitor may be suspected. Check this by bridging it with another capacitor of the same value. Also, if any one of the resistors changes in value, the bias of the transistor will also change and will affect the output. Resistors can be checked by measuring their values out of circuit, and if the resistors are not in tolerance, they should be replaced. However, removing them from the circuit to test them is a time-consuming task, and generally a resistor will

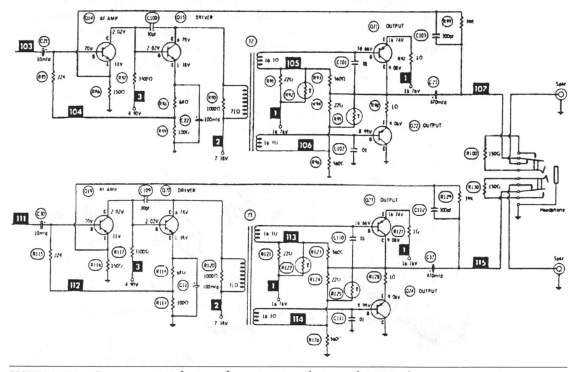

FIGURE 6.25 Component values and operating voltages of a typical stereo system. (*Courtesy Howard W. Sams & Co., Inc.*)

not change unless it has been overheated by too much current, in which case either it will often look damaged or the board around it will appear scorched.

If no signal enters the af amplifier stage, an open coupling capacitor may be suspected. Transformer T2, which acts as a splitter, will cause this channel to go dead if it is shorted. Often, a shorted transformer will have a charred appearance and will easily be identified by its "burned" smell. Using an ohmmeter, you should read 16.3 ohms (Ω) across each half of the secondary coil. A shorted coil will read near zero.

If there is low volume and high distortion in this channel, one might suspect Q21 and/or Q22 to be defective. If only one of these transistor outputs is found defective, replace both with a matched pair. It is important that these transistors be properly balanced for best signal reproduction. Remember, when you replace power transistors, always use a heat sink and insulating mica. Also, use heat-sink silicon compound if necessary, as shown in Fig. 6.26.

Anytime a radio receiver requires bench servicing, correctly connect the radio to a well-filtered power supply. Use the correct voltage and correct polarities. If it is an automobile radio, remember to apply a voltage of 12 V dc and not less, as the automobile electrical system typically operates between 12 and 13 V. Also, remember to arrange all wires neatly. Shorted speaker wires can blow out the output transistors and can cause static. Never operate the receiver without having speakers hooked up, since this could have the same effect.

Figure 6.27 shows a typical three-stage amplifier commonly used in a radio receiver. If Q3 draws excessive current, this will cause Q1 to conduct more heavily. The collector voltage at Q1 will decrease, because Q1 draws more current. As the collector voltage of Q1 drops, this will in turn reduce the conduction of Q2 and Q3. Together, this circuit will have low output and high distortion. Replacement of output transistor Q3 should solve the problem.

There are a number of problems that can occur with digital versatile disc (DVD) recorders and players. Some of the typical problems associated with them

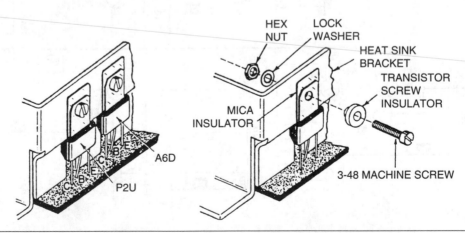

FIGURE 6.26   Example of proper installation of an output transistor. (*Courtesy Motorola, Inc.*)

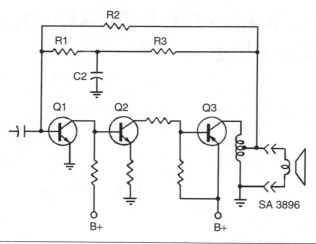

FIGURE 6.27   Typical three-stage amplifier. (*Courtesy Motorola, Inc.*)

include frozen screens, inability to play a disc, sound issues, noise and distortion, picture issues, and poor resolution. For example, if the recorder does not play a disc, some of the remedies might include a dirty or damaged disc, moisture or condensation in the unit, or electronic or mechanical malfunction of the apparatus. When troubleshooting sound and power issues, start with the basics. Always ensure that connections are properly secure, and switching to the correct soundtrack such as Dolby or PCM mode on the disc menu. Also, data files may be mixed such as an MP-3 file on the same disc which can cause noise or sound distortion. When servicing picture problems, consider cleaning the unit, resetting the color system of the player, ensuring the television source is properly set, ensuring a channel scan has been conducted when using a cable box, and checking all video input and output connectors.

When cleaning dirty discs, there are several commercial disc-cleaning kits available. The microfiber cloth is excellent for cleaning dirt, fingerprints, and smudges on DVDs, Blue-rays, and CDs. Scratches are very difficult to repair. However, it is possible that burning another disc in the hard drive may help. When cleaning a DVD, gently wipe the disc in straight lines from the center. Avoid using a circular motion. Sometimes moistening the cloth with distilled water or isopropyl alcohol can help remove stubborn smudges. Make sure everything is dry before trying the disc again. Also, damaged discs might be repaired through use of polish cleaners used by opticians or a gentle detergent. Avoid scrubbing the disc too hard, which ultimately may cause more damage.

It is often necessary to follow the manufacturer's diagnostic troubleshooting charts when available. For example, a typical diagnostic chart will give a step-by-step procedure covering such items as the software, drive analysis, video and audio settings, data cable analysis, and mechanical and connection coverage.

Black-and-White CRT TV Monitor Troubleshooting

While black-and-white CRT TV monitors are virtually nonexistent today, they are still used in a variety industrial, security surveillance, computer monitors, and special visually impaired applications. Therefore, although it may be cheaper to simply replace the defective television, in some cases, it can still be worthwhile to repair some of the easier defects. Televisions and computer monitors are similar in circuit design. Because each stage of a TV receiver or computer monitor has a specific function to perform, it is likely that certain symptoms can easily be diagnosed as a problem in a specific stage(s). The video and audio qualities can be used to track down the defective stage(s). You will have to use a voltmeter, an oscilloscope, or both, and at times other special TV testing equipment. The voltages and signal at critical stages are found on the manufacturer's schematic.

If many stages are involved, use the half-splitting method. Examine the signal at a stage that is halfway between a known good stage and the output. Should the results be correct, then move forward to another stage that is halfway between the previous test point and the output. However, if the first test indicated a defective signal, you can use the half-split method in the opposite direction until the defective stage is isolated. This technique can minimize the number of test measurements.

Weak Picture and Sound

If both the sound and the picture are weak and distorted, a possible cause is a defective signal input—antenna system, computer cable connection, or television cable hookup. For example, the antenna system may have a bad antenna, loose connection, bad cabling, or an improperly oriented antenna. Check for the proper signal, using a signal level meter or substituting another TV. Should the antenna system prove good, the problem is likely to be in the tuner section. On old mechanical tuners, selection of channels is accomplished by changing frequencies of the oscillator by rotating a different coil into position. You should check it for proper alignment and clean all the contacts. Modern tuners are all solid-state and electronically selected. There is usually nothing that you can do in the field to repair these tuners.

Rotate the contrast and brightness controls to verify the absence of the signal. If the tuner appears to be operating properly, the problem may be in the i.f. circuit, video detector, or AGC. All these stages are common to both the sound and the picture; therefore, a bad transistor, a change in resistive value, or low voltage supply can contribute to the problem.

Figure 6.28 illustrates an older bipolar transistor i.f. amplifier circuit. The input consists of three i.f. stages: sound trap and detector, a video detector, and traps to eliminate carrier signals of adjacent channels. When you test a TV, examine the schematic and then check the direct-current (dc) voltages and signals as indicated. For example, if an abnormal collector-to-emitter voltage reading is found, it indicates either a defective component in the collector circuit or a change in transistor current conductance. If we assume that the transistor and collector circuit is normal,

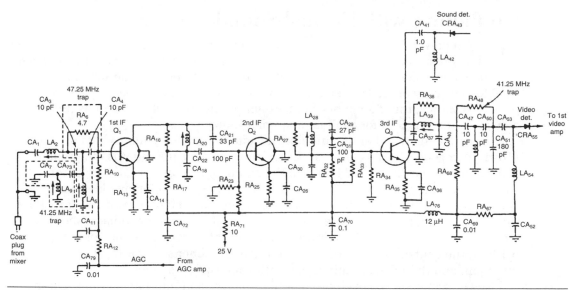

FIGURE 6.28 Typical i.f. amplifier circuit. (*Courtesy Howard W. Sams & Co., Inc.*)

there has been a shift in the base-to-emitter voltage bias. That bias shift can be the result of a resistor value change or leaky capacitors in the base or emitter circuit.

Notice that the AGC circuit controls the bias of the first i.f. stage. If the AGC circuit is defective, it could reduce the gain of the i.f. stages and create a weak picture.

Good Picture, Weak Sound

If the picture is good but the sound is weak and distorted, chances are the audio i.f. amplifier, FM detector, af amplifier, or speaker is causing the problem. The FM detector is the most likely cause. Check the FM detector for proper voltage and signal first. If the test voltage and signal match the manufacturer's specification, the problem is the af amplifier or speaker. An easy way to check the audio stage is to increase the volume and note the noise. If there is noise as the volume increases, the af amplifier output and the speaker are working; therefore, the problem must be in the signal or the previous stage. If not, the problem is in the FM detector or audio i.f. amplifier.

Weak Picture with Normal Sound

If the picture is weak but accompanied by normal sound and a bright screen (raster), the probable stages that can affect the picture are the antenna system, rf amplifier, converter, local oscillator, i.f. amplifier, video detector, AGC system, or, most likely, the video amplifier. The sound could be perceived as normal after leaving the video detector if the audio section has a number of amplification stages which would amplify a weak signal. Once again, a decrease in voltage from the low-voltage power supply can be the source of the problem. Another possible cause is the video detector. The video detector is the diode CRA55 at the output of the i.f. amplifier shown in Fig. 6.28.

No Picture with Normal Sound

If the TV receiver does not produce a picture but has a raster, or illuminated picture tube, the defect exists in the stages before the sound pickoff. However, it is possible that the video amplifier circuit is at fault. If there is no snow on the screen, the trouble is most likely in the video detector or the i.f. amplifier stage. However, if the picture is snowy, the rf amplifier in the tuner or antenna/cable system is probably defective. See Fig. 6.29 for an illustration of a snowy screen.

To determine whether the tuner or the antenna/cable system is at fault, simply substitute a good TV in its place. If the snow disappears, the tuner is the problem. If the snow remains, the problem is in the antenna or transmission line. Often, a defective rf amplifier causes the picture to be snowy. Also, many older tuners contain silver contacts that have a tendency to tarnish and become dirty. If rocking the tuner channel switch causes the picture to shake, the tuner contacts should be cleaned with a commercial tuner wash. Take off the tuner cover and spray the tuner wash on the contacts of the tuner as you rotate the tuner selector five times in one direction, and then five or more times in the opposite direction.

Newer sets incorporate circuit modules that contain multiple sections. Often, the complete module is simply replaced with a new one. Also, unlike old-fashioned rotary tuners, pushbutton control modules can become dirty, sticky, worn, or defective, requiring replacement. Do not overlook the possibility of cable defects. Check cable for possible shorts and opens; if in doubt about its condition, replace it.

Sound Normal But No Raster

If the receiver lacks a raster, the fault could exist in the high-voltage power section. There could be a problem in the horizontal deflection stage, such as the flyback transformer or damper. Check the high voltage with a high-voltage dc probe to determine whether there is high voltage at the anode of the CRT. Be careful when you check this voltage, because dangerous arcs can be drawn. If the voltages match the manufacturer's specifications, the picture tube may be at fault. However, if there is a lack of dc voltage, check to see whether there is ac voltage from the flyback

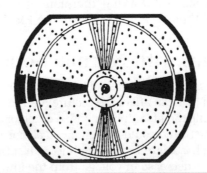

FIGURE 6.29   Snowy, weak television reception.

transformer. Newer circuitry may have only dc voltages at this stage. Compare the positive and negative dc voltages in the circuit with the values indicated in the service manual schematic to isolate the problem. A blue arc indicates ac voltage; dc voltages produce a white arc. If an ac arc can be drawn at the flyback transformer, the high-voltage rectifier is defective. Lack of an ac arc could indicate a faulty flyback transformer or horizontal circuit.

Another possibility for the lack of a raster is a faulty tuner or i.f. stage that is sending only black signals; therefore, the screen is dark.

The picture tube, like any other kind of tube, works on the principle of thermionic emission. Weak emission causes the picture tube to become out of focus and develop a silvery tint. An easy way to identify a weak picture tube is to turn up the brightness control. If the picture becomes silvery and out of focus, compared to when it is turned down, then the picture tube can be considered to be bad or going bad. If the high voltage is absent, the trouble is in the high-voltage rectifier, damper, horizontal output, or horizontal oscillator.

Figure 6.30 shows a simplified schematic of a horizontal system. Notice the horizontal oscillator, horizontal output, flyback transformer, and damper diode. Before you check the high-voltage rectifier or flyback transformer, the picture tube must be discharged. Using an alligator clip lead, clip one end to the chassis and the other end to the shank of a screwdriver. The anode is the conductor under the rubberlike grommet. Isolate the rectifier or flyback transformer and check for open circuits. Check the condition of the rectifier, or simply use a substitute diode. Check the flyback transformer by making a resistance test with an ohmmeter. Keep in mind that the horizontal oscillator must be running to get high voltage; therefore, check the horizontal oscillator output.

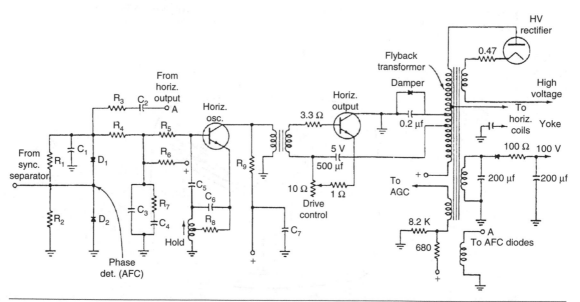

FIGURE 6.30   Typical horizontal system. (*Courtesy Howard W. Sams & Co., Inc.*)

Also, never exceed the proper high-voltage rating, or possible undesired X-radiation can be emitted by the picture tube. While degaussing shields and other protective devices are often used to help shield X-rays, leakage can still occur. Always check the high voltage with an accurate meter. Some troubleshooters will check the high voltage while turning the brightness control up and down to make sure that the high-voltage value is not exceeded and that the circuit has proper regulation.

Sound Normal, Picture out of Sync

If the sound is normal but there are heavy slanted streaks across the screen, the problem is that the horizontal deflection is out of synchronization. Figure 6.31 illustrates the horizontal synchronization problem, which is commonly called *horizontal tearing.* Check the horizontal deflection control to verify that it is set properly. If it is, the fault exists in the horizontal oscillator. If it stops running, there will be no pulse to drive the horizontal output or deflection coil in the yoke.

Sound Normal But Picture Tearing, Reduced Width

If your picture has tearing or heavy slanted streaks across the screen as well as vertical roll, first check for proper setting of the horizontal and vertical controls. If these are properly set, probable causes are defects in the sync separator or sync amplifier stages. There is a possibility that defects can exist in both horizontal and vertical deflection systems. Figure 6.32 illustrates a typical vertical deflection system. It consists of a vertical oscillator, a vertical driver, and a vertical output stage that is tied to the yoke.

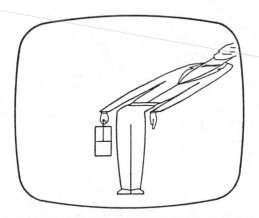

FIGURE 6.31   Horizontal tearing.

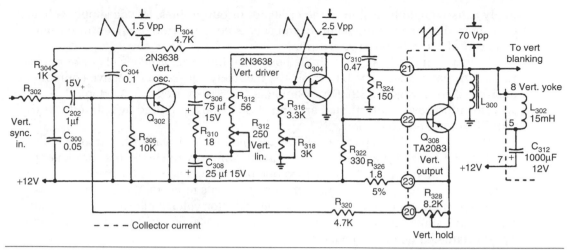

FIGURE 6.32 Typical vertical deflection system. (*Courtesy Howard W. Sams & Co., Inc.*)

Sound Normal But Picture Rolls and Folds, with Reduced Height

Other vertical system problems are shown in Fig. 6.33. If the picture rolls vertically, a probable cause is a defect in the vertical oscillator. If the oscillator stopped oscillating, there would be no vertical deflection, and all that would be seen on the screen would be a bright horizontal line. If you experience reduced height in the picture, the problem is a weak vertical output. As indicated in Fig. 6.32, several

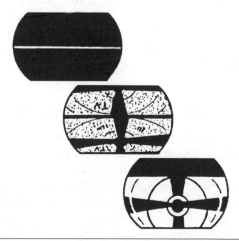

FIGURE 6.33 Other vertical system problems.

likely causes could be a shift in the oscillator or output bias, low dc supply voltage, or a shorted or open component. It probably is not a shorted oscillator or output transistor if there is a partial picture.

However, if there is only a single horizontal line, a shorted oscillator or output transistor may be suspected. In this case, the likely cause is a defective component, such as C306, C308, or R310. For example, if C306 is shorted, the sawtooth-waveform process is interrupted and the bias of Q302 is shifted, which reduces the amplification and oscillation.

If C306 is open, the picture is likely to "fold over" with a white bar and compressed lines at the bottom of the screen. One way to check this component is to bridge capacitor C306 with a good one (or to use a capacitor substitution box) while the set is on. If the picture height returns to normal, the problem has been identified. Incidentally, checking the collector voltage of Q302 in this circuit would have shown it to be low because capacitor C306 was open and not placing appropriate charge on the collector.

Should the screen have a reduced height as well as trapezoidal distortion, the likely cause is a defective yoke or trouble in the pincushion correction circuit in a color receiver. The problem is not in the vertical oscillator or output.

Often, when the troubleshooter needs to determine whether vertical roll or horizontal tearing is caused by the oscillator or sync stage, a simple test can be performed. If the picture can be made to hold when the oscillator control is turned but fails to stay held, the problem lies in the sync circuit. A defective diode in the horizontal AFC often causes the problem of horizontal tearing. See Fig. 6.30 for an example of AFC diodes in a horizontal circuit. In Fig. 6.30, if C5 is open, the horizontal voltage is decreased, which reduces the picture width. If the horizontal oscillator transistor is shorted or the 3.3-KV resistor is shorted, there will be no horizontal voltage. On new models, the entire module or circuit board may simply need to be replaced.

Lack of vertical deflection could be caused by a bad oscillator, output transistor, or IC. An open emitter bypass capacitor or resistor could also cause insufficient gain which would decrease the height in the picture. Keep in mind that trapezoidal distortion is usually caused by a defective deflection yoke and not the vertical circuit. A "slanted picture" can be corrected by simply loosening the deflection yoke and turning it in the correct direction. Remember, never overtighten the deflection yoke, as the picture tube neck can be easily broken.

Normal Picture But Poor Sound

If the picture is normal but the sound is missing, check the i.f. amplifier, FM detector, or af amplifier stages. There could also be a defective speaker coil. A weak sound suggests improperly adjusted fine-tuning control or a shift in the local oscillator due to changing component values. As multiple stages are involved, use the half-splitting troubleshooting technique to quickly identify the defective stage. Check for a defective transistor, IC, or module in the sound section. A change in a component value or defective cable in an audio stage can affect the gain of an amplifier. Sound distortion can be caused by a defective interstage coupling capacitor.

Dead Set

Like the radio, if a TV set is dead, check the power supply. Possible causes include a blown fuse, tripped circuit breaker, open line cord, shorted or open cable, or a defective component—on/off switch, transformer, diode or rectifier, thermistor, or power supply IC module.

Color CRT TV Monitor Troubleshooting

Like the old black-and-white televisions, the color CRT TV monitor is virtually nonexistent today, but it is still found in many homes, and special applications such as industrial monitoring, computer monitors, security and surveillance, and process operations. The basic CRT television uses three separate cameras, with each camera being sensitive to only one of the three primary colors (red, blue, and green). In addition to these primary colors, complementary colors such as yellow, orange, cyan, and magenta can be produced. Various combinations of these primary colors will produce any color to which the human eye is sensitive. The hue or tint is the distinctive color itself. Saturated colors are vivid, strong colors. Lack of saturation in a color results in a pale, weak color. Chrominance refers to the combination of the hue and saturation. Luminance refers to the amount of brightness perceived.

The three cameras scan the scene in unison. The primary colors of red, blue, and green are fed into a matrix at the transmitter, which creates a Y, or luminance, signal and chrominance, or color signals I and Q. The Y signal contains the proper proportions of red, blue, and green content so it can re-create a normal black-and-white signal. This signal is used to modulate the carrier. The I and Q chrominance signals are used to modulate a 3.58-MHz color subcarrier that is suppressed by the modulation process. The composite signal has a carrier, the Y (or luminance), and the I and Q (or chrominance signals), along with the FM audio.

If the receiver is a black-and-white (monochrome) set, only the Y signal is detected and processed. The chrominance signals I and Q cannot be detected and processed because the receiver lacks the 3.58-MHz oscillator needed to recover the I and Q signals. Thus, a color receiver requires a 3.58-MHz oscillator to enable the detection of the I and Q signals.

A block diagram of the color section of a television receiver is shown in Fig. 6.34. The color signal comes from the video amplifier into the chroma amplifier, where the signal is amplified. Notice that after the video amplification, the Y signal is immediately available with the exception of a delay of 1 microsecond (ms) so that the Y signal and the I and Q signals arrive at the CRT at the same time. It requires about 1 ms for the I and Q signals to undergo additional signal processing.

After the chroma signal is amplified, it is sent into a bandpass amplifier of 2 to 4.2 MHz, which separates the I and Q content from the Y content and is then sent to the I detector and to the Q detector. These detectors have input from the 3.58-MHz crystal oscillator that aids the detection. Notice that the 3.58-MHz oscillator is phase-shifted 90° at the transmitter to separate the various signals.

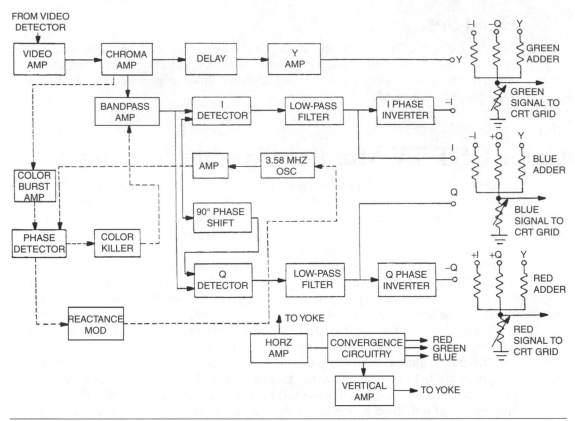

FIGURE 6.34 Block diagram of a TV color section.

Once the I and Q signals are detected, they are sent to their respective low-pass filters and then processed by a phase inversion for both positive (+) and negative (−) chroma signals. The positive and negative chroma signals are formed by

$$\text{Green 5 } 2I2Q1Y$$

$$\text{Blue 5 } 2I1Q1Y$$

$$\text{Red 5 } I1Q1Y$$

The I, Q, and Y signals are summed in the three-color adder circuits, with resistor values providing the proper proportion of each signal. Each color signal is then sent to the appropriate CRT grid to control the beam's intensity. There is a rheostat at each adder circuit to allow the intensity of each color to be varied proportionately to the other colors.

The subcarrier crystal oscillator is not precise enough to permit proper detection of the chroma signals. Therefore, at the transmission of the color signal, a sample of it is placed on the so-called back porch of the horizontal blanking

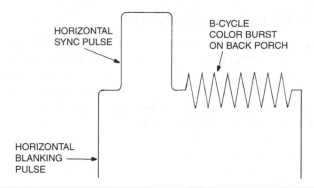

FIGURE 6.35   Horizontal blanking pulse with back porch.

pulse, as shown in Fig. 6.35. The color burst amplifier receives a portion and sends it to the phase detector. The phase detector compares the 3.58-MHz signal to the color burst; if the two are not equal, a dc signal is sent to the reactance modulator, which "pulls" the two signals into precise synchronization.

The phase detector also sends a signal to the color killer. The purpose of the color killer is to prevent any signals from the chroma circuits during a monochrome broadcast. As long as there is a color burst present, the color killer is off. However, when the color burst is absent, as it is during monochrome broadcasting, the phase detector sends out a dc signal that "kills" the bandpass amplifier.

Correct color reproduction requires undistorted chrominance and luminance signals. Any defects that alter a black-and-white picture will have an effect on the color picture. A general localization of color troubles can be made in terms of the raster and monochrome picture quality. A good raster indicates normal dc voltages, and a good monochrome picture indicates normal Y signal and correct dc voltages for the picture tube.

Absence of Color

If there is an absence of color, look for a defective IC or module in the I and Q signal processing stages. The color killer control may be improperly set or defective. Also, the bandpass amplifier stage may be inoperative. Check for the presence of the 3.58-MHz signal and the color burst signal at the oscillator and subcarrier.

Weak or Faded Color

Weak or faded color can result from an improperly adjusted bias control on the picture tube or improperly adjusted screen and drive controls. A weak transistor or IC in the I and Q signal processing stages or defects in the bandpass amplifier may also cause the problem. The tuner and i.f. stages may be slightly out of alignment and could contribute to the problem. Check for the presence of the 3.58-MHz signal and the color burst signal at the oscillator and subcarrier amplifier.

Screen Dominant Color

Should the receiver have a dominant blue color, the probable cause is an improperly adjusted green and red drive or misadjusted screen controls. However, if the receiver has a dominant red color, check for improperly adjusted blue and green drives. There is also a possibility that the picture tube is defective.

If some colors are more vivid than others, generally the problem is a misadjusted screen or color drive control.

Color Killer

The color killer cuts off the chroma amplifier during a black-and-white transmission. A defective color killer results in color noise, called *confetti*, which looks like snow but with larger color spots. If confetti is present during a monochrome reception, check the adjustment or circuit of the color killer.

Color Bars

Another typical problem in color sets is the presence of color bars in the picture. Usually the reactance transistor, the automatic frequency phase control, or a defective color burst is the cause of the problem. Many of these circuits are contained in an IC called a *chroma processor*; therefore, the IC or entire module must be replaced.

Other Color Problems

A picture tinted black and white suggests that the purity is out of adjustment or that the picture tube needs degaussing. Most TVs and computer monitors have automatic degaussing. However, should the flesh tones vary with image position on the screen, check for incorrect purity adjustments.

If the colors are not properly registered, check the convergence for proper adjustment or operation.

Poor focus suggests a defective focus rectifier or focus circuit. Check the rectifier first, then the other circuit components. Consider defective static control panel assembly, cable, pin connectors, and sweep.

If the colors are smeared, check for loss of the Y signal or a fault in the video amplifier system.

Total Convergence Setup

The need for making convergence setups is declining, as most modern sets have a preset convergence circuit. When a new picture tube is installed in a television set, a total convergence setup may be necessary. The standard procedure can be found in the television service manual. Generally, a total convergence setup consists of first adjusting the proper picture size, focus, and linearity. The brightness should also be adjusted to the typical viewing level. A dot, bar, or crosshatch generator is generally used in the setup procedure.

Purity Adjustment

Bias the blue and green guns to cutoff, and slide the deflection yoke forward. Adjust the purity magnets until the red focuses exactly in the middle of the picture tube. Now, push the deflection yoke backward until the raster is completely red.

Static Convergence

Turn on the green gun. Adjust the red and green static convergence magnets to merge two colors in the center of the picture tube until you get one yellow dot. Now, turn on the blue gun and merge all three colors in the center of the picture tube until the color white is formed.

Dynamic Convergence

Adjust each dynamic control to converge the top, bottom, and sides of the picture tube. Before you complete the total convergence, it is important to adjust the picture for the best gray tracking. Gray tracking ensures the best possible black-and-white balance. While the TV is on and the color turned down, adjust the red, blue, and green drive controls until the best overall gray raster is obtained.

The last step is adjustment of the screen controls. Set the service switch in the service position. This will display a horizontal line across the center of the screen. Now, turn all three color screen controls counterclockwise, and then slowly adjust each one until the color is just visible.

If the convergence does not seem to hold during the operation of the TV, a defective convergence rectifier may be the reason.

Servicing Late-Model Television Sets

The material in this chapter has dealt with overall television theory of operation and directed you in isolating problems to the most likely section. Most of the schematics shown are from older television receivers. In modern television sets, there are fewer individual semiconductor components. Entire sections of the television such as the tuner, i.f. and video amplifiers, sweep, and audio output are handled by ICs. Figure 6.36 shows the IC functions of a modern receiver. ICX1200 contains the i.f. amplifiers, video amplifiers, video detector, FM audio detector, and vertical and horizontal sweep generation, all in a single IC! This is not to say that the IC is the only thing that can go wrong. Each IC has many peripheral components connected to it. In addition to the ICs shown, a microprocessor is used to control the entire system.

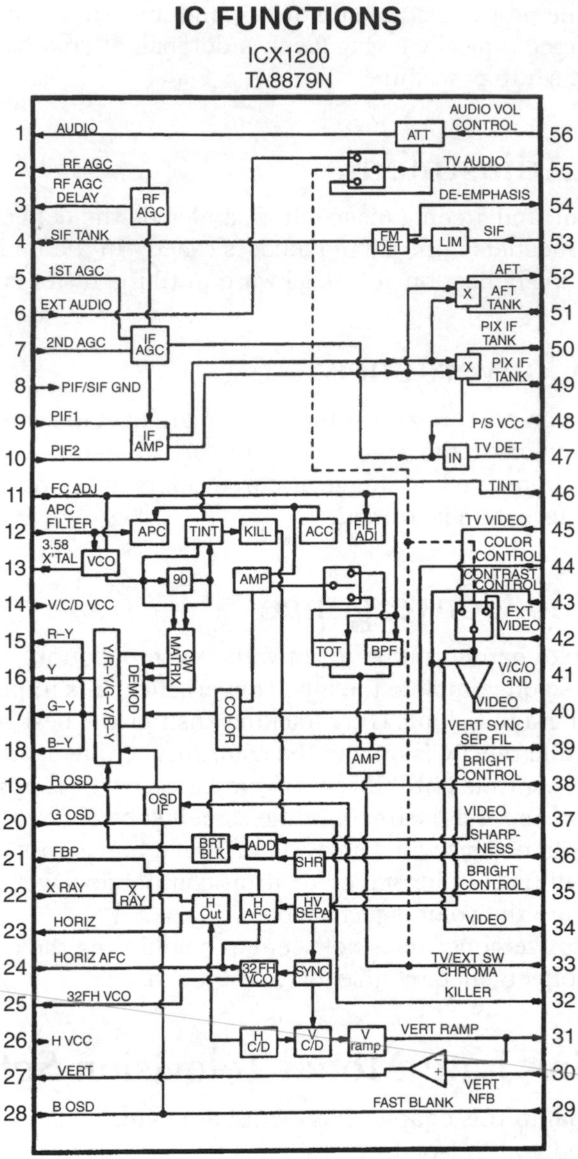

IC FUNCTIONS

ICX1200
TA8879N

FIGURE 6.36 A television component block diagram. (*Courtesy SAMS Photofacts.*)

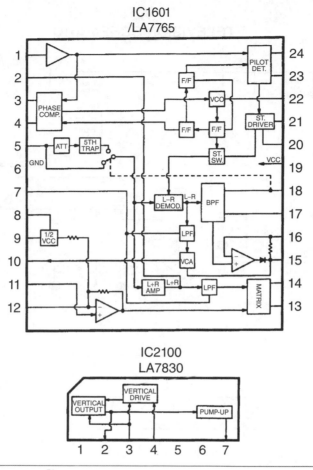

FIGURE 6.36   (*Continued*)

The schematic of Fig. 6.37 was supplied by SAMs Photofacts. For over 50 years SAMs has provided technical documentation of consumer electronic appliances for professional technicians and electronic hobbyists who enjoy repairing their own equipment. Notice how various functions of this television receiver are distributed on the schematic diagram. Expected waveforms and dc voltage levels are provided at all the critical points in the circuit. Most of the small signal-processing stages are contained in the single IC whereas power amplification and signal buffering between stages are provided by discrete transistor circuits. Notice that the vertical driver is a separate IC whereas the horizontal output is made up of a two-stage power transistor circuit. Troubleshooting a modern television involves using your knowledge of signal flow to predict which section would be likely to cause the problem. Within the likely section, an oscilloscope can be used to determine which (if any) waveforms are incorrect. Once the problem area is localized, further measurements with a dc voltmeter can often identify the defective component.

A
TELEVISION SCHEMATIC

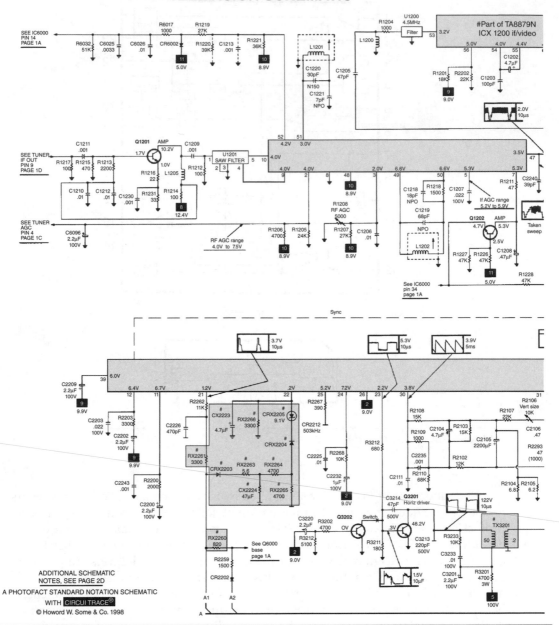

ADDITIONAL SCHEMATIC
NOTES, SEE PAGE 2D
A PHOTOFACT STANDARD NOTATION SCHEMATIC
WITH CIRCUI TRACE®
© Howord W. Some & Co. 1998

FIGURE 6.37   The i.f. and video section schematic with partial sweep circuits. (*Courtesy SAMS Photofacts.*)

B
TELEVISION SCHEMATIC

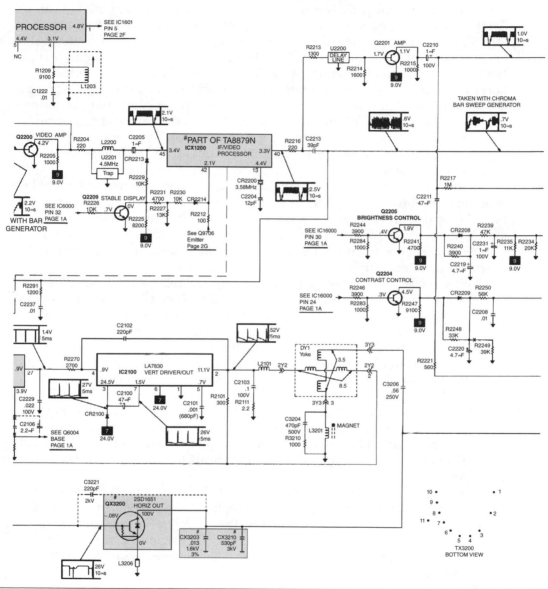

FIGURE 6.37   (*Continued*)

A common problem in modern television repair involves the power supply. Older sets had a simple linear power supply with several voltages regulated and sent to various sections of the receiver. Newer sets use a conventional dc power supply circuit to generate a few primary supply voltages that are used to get the horizontal sweep section oscillating. These oscillations are picked off of the secondary windings of the flyback transformer and used to generate other voltages for the television's circuits. Figure 6.38 shows an example of such a power supply. Notice the transformer windings of TX3200 (horizontal output transformer) in the lower left-hand corner. This presents a problem for the troubleshooter when a set is dead. The oscillator needs the power supply in order to work, but the power supply needs the oscillator in order to work. Finding the defective component in a closed-loop system like this one is not an easy task! Every set is a little bit different, but generally the service manual will describe a procedure for isolating power supply problems.

Most service manuals for analog CRT televisions use a combination of block diagrams and flowchart systems to aid the troubleshooter. These diagrams provide a quick and convenient means to locate faulty components through circuit illustrations and troubleshooting systematic checkpoints. For example, in troubleshooting a video projection system, Fig. 6.39 illustrates a troubleshooting flowchart of a dead set.

Follow the start of the flowchart by checking presence of 120 V ac at the pins 1–3. If there is no voltage, proceed to check the fuse. If voltage is present, continue to pins 6, 4, 3, and 1 with voltage checks. When abnormal measurements are discovered, repair procedures are performed, such as replacing an IC, cables, IC boards, or assemblies or correcting poor connections.

A schematic or block diagram is another common aid found in troubleshooting manuals (Fig. 6.40). This guide allows the troubleshooter to check specific measurements at designated points visually referenced on enhanced block diagrams. These schematics or block diagrams typically have an accompanying flowchart to guide the troubleshooter through a set of logical sequences in isolating a fault. Most television manufacturing companies produce detailed technical training and servicing manuals that are essential in repairing televisions and other electronic products.

The tuner section of most modern sets is made up of a microprocessor and digitally controlled rf processing ICs with peripheral components that make up *RLC* (resistor-inductor-capacitor)–tuned circuits. The rf processing unit is not normally repairable but can be replaced if it is determined to be defective. It is usually a separate module enclosed in a shielded cage. The microprocessor also must be replaced if defective.

These television sets are still being manufactured in a manner that allows them to be repaired. The printed-circuit boards are the standard "through-hole" type that have been used for the past 30 years. ICs are in dual in-line packages that allow them to be replaced as described later in Chap. 7. Other components such as capacitors, transistors, and diodes can also be easily removed and replaced. In other words, televisions are still relatively serviceable. Early in the 21st century, digital

POWER SUPPLY SCHEMATIC

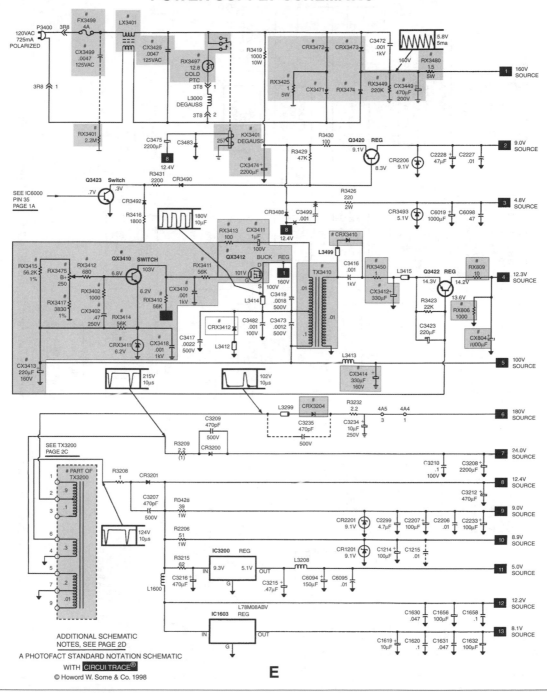

FIGURE 6.38   Typical television power supply. (*Courtesy SAMS Photofacts.*)

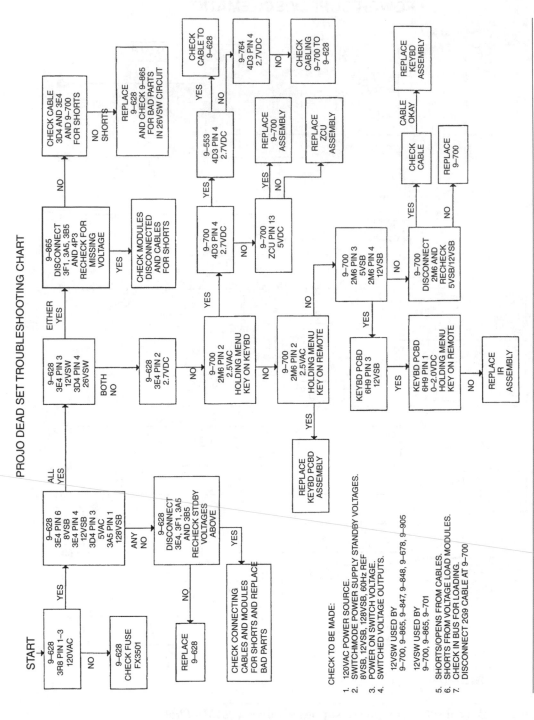

FIGURE 6.39 A troubleshooting flowchart diagram. (*Courtesy Zenith Electronics Corporation.*)

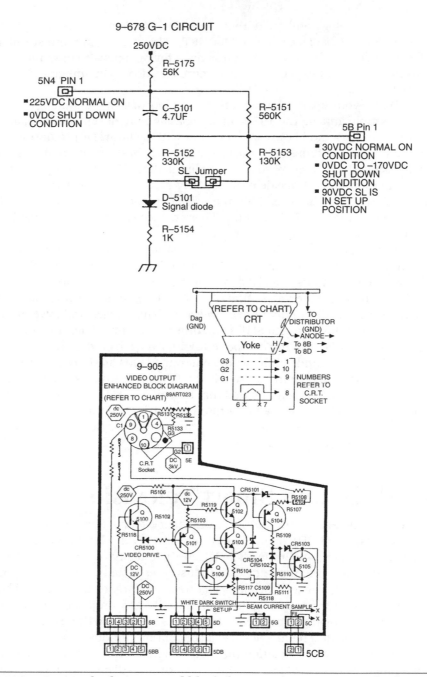

FIGURE 6.40 A typical schematic and block diagram. (*Courtesy Zenith Electronics Corporation.*)

HDTV technology will make these sets obsolete. You can also expect new display technology to displace the need for CRTs. This means no more strong magnetic fields to deflect the beam. With less need for high-power components to drive the displays, you can expect more and more circuitry to be integrated into special-function ICs.

When you repair components, remember that the technology is advancing. For example, during the 1970s, many chassis were designed with stamped metal bases and lugs and hand-wired together with 200-watt (W) soldering irons using combinations of tubes and transistors called *hybrid circuits*. Solid-state devices then necessitated use of 40-W soldering irons as these ICs and components were very closely mounted on plastic boards. Many modern-day circuits use miniaturized surface-mounted devices (SMDs) or chip components, which are tiny blocks soldered to metallic-foil modules. These devices are extremely fragile and can be damaged by movement or overheating.

Resistors and capacitors are also miniaturized and are often made with tiny ceramic bases of film materials. As with SMDs, ICs, transistors, diodes, and other delicate components, special heat-sensitive tools and equipment are necessary. A controlled heat soldering station of 25 W with rosin core solder as small as 0.3-mm diameter is needed. Tiny tipped tweezers, dental picks, and jeweler's screwdrivers are useful in dealing with SMDs and other miniaturized components. When you remove an SMD, grip the component with tweezers and gently rock it as you heat the solder at the ends of the leads. When the leads are loose, gently break the SMD away by twisting it from its glued body. Once these miniaturized components are removed, it is not recommended that they be reused. Also, when you remove tiny ICs, gently heat the pins and lift the IC up, using a dental pick, as soon as the solder melts. Likewise, when you replace components, position them with tweezers and gently apply solder to the leads.

HD, UHD, and Smart Televisions

There are several HDTV formats in existence today. Some of these include the liquid crystal display (LCD), digital light processing (DLP), light-emitting diode (LED), plasma, ultra high-definition TV (UHDTV), and smart televisions. These televisions have revolutionized the television industry, especially since the original CRT, by allowing clearer images, brighter pictures, and larger screen sizes. They also are significantly thinner and lighter than the CRT and projection televisions.

The LCD operates by means of a series of cathode florescent lamps at the base of the television screen. They use a concept of millions of separate LCD shutters composed of a grid formation that controls light emission. The degree of light passing through the shutters creates a color tone displayed in sub-pixels that provide the image. The liquid crystals are integrated and layered in distinct directions to create light controls. LCDs generally have excellent brightness control, are lightweight, reliable, and produce clear pictures. LCDs have been used for not only television entertainment but as computer monitors because of their high clarity and light weight. Some traditional drawbacks of them have been the reduction in picture clarity when observed at an angle, some difficulty with black images, high motion resolution, and expensive production of large-screen televisions.

The quality of LCD televisions has been enhanced through the use of LED technology. These LED televisions are essentially an LCD framework with embedded LEDs. The LEDs are mounted at the base and illuminate the LCD panel, which allows for a brighter, sharper, and high-quality picture. It also provides greater energy efficiency, thinner display, and lighter weight.

The DLP television works on the principle of a series of mirrors that refract light by a matrix configuration that creates a picture image. Each single pixel is equivalent to a mirror and the entire system can be composed of literally thousands of these mirror formations. The quality of the picture image is dependent on the composition and amount of these mirrors. LDP televisions have traditionally been very affordable as compared to other formats and they produce a relatively good image and brightness. However, they tend to be larger in size than LCD and plasma televisions, and therefore are not as suitable for wall mounting. Unlike LCDs the plasma tends to have a better-quality picture when viewed at an angle. Some older models have also been criticized for creating slight streaks of colors when fast images are displayed. The biggest issue with DLP includes the need for replacing the light bulb contained within the video projection box. While these bulbs are easily replaced, they need a higher degree of maintenance as compared to LCDs and plasma televisions.

The plasma television works on a principle of displaying varying minute cells consisting of electrically charged ionized gasses. This composition essentially creates a florescent light or plasma. There are thousands of tiny cells protected by a dielectric material. When current flows through the gas within a cell, ultraviolet photons are released and interface with the florescent material, producing visible light. There are different colored florescent materials composed of each pixel, which produce red, green, and blue primary colors. Combinations of these colors can be produced by the amount of current. Plasmas have traditionally experienced long life, good-quality images, excellent brightness, and contrasts in size that offer flexible wall mounting.

However, some earlier plasma televisions have experienced an issue of "burn-in." Because of the construction of fluorescent material, traces of images can be burned into the display when operated for long periods of time. For example, if the plasma television has been operating for an entire day, it is possible that an image, such as a company logo, could be permanently burned into the display. Therefore, plasmas tend not to be used as computer monitors. Also, though plasmas tend to create very vivid colors, some models have been criticized for low brightness. However, there may be some limitations to using plasmas in highly lighted rooms or in direct sunlight. Also, given that plasmas use natural gasses, altitude may be an issue and impact the performance and quality. Moreover, plasma televisions tend to be more fragile and the mere transporting or storage of these plasmas on their side can cause harm.

Ultra HDTVs (8k UHDTVs and 4320p) offer 16 times the pixels of the 1080 HDT televisions with super surround sound capabilities. These televisions can create images using millions of pixels with resolutions of 7680 × 4320 or 32.2 megapixels. The UHDTVs range in various sizes, typically 55 to 84 inches. They use liquid-crystal-display panel technology that requires higher precision than HD televisions. Other features include gyrostabilized ultra HD aerial camera systems, external memory storage, and medical and other professional, precision video equipment.

Smart televisions are basically any sophisticated television that has the ability to connect to the Internet. These televisions are a more advanced generation television than traditional HDTV that has capabilities of using applications (apps) and web browsing. They often have functionality of connecting to entertainment networks, streaming video, social applications, and 3D capabilities. They generally require a setup box and other Internet connection devices for full range of entertainment modes.

HD, UHD, and Smart Television Troubleshooting

The troubleshooting of HD, UHD, and smart televisions requires the use of the manufacturers' service manuals by trained company certified technicians. For example, one of the basic service needs of these televisions is often calibration. Most high-density televisions are calibrated at the factory. However, the picture settings may need to be adjusted by a skilled consumer or troubleshooter. Calibration generally includes measuring and determining the desired amount of color intensity, contrast and brightness, color details, and temperature. These adjustments are based on the manufacturer's equipment using the factory service manual.

When troubleshooting HD, UHDT, and smart televisions, you can obtain manufacturer's service manuals with a service contract agreement. Most manufacturers have an authorized website where these manuals are filed and can be obtained. The typical service manual starts with safety precautions listing model numbers, safety notices, specifications, current leakage tests, and needed equipment. The specifications generally outline temperature, humidity, power, and performance considerations along with testing methods and environmental operating data. In addition, specifications for chroma, brightness, luminance, contrast, video inputs, and RGB specifications are included. For example, typical RGB specifications include resolution, horizontal frequency, vertical frequency, and pixel clock.

An adjustment instruction section is typically included, which contains detailed instructions on application, specifications, PCB check process, USB downloads, connect conditions, download steps, and a myriad of adjustment factors. Detailed screen shots along with an exploded view of the device and schematic diagrams are usually included. The most useful part for the troubleshooter is the step-by-step diagnostic diagrams. For example, a typical troubleshooting diagram begins with the contents of the television associated with the specific type of error code and the page number where the diagram can be found. An example of the typical symptoms and error categories is outlined as follows:

Error Symptom	Error Category
Audio Problems	No sound Volume issues Noise in sound No audio but normal video Intermittent audio
Video Problems	Picture noise Picture distortion Picture not in full screen Static or noise within picture display Poor video resolution Color issues Video lag No video but normal audio No video and no audio Vertical or horizontal bars
Function Errors	Remote control problems Key errors Recording errors Memory errors
Power Errors	No power Intermittent power Power on when indicator is off Power indicator is on but no power Automatic power error

The general steps in troubleshooting televisions consist of following the step-by-step diagnostic troubleshooting charts, using the proper equipment, while referencing the block diagrams. Figure 6.41 illustrates a typical flowchart for troubleshooting procedures for an eight-blink error code failure. The flowchart begins with the troubleshooter checking all cables between the SS board and the panel. If, for example, the monitor is able to turn on and stay on, the decision tree would direct the troubleshooter to unplug the monitor and check for continuity between pins 1 and 2 of connectors SS61 and SS64 on the SS board. If, for example, the monitor does not stay on, the troubleshooter would be directed to check the connections between the SS, C2, and A boards. Last, if continuity is good in all the connectors, the troubleshooter would be advised to replace the SS board. If not, the connections between the SS board and the panel should be checked and, if necessary, the panel may need to be replaced.

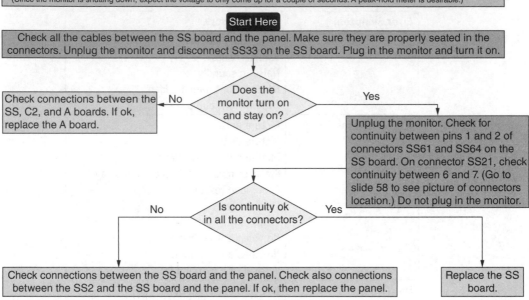

Cautions: Disconnect the AC power prior to making any disconnection or connection.
When taking voltage reading, place the voltmeter probe at the test point, component, or connector's pin indicated before connecting the monitor to the AC line. This will ensure voltage reading accuracy before the monitor shuts down.
(Since the monitor is shutting down, expect the voltage to only come up for a couple of seconds. A peak-hold meter is desirable.)

Start Here

Check all the cables between the SS board and the panel. Make sure they are properly seated in the connectors. Unplug the monitor and disconnect SS33 on the SS board. Plug in the monitor and turn it on.

Does the monitor turn on and stay on?

No — Check connections between the SS, C2, and A boards. If ok, replace the A board.

Yes — Unplug the monitor. Check for continuity between pins 1 and 2 of connectors SS61 and SS64 on the SS board. On connector SS21, check continuity between 6 and 7. (Go to slide 58 to see picture of connectors location.) Do not plug in the monitor.

Is continuity ok in all the connectors?

No — Check connections between the SS board and the panel. Check also connections between the SS2 and the SS board and the panel. If ok, then replace the panel.

Yes — Replace the SS board.

FIGURE 6.41 Troubleshooting procedure for an eight-blink error code failure. (*Courtesy Panasonic Corporation of North America.*)

The diagnostic chart is essentially a decision-making tree. For example, if there is a video error code, the decision-making tree will guide the troubleshooter through each step depending on the functioning of the television. If there is no video but normal audio, then a different section will be checked versus if both video and audio are gone. The instructions will guide the troubleshooter to check power boards, specific circuits, modules, as well as ancillary equipment such as cables and connections. If, for example, there is a picture issue, the diagnostic chart may begin with checking if there is a sufficient RF signal available. This would be done using a digital signal level meter. Depending on the level of signal, reconnections or installation of booster may be considered.

When flowchart diagrams for troubleshooting are used, the inclusion of circuit board pictorials is generally included to assist the troubleshooter. These pictorials are actual pictures of the components and circuit boards of the television, and allow the troubleshooter to visually see the devices that are referenced by the flowcharts. Figure 6.42 illustrates examples of an eight-blink error code detect circuit concept for a plasma television.

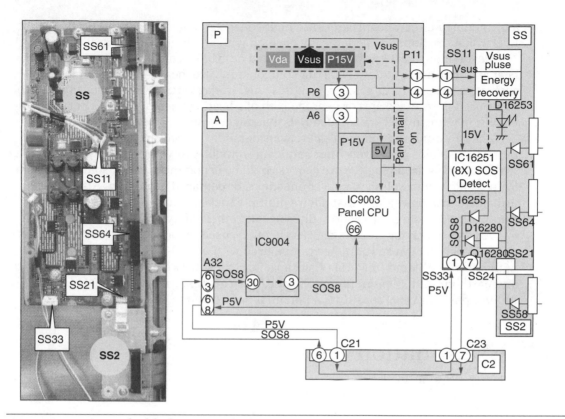

FIGURE 6.42   An eight-blink error code detect circuit concept for plasma TV. (*Courtesy Panasonic Corporation of North America.*)

Preventive Maintenance

Sound equipment breakdowns can often be directly attributed to owner abuse. Stereos should not be played at full volume or for an excessive length of time. Heat can cause premature breakdown of the speakers and amplifier outputs. Never play a stereo with the speakers removed or with the incorrect number of speakers. A mismatched impedance can break down the unit. Therefore, always match the impedance of the speakers with that of the output amplifiers.

Television sets, like other types of receiver, should be operated with care. Never abuse the channel selector or other controls. Maintain proper adjustment of controls and replace any broken parts.

Periodic cleaning of the receiver can often prevent a breakdown. A dirty circuit can increase the heat in the circuit, eventually wearing down the components. Also, a dirty, dusty circuit board can cause high-voltage arcing and a lack of brightness. Every chassis should be periodically cleaned with an air gun or vacuum cleaner.

Never allow the TV to become overheated or to be in contact with moisture. Both heat and moisture can destroy circuit components. Also, heavy moisture in the air, such as in basements or porches, can cause high-voltage arcs. Do not install a television in a closed-in recessed area or near the path of a heating system. Also, do not cover the television with any decorative materials that might obstruct proper ventilation.

Keep in mind the common safety considerations. Always reinstall all safety and protective shields and covers after servicing a product. Never replace a polarized plug with a nonpolarized plug. Severe shocks could be experienced. After soldering, look for possible solder splashes, cold connections, or damaged insulation. Also, always use an electric power strip that allows multiple sockets, energy-saving features, standby modes, surge protection, filtering, and other safety and operation features.

Always check for possible ac leakage on exposed metallic parts, such as cabinets or tuning knobs. An ac voltmeter can be used with a 1500-V 10-W resistor in shunt with a 0.15-millifarad (mF) 150-V ac capacitor to test between the metal product and earth ground. Measurements above 0.75 V rms (root mean square) [or 0.5 milliampere (mA) ac] constitute a potential hazard and need to be corrected.

Self-Examination

Select the best answer:

1. An audio frequency is said to be within
   A. 400 to 40,000 Hz
   B. 20 to 20,000 Hz
   C. 100 to 10,000 Hz
   D. None of the above
2. A modulated rf wave consists of
   A. An rf wave and cw
   B. An rf wave and a radio wave
   C. A cw wave and an rf wave
   D. None of the above
3. Which of the following demodulates an rf modulated wave?
   A. An rf amplifier
   B. An i.f. amplifier
   C. A mixer
   D. A detector
4. In most mixers, the oscillator frequency is _____ the carrier frequency of the input signal.
   A. Lower than
   B. Higher than
   C. The same as
   D. None of the above

5. A push-pull amplifier requires
   A. A splitter at 180° phase shift
   B. An inverter at 90° phase shift
   C. Both (A) and (B)
   D. None of the above
6. A suspected shorted Zener can be checked by
   A. Bridging
   B. Lifting one end of the diode
   C. Shorting across the diode
   D. None of the above
7. What receiver stage is used to reduce signal fading and keep a constant volume?
   A. Detector
   B. Intermediate-frequency amplifier
   C. An af amplifier
   D. An rf amplifier
   E. AGC
8. In FM multiplex reception, a monaural receiver uses
   A. Only the 19-kHz pilot signal
   B. Only the L2R sideband
   C. Only the L1R carrier signal
   D. All the above
9. Hum is most likely caused by a defective
   A. Diode
   B. Transistor
   C. Filter
   D. None of the above
10. Thermally intermittent components can often be checked by
    A. Heating and freezing
    B. Tapping
    C. Bridging
    D. None of the above
11. When a dead set is examined, which item(s) should be checked?
    A. On/off switch
    B. Power supply diodes
    C. Fuse
    D. Open filament
    E. All the above
12. The decibel reading for proper separation between channels of a multiplex unit is
    A. 5 dB
    B. 10 dB
    C. 20 dB
    D. 40 dB

13. A dirty tape head should be cleaned with
    A. Gasoline
    B. Etching solution
    C. Isopropyl alcohol
    D. Any of the above

14. If only one of the transistors in a push-pull amplifier is bad,
    A. Replace only the bad one.
    B. Replace both of them.
    C. (A) or (B)
    D. None of the above

15. A shorted capacitor can be checked by
    A. Bridging
    B. Substitution
    C. Both (A) and (B)
    D. All the above

16. A common problem that causes low volume, lack of treble control, and distortion is
    A. Defective AGC
    B. Defective volume control potentiometer
    C. A dirty tape head
    D. Defective motor and drive mechanism

17. An oscilloscope is an effective instrument for locating defective stages because it can display
    A. The waveform of the signal
    B. The frequency response of a stage
    C. Noise on a signal
    D. All the above

18. If a receiver squeals, howls, or motorboats, the most likely cause is a
    A. Transistor
    B. Filter capacitor
    C. Weak battery
    D. Resistor

19. When you are resistance-checking a quasi-complementary amplifier,
    A. There should be a low-resistance reading between common and V_{CC}.
    B. There should be a low-resistance reading between common and the ground.
    C. There should be a difference between resistance measures, between the common and ground, and the common and V_{CC}.
    D. There should be no difference between resistance measures, between the common and ground, and the common and V_{CC}.

20. When you use an ohmmeter to identify a short, the ohmmeter reading should indicate
    A. Zero
    B. Infinite
    C. 100 kV
    D. 1 MV

21. If an FM receiver signal drifts, a possible cause is
    A. A defective AFC circuit
    B. A defective oscillator circuit
    C. A shorted or open component in the AFC
    D. All the above
22. Which of the following stages are common to both AM and FM receivers?
    A. Tuner, local oscillator, detector, af amplifier
    B. The rf amplifier, mixer, i.f. amplifier, af amplifier
    C. Local oscillator, rf amplifier, frequency discriminator, detector
    D. Tuner, i.f. amplifier, detector, af amplifier
23. The AM detector performs two basic functions in the receiver:
    A. Amplifier and filter
    B. Buffer and amplifier
    C. Buffer and detector
    D. Rectifier and filter
24. One set of 262½ scanning lines represents
    A. A field
    B. A frame
    C. A cycle
    D. An interlace set
25. A total of 525 scanning lines represents
    A. A field
    B. A frame
    C. A cycle
    D. An interlace set
26. The number of frames per second is
    A. 20
    B. 30
    C. 60
    D. 120
27. The vertical oscillator runs at a frequency of
    A. 20 Hz
    B. 30 Hz
    C. 60 Hz
    D. 120 Hz
28. The horizontal oscillator runs at a frequency of
    A. 60 Hz
    B. 15,750 Hz
    C. 3.58 MHz
    D. 45.75 MHz
29. The sound and video signals are separated at the
    A. Intermediate-frequency stage
    B. Video detector
    C. Video amplifier
    D. Burst separator

30. The vertical and horizontal pulses are separated at the
    A. AFC
    B. High voltage
    C. Sync separator
    D. AGC
31. If both sound and picture are weak and distorted, the problem is most likely in the
    A. AFC
    B. Tuner
    C. Sound
    D. Video
32. Lack of a raster often indicates
    A. No television signal
    B. No video signal
    C. No AGC
    D. No high voltage
33. A signal horizontal line across the middle of the screen indicates trouble in the
    A. Tuner section
    B. Vertical section
    C. Horizontal section
    D. Video section
34. Slowly rising white "hum bars" in the television are caused by a bad
    A. Rectifier
    B. Picture tube
    C. High-voltage transformer
    D. Filter
35. A silvery, out-of-focus picture usually indicates a bad
    A. Rectifier
    B. Picture tube
    C. High-voltage transformer
    D. Filter
36. An overloaded picture can often be corrected by adjusting the
    A. Vertical oscillator
    B. Horizontal oscillator
    C. AGC
    D. Color killer
37. Vivid, strong colors are often referred to as
    A. Hue
    B. Luminance
    C. Saturation
    D. Chrominance
38. The amount of brightness perceived is referred to as
    A. Hue
    B. Luminance
    C. Saturation
    D. Chrominance

39. The alignment of all three color guns to a common point is referred to as
    A. Demodulation
    B. Confetti
    C. Blooming
    D. Convergence
40. Color confetti can often be eliminated by adjusting the
    A. Color chroma amplifier
    B. Color detector
    C. Color killer
    D. Color oscillator
41. The presence of color bars often indicates trouble in the _____ circuit.
    A. Reactor
    B. Horizontal
    C. Burst separator
    D. Chroma amplifier
42. Before a convergence setup is performed, one should first perform
    A. Gray tracking
    B. Degaussing
    C. Screen setting
    D. Alignment
43. The initials of UHDTV stand for
    A. Unit high-plasma television
    B. Ultra higher-distance television
    C. Ultra high-definition television
    D. Ultra high-diode television

Questions and Problems

1. Describe the development and characteristics of a modulated radio wave.
2. What is a crystal detector?
3. Draw a block diagram of a superheterodyne receiver.
4. Why is a splitter needed in a push-pull amplifier?
5. Explain the procedure of repairing a "dead" radio receiver.
6. Generally, what could cause a receiver to squeal, howl, or motorboat?
7. How can thermally intermittent components be located in a receiver?
8. What is a head demagnetizer, and why is it used?
9. What is a record stroboscope?
10. Explain the procedure of correcting a slowly turning record turntable.
11. Explain one procedure used to check a suspected shorted Zener diode.
12. Explain how to clean a tape recorder head.
13. What is a noise generator?
14. List some typical noise suppression equipment.
15. What is the difference between the methods of bridging a capacitor and substituting a capacitor?

16. Draw a block diagram of a black-and-white television from memory.
17. Explain the basic function of each circuit in the black-and-white television.
18. Explain common problems associated with a power supply in a television receiver.
19. Explain what happens to the overall picture of a CRT when its emission becomes weak.
20. How can the picture be improved when the tube has weak emission?
21. How can a service technician tell if vertical roll is being caused by the vertical oscillator or the vertical sync?
22. What stage of the television receiver is probably at fault when the picture has trapezoidal distortion?
23. What is confetti?
24. What are the different colors between ac and dc high voltage?
25. Explain how the service technician can tell whether the snow is being caused by the tuner or by the antenna?
26. What do the terms *hue*, *saturation*, *chrominance*, and *luminance* mean?
27. Explain the purpose of each section of a color television.
28. If color bars are present in the picture, what section should be investigated?
29. What section would most likely be causing a problem of weak color?
30. Explain the precautions to observe in removing surface-mounted devices (SMDs).
31. Explain typical problems with digital versatile discs (DVDs).
32. Explain the difference among the HD, UHD, and smart televisions.
33. What is the difference between liquid crystal display (LCD) and digital light processing (DLP) televisions?
34. Explain the difference between the light-emitting diode (LED) and plasma televisions.
35. Explain the basics of a smart television.

Chapter 7

Troubleshooting Digital Circuits

All electronic circuits are made up of devices that process either analog or digital signals, or both. An analog signal has an electrical waveform that has continuous values throughout time, whereas a digital signal is usually discrete. Continuous means that the signal has electrical values, such as voltage or current, that vary smoothly between one level and another. Discrete signals can only take on a very limited number of values. In most digital systems, only two values are used to represent an ON (1) state and an OFF (0) state. A few digital systems use four or eight different states or levels of voltage, current, frequency, or another electrical quantity. In any case, the number is limited and in-between values are not used.

Analog circuits are used to amplify process and filter analog waveforms that later may be connected to digital signals. As a result, the real-world and digital electronic devices can communicate with each other. Traditionally most circuits were analogs that were composed of discrete electronic components covered in Chap. 1. Digital electronics uses waveforms that have quantities with discrete values. Modern equipment such as computers, TVs, and cell phones use microprocessors that are made up of digital circuits.

In the early 1900s, predecessors to transistors called "thermionic valves" were created, which later became known as "vacuum tubes." The first electronic digital computer, built with vacuum tubes, was invented in 1947 by J. Mauchly and J. P. Eckert for the United States Army and it was called Electronic Numerical Integrator and Computer (ENIAC). Also in 1947, the first transistor was invented by Bardeen, Brattain, and Shockley at Bell Labs. Transistors are three-terminal devices, which are used for switching and amplification.

Since the 1950s, billions of transistors have been manufactured individually and as part of integrated circuits (IC). Later developments were large-scale integration (LSI), very-large scale integration (VLSI), and most recently, ultra-large-scale integration (ULSI) technologies. Each development contained increasing number of transistors.

Analog and Digital Circuits

Radio communications is a good example for analog signals. As discussed in Chap. 6, when a transmitter sends radio waves modulated by sound waves through the air, the receiver picks up the signals and then plays them through the speaker. Other old analog communication examples are reel-to-reel tapes, cassette tapes, and records. An example of new analog technology is a switched capacitor circuit that is used in power supplies. This circuit is constructed out of operational-amplifier (op-amp) comparators. Op-amps are manufactured from transistors. The switched capacitor circuits handle voltage differently than conventional analog ones. As a result, greater power efficiency is achieved.

An analog signal is an electrical waveform that has continuous amplitude values throughout time. For example, a human voice is a continuous signal with different amplitudes and frequencies. Figure 7.1 shows a sample of human voice. Amplitude is the amount of voltage or current variation of a waveform. Frequency is defined as the number of cycles per second for a periodic alternating-current (ac) signal, which can be a sinewave, square, or triangular waveform. Besides voice, actual physical variables such as light, pressure, and temperature are examples of analog signals.

Digitization occurs by taking samples of the analog waveform at equal time intervals and converting them to digital signals through analog-to-digital converters (ADC). For example, in order to record a signal on a compact disc (CD), voice needs to be converted to digital codes first. The accuracy of conversion is called *resolution* or *quantization*, which is the number of binary bits used to represent an amplitude at a given time.

The formula to calculate the number of quantized levels for resolution is 2^n, where n = number of bits used to represent an amplitude. For example, if 8-bit ADC is used, the analog signal is quantized into $2^8 = 256$ discrete levels. The more bits are used, the higher the resolution is and the better and more reliable digital signal is produced. In order to playback the CD, the digital waveform is converted back to analog through a digital-to-analog converter (DAC), and after it gets amplified through a linear amplifier, the analog voice is heard through the speakers.

As technology developed, smaller size, more reliability, and better sound quality have been achieved through the development of digital sound devices. CDs were invented in 1982; they were followed by the inventions of MP3s and iPods in the late

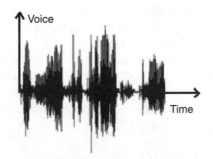

FIGURE 7.1 Sample waveform of a human voice.

FIGURE 7.2   Analog example—record player.

1990s and early 2000s. Figures 7.2 and 7.3 show the old (analog) and new (digital) technologies for audio recording.

Digital computers have also become smaller over time and more reliable, and electrical noise has been minimized. Technology has advanced by leaps and bounds since one of the oldest examples in digital communications, namely telegraph operating with the short and long beeps of Morse code. Telegraph signals are

FIGURE 7.3   Digital example—MP3.

transmitted and received through the wires and each beep represents a logic 1 or 0, explained in the next section.

Binary Code

Digital electronics uses binary numbers to represent a quantity. Binary numbers are 1 and 0, which are called *bits*; and groups of bits form *codes*. Each code represents an alphanumeric character, which is a letter, number, or a symbol. Figure 7.4 shows the comparison of binary and decimal numbers from zero to seven. With 3 bits we can count up to $2^3 = 8$ bits, which are binary numbers 000 to 111.

The leftmost digit in the binary code is called the most significant bit (MSB) and the rightmost digit is called the least significant bit (LSB). For example, for 011, the MSB is 0 and the LSB is 1. Each digit in the binary number has a weighting factor. For a three-digit binary number, the weighting factors will be multiples of 2s as follows: $2^2\ 2^1\ 2^0$, where $2^0 = 1$ is the weighting for the LSB, $2^1 = 2$ is the weighting for the next bit to the left of LSB, and $2^2 = 4$ is the weighting for the MSB. The value of the digital code is decided by adding all the digits multiplied by their weighting factors.

For the binary code 101, the LSB 1 is multiplied by its weighting of $2^0 = 1$ (1×1); the bit left to LSB, which is 0, is multiplied by its weighting of $2^1 = 2$ (0×2); and the MSB = 1 is multiplied by its weighting of $2^2 = 4$ (1×4). The final conversion comes out to be

$$(1 \times 4) + (0 \times 2) + (1 \times 1) = 4 + 0 + 1 = 5$$

We conclude that decimal equivalent of 101 is 5.

Digital signals are made up of discrete signals, which are composed of logic 1s and 0s. Logic 1 can be represented by

- A 5-volt (V) pulse on a copper wire using transistor-to-transistor logic (TTL) technology
- A light pulse on a fiber-optic cable
- A sinewave signal at a high frequency transmitted through airways

BINARY			DECIMAL
0	0	0	0
0	0	1	1
0	1	0	2
1	0	0	4
1	0	1	5
1	1	0	6
1	1	1	7

FIGURE 7.4 Binary to decimal conversion.

Logic 1 can also be identified as High, On, or True. Logic 0 can be represented by

- A 0-V signal on a copper wire using TTL technology
- No light pulse on a fiber-optic cable
- A sinewave signal at a low frequency transmitted through airways

Logic 0 can also be identified as Low, Off, or False. "On" can also be represented by a closed electronic switch, and "Off" represents an open switch.

Logic Gates

Logic gates are IC chips that are composed of basic electronic parts, including diodes, transistors, and resistors. Logic gates can make logical decisions to operate electronic components or equipment. For example, a logic gate can be implemented in a home security alarm system. Basic logic gates are AND, OR, and NOT gates. NAND, NOR, EXCLUSIVE-OR (XOR), and EXCLUSIVE-NOR (XNOR) gates are constructed out of AND, OR, and NOT gates. For simplicity we concentrate on 2-input gates.

The schematic symbol for an AND gate is shown in Fig. 7.5, where A and B are the inputs and C is the output. The truth table in Fig. 7.6 illustrates all combinations of possible input signals and shows which output each will produce. Each one of the input and output signals is either logic 1, 0, or undefined. For simplicity we cover logic 1 and 0 primarily.

The truth table shows us that the only way we can get a 5-V (1, True or On) output is by applying 5-V signals to both inputs A AND B. Any other combination of inputs will provide a 0-V output. Figure 7.7 shows a simplified example of an AND gate circuit for a house with a security system for one door and one window. In order to secure the house completely, the system must be installed in every window, door, and opening. Each opening or window and door must have a switch that should be closed before the owner leaves the house. If the light bulb

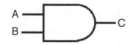

FIGURE 7.5 AND gate.

A	B	C
0	0	0
0	1	0
1	0	0
1	1	1

FIGURE 7.6 AND truth table.

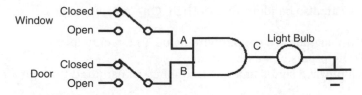

FIGURE 7.7 AND gate example.

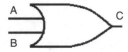

FIGURE 7.8 OR gate.

goes dark, an alarm will be activated, which is not shown in the picture. Under normal circumstances, once all openings (switches A and B) are securely closed, the security system is activated and the light is on. If an intruder breaks through a window or a door, the switch A or B will be open, which will sound the alarm system as the light goes off.

Figure 7.8 demonstrates the OR gate where A and B are the inputs and C is the output. The truth table for this OR gate is shown in Fig. 7.9. The truth table shows us that the only way we can get a 0-V (logic 0, False or Off) output is by applying a 0-V signal to both A and B inputs.

Any other combination of inputs will provide logic 1 output. An example of an OR gate is shown in Fig. 7.10, which demonstrates a security system where each one of the two inputs is connected to a switch that becomes a 1 if the door or window

A	B	C
0	0	0
0	1	1
1	0	1
1	1	1

FIGURE 7.9 OR truth table.

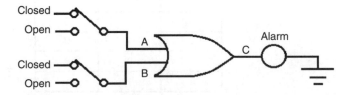

FIGURE 7.10 OR gate example.

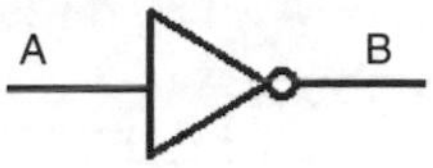

A ▷ B

FIGURE 7.11   NOT gate.

A	B
0	1
1	0

FIGURE 7.12   NOT truth table.

is opened. The output of the gate is connected to an alarm. If any of the switches connected to A, B, or both are on, the alarm will turn on. If both switches are off, the alarm will turn off.

Figure 7.11 is the schematic symbol for a NOT gate. The NOT gate has only one input and one output. The bubble at the output is called "active low" signal, which designates inversion or complementing. In the alarm with the AND gate done earlier, a NOT gate would be useful to trigger the alarm when the signal to the light is NOT on. The truth table for a NOT gate is shown in Fig. 7.12. The NOT gate inverts the logic 0 to 1 and logic 1 to 0.

The schematic symbol and the truth table for a NAND gate are shown in Figs. 7.13 and 7.14. NAND gate is the combination of AND and NOT gates. For a two-input NAND gate, output C is logic 1 when either A or B are logic 0, or when both A and B are logic 0; output C is logic 0 only when both inputs A and B are logic 1. An example of troubleshooting a NAND gate is in Fig. 7.15, where each one of the two inputs is connected to a switch. Output of the gate is connected to a light bulb.

FIGURE 7.13   NAND gate.

A	B	C
0	0	0
0	1	0
1	0	0
1	1	1

FIGURE 7.14   NAND truth table.

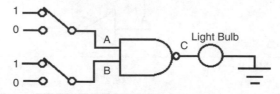

FIGURE 7.15 NAND gate troubleshooting example.

FIGURE 7.16 NOR gate.

A	B	C
0	0	1
0	1	0
1	0	0
1	1	0

FIGURE 7.17 NOR truth table.

If any of the switches connected to A, B or both of them are on, the light will turn on. If both switches are off, the light will turn off.

The schematic symbol and the truth table for a NOR gate are shown in Figs. 7.16 and 7.17. NOR gate is the combination of OR and NOT gates. For a two-input NOR gate, output C is logic 1 only when both A and B inputs are logic 0; output C is logic 0 when either input A or B is logic 1, or when both inputs A and B are logic 1.

Troubleshooting example of a NOR gate is shown in Fig. 7.18 where the output light will turn on only when both switches connected to A and B inputs are 0 at the same time. If one of the switches is 1, the light will turn off.

The schematic symbol and the truth table for an EXCLUSIVE-OR (XOR) gate are shown in Figs. 7.19 and 7.20. For a two-input XOR gate, output C is logic 0 when

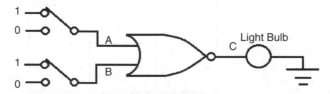

FIGURE 7.18 NOR gate troubleshooting example.

FIGURE 7.19 EXCLUSIVE-OR (XOR) gate.

A	B	C
0	0	0
0	1	1
1	0	1
1	1	0

FIGURE 7.20 XOR truth table.

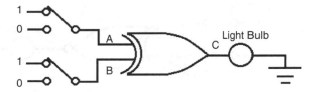

FIGURE 7.21 XOR gate example.

inputs A and B both are logic 1, or both are logic 0s. Output C is logic 1 when input A is logic 0 and input B is logic 1, or when input A is logic 1 and input B is logic 0.

Figure 7.21 shows an example of an XOR gate, where the output light will turn on only when only one of the switches connected to A or B inputs is on. When both switches are on or off at the same time, the light will be off. An application of an XOR gate is the circuit troubleshooting example shown in Fig. 7.22. Two similar circuit outputs are connected to the inputs of the XOR gate in order to be tested. If these two circuits function as expected, their outputs are the same, either logic 1 or 0. As these identical outputs of the circuits are applied to the inputs of the XOR gate, the output of XOR gate will be low, which indicates no failure. If the outputs of the identical circuits do not match, the XOR gate will turn on a light flagging a fault in one of the circuits.

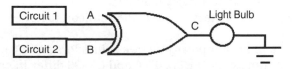

FIGURE 7.22 Troubleshooting XOR gate example.

FIGURE 7.23   EXCLUSIVE-NOR (XNOR) gate.

A	B	C
0	0	1
0	1	0
1	0	0
1	1	1

FIGURE 7.24   XNOR truth table.

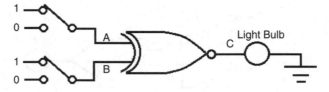

FIGURE 7.25   Troubleshooting XNOR gate example.

The schematic symbol and the truth table for an EXCLUSIVE-NOR (XNOR) gate are shown in Figs. 7.23 and 7.24. For a two-input XNOR gate, output C is logic 1 when A and B inputs are both logic 1 or both 0; output C is logic 0 when A is logic 0 and B is logic 1, or when A is logic 1 and B is logic 0.

An example of troubleshooting an XNOR gate is shown in Fig. 7.25, where the output light will turn on only when both switches connected to A and B inputs are on or off at the same time. If the switches connected to A or B inputs have different values, the light will turn off.

Digital Technologies

Two major digital technologies are used to manufacture logic gates:

- TTL (transistor-to-transistor logic)
- CMOS (complementary metal-oxide semiconductor)

TTL is made up of bipolar junction transistors (BJTs) and CMOS is made up of field-effect transistors (FETs). TTL and CMOS differ in characteristics such as switching speed, noise immunity, and power dissipation. TTL is the classic digital technology, and since it is not sensitive to electrostatic discharge (ESD), it is easier to use than CMOS technology. TTL has faster switching speed, shorter propagation delay, and lower power dissipation than CMOS technology.

TTL and CMOS chips differ in the letter designations that follow the first set of two numbers. For example, 74ACT04 designates a CMOS chip and 74LS04 designates a TTL chip. If the first two numbers are 74, it shows a commercial product. If the first two numbers are 54, it shows a military product. The letters are for IC chip descriptions and the last two digits indicate the type of gate. Examples of TTL IC chips are as follows:

```
74 (without letters)—Standard TTL
74S—Schottky TTL
74AS—Advanced Schottky TTL
74LS—Low-power Schottky TTL
74ALS—Advanced low-power Schottky TTL
74F—Fast TTL
```

Examples of CMOS chips are as follows:

```
74HC—High-speed CMOS
74AC—Advanced CMOS
74AHC—Advanced high-speed CMOS
74LVC—Low-voltage CMOS
74ALVC—Advanced low-voltage CMOS
```

The following list shows some examples of types of IC chips designated by the last two or three digits. Quad means that there are four gates in the chip. Triple means there are three gates and hex means there are six gates in the package.

```
74LS00—2-input quad NAND
74LS02—2-input quad NOR
74LS04—Hex inverter
74LS08—2-input quad AND
74LS10—3-input triple NAND
74LS11—3-input triple AND
74LS20—4-input dual NAND
74LS21—2-input AND
74LS27—30-input quad OR
74LS30—8-input single NAND
74LS32—2-input quad OR
74LS86—Quad XOR
74LS266—Quad XNOR
```

Figure 7.26 shows the pin-out diagram and manufacturer's data sheet for 74LS04.

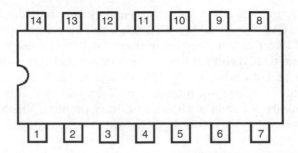

FIGURE 7.26   74LS04 pin-out diagram.

Voltage Specifications

Figure 7.27 indicates that a 74LS04 input is interpreted as logic 1 when the input voltage V_{IH} is 2 V or more and is interpreted as logic 0 when the input voltage V_{IL} is 0.8 V or less. The conditions for a 74LS04 output voltage V_0 are different. When its output V_{OH} is a logic 1, its voltage should be at least 2.7 V, and when its output V_{OL} is logic 0, its voltage should be 0.5 V or less. V_{OH} is a minimum acceptable value. A lower value than the minimum V_{OH} indicates a failure or damaged gate. V_{OL} is a maximum value. A higher value than the maximum V_{OL} also indicates a fault in the device. The abbreviations are explained as follows:

> OL = Output low
> OH = Output high
> IL = Input low
> IH = Input high
> L = Logic 0
> H = Logic 1

Figure 7.28 demonstrates the pin-out diagrams of typical IC chips, including 74LS00, 74LS02, 74LS04, and 74LS08.

Voltage	Minimum	Maximum
V_{OL}		0.5 V
V_{OH}	2.7 V	
V_{IL}		0.8 V
V_{IH}	2.0 V	

FIGURE 7.27   MIN and MAX input and output voltages.

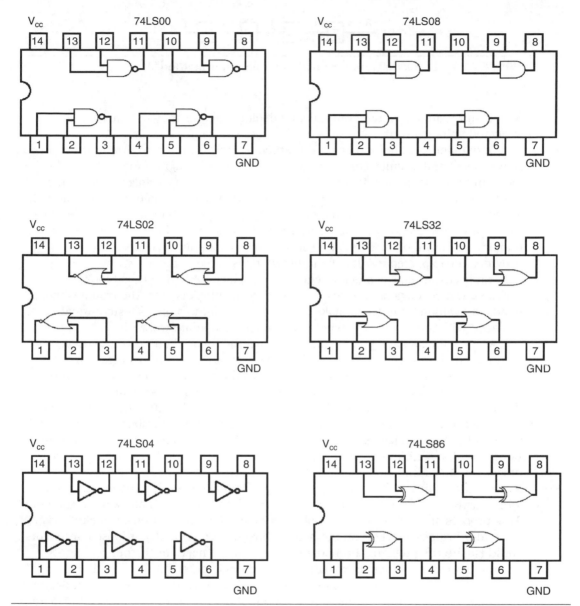

FIGURE 7.28   Pin-out diagrams for typical IC chips.

Troubleshooting Techniques

Troubleshooting is the process of finding the faults that cause the digital components or circuits to malfunction. It is advisable for you, as a troubleshooter to learn the theory behind the component or the circuit before you attempt to troubleshoot.

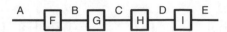

FIGURE 7.29   Block diagram for divide-and-conquer method.

As a result, you can test the outputs and decide whether they operate at the expected voltage levels.

Divide-and-conquer is a general method to apply in order to troubleshoot analog as well as digital circuits. For example, given the block diagram Fig. 7.29, the best location to test is the middle of the block, point C. Let's say voltage measurement at point C did not agree with the expected value; then you divide the circuit into half to the left of point C which would be point B. If the point B measurement tests well, we conquer the diagram to the right of B and divide it in half. We have a decision to make in selecting point C or D to test next. The best point to choose is D, and if it provides the expected results, the fault will be isolated to block I.

If it is cost-effective, block I needs to be replaced. Weighing the block replacement cost against the cost of repairing the block is what the troubleshooter needs to evaluate in order to make a decision. Once the block is tested, repaired or replaced with a new one, the whole circuit should operate well.

As a troubleshooter, you should also learn the theory behind what you are testing, which in turn will help you learn why a certain circuit voltage, current, or resistance measurement should provide results according to the specifications. You should order the service manual online or through regular mail for the device, block, circuit, or the equipment. For devices such as ICs, the name of the manufacturer will be on top of the package along with the number of the part. A search of the manufacturer's website will usually lead to data sheets for those parts. Being familiar with the manual will help the troubleshooter know what to expect out of the test measurements.

The first thing to do in troubleshooting is to make sure that the power is on while testing the circuit and the power is off when replacing a block or component. This process also helps the troubleshooter take safety precautions. Checking visually for solder bridges, crack(s) on the boards, burned components, or loose connectors could isolate the problems immediately without the need for further testing. Testing instruments such as "logic probe" and "logic pulser" are covered later in this chapter. Using your other senses such as listening to an unexpected and undesirable sound, or the "sniff test" for the telltale smell of a burned component, can also help troubleshoot the circuit as well.

Safety

For the safety of the components, the troubleshooter should wear grounding straps while handling electrostatic-sensitive devices such as CMOS chips. Figure 7.30 shows a grounding strap. After touching the metal piece (chassis) of equipment, the technician can touch the device and if there are any static electric charges, they will

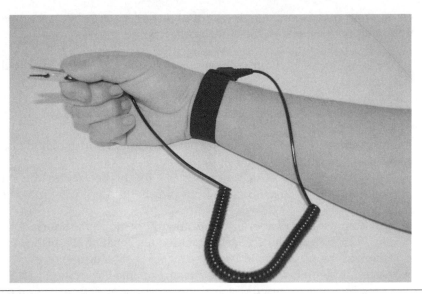

FIGURE 7.30   Grounding strap.

be discharged through the body, the ground strap, and the common circuit ground. This process is called *electrostatic discharge* (ESD).

For the safety of the troubleshooter, always power down the circuit before trying to remove or change components. As explained in Chap. 1, there are many kinds of faults with electronic devices; some of the more common ones in digital circuits include dirt and corrosion of copper components, moisture, high amperage, overheating of devices, smoke, and low voltage due to high currents caused by shorts.

Lightning could damage the circuit to a point where it cannot be repaired. Voltage spikes can occur as a result of printers, motors, relays, or any other electrically noisy equipment being attached to the same power outlet. These voltage spikes can cause logic 1 to become a 0, or logic 0 to become a 1 or undecided. Moving the circuit away from such equipment usually cleans up the noise on the digital signals and solves the problem when plugged into a different outlet or circuit farther away. Overheating can be minimized or prevented by installing heat sinks and/or fans around the components such as microprocessors, which can heat up excessively. If overheating is the problem, check for items piled up on top or on the sides of the enclosure that might be blocking the cooling vents.

Digital Circuit Troubleshooting Examples

Logic probes and logic pulsers are the two kinds of tools that can be used to test digital circuits. Figures are shown in Chap. 2 to demonstrate these instruments. A logic pulser is the tool that injects logic 1 and 0 pulses so that the logic probe can test the outputs. Contacting the point of test on a circuit with the tip of the logic probe will provide one of the three scenarios:

1. The indicator light on the logic probe will be bright indicating that the logic level is 1.
2. The indicator light will be off indicating logic 0.
3. The indicator light will be dim indicating undecided or gray region (not logic 1 or 0).

The following are scenarios causing possible troubles with digital circuit boards:

- An IC chip can be inserted incorrectly on the board causing shorts or opens. For example, in Fig. 7.31, the bent pin that did not plug into the socket caused an open.
- Too much solder (solder bridge) at a joint could cause shorts. For example, in Fig. 7.32, due to poor soldering techniques, a ball and a bridge of solder were formed which connected two pins together causing a short.
- A circuit board can be cracked, causing opens. For example, in Fig. 7.33, the crack in a board caused discontinuities or opens across several copper traces.

Electrical shorts could cause too much current flow through the circuit. Consequently, components could overheat and burn that would cause opens in the circuit. For example, in Fig. 7.34, the component is dark, which shows a good possibility of component burnout causing opens.

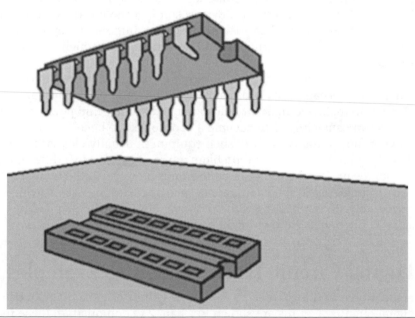

FIGURE 7.31   Bent pin.

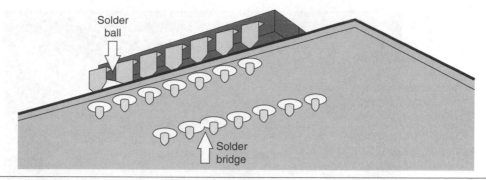

FIGURE 7.32   Solder bridge and solder ball shorts.

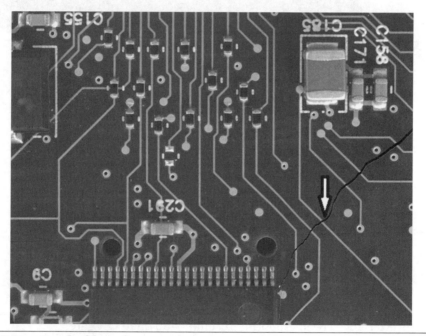

FIGURE 7.33   Cracked board.

The IC chip 7432 quad OR gate in Fig. 7.35 represents an example of digital component troubleshooting. In order to troubleshoot the chip, the first step is to disconnect it from the rest of the circuit. After disconnect, apply certain conditions at the inputs of one of the OR gates and check the output to observe whether it provides the expected results. For example, in Fig. 7.35, connect the logic pulser at input pin 5, logic probe at output pin 6, power supply at pin 14, and common ground to pins 4 and 7.

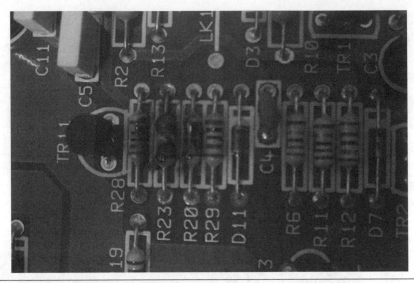

FIGURE 7.34   Burned components.

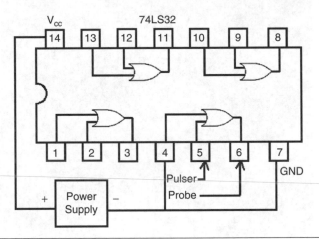

FIGURE 7.35   Troubleshooting IC chip 7432 quad OR gate.

If the gate is working properly, the probe will have light on and off at the same time as the pulser. Because pin 1 is connected to ground (logic 0) and also connected to OR gate input at pin 1, this procedure will test one of the four OR gates. The rest of the OR gates can be tested following the same procedures at their appropriate pins.

In order to troubleshoot the IC chip 7408 quad AND gate, first disconnect the chip from the rest of the circuit. In Fig. 7.36, connect the pulser at input pin 2, the probe at output pin 3, power supply at pins 14 and 1, and common ground at pin 7.

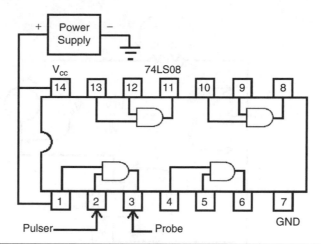

FIGURE 7.36 Troubleshooting the IC chip 7408 quad AND gate.

If the gate is good, the probe will have the light on and off at the same time as the pulser. Because pin 1 is connected to power supply at 5 V and it is at logic 1 level, the probe at output pin 3 will light up whenever logic 1 is injected at pin 2. This procedure will test one of the four AND gates. The rest of the AND gates can be tested following the same procedure at the appropriate pins.

Opens and Shorts

The topic of opens and shorts is essential to understand in order to perform troubleshooting. Often, circuits fail due to opens and shorts; they need to be troubleshot before repairing to resume normal operations.

Open Circuit

Whenever a complete break in a current path occurs, it results in an open circuit. In an open circuit there is no current flow. One of the most common sources of open circuits in circuit board systems happens when a connector to the edge of a board is loose. A broken wire would have the same effect. In Fig. 7.37, the normal current flow through the 10-kilohm (kΩ) resistor and the digital circuit is disrupted. Disruption occurs because the wire connecting a to b has been removed or disconnected and there is now a very high resistance between those two points. An ammeter measuring current through the 10-kΩ resistor or the digital circuit would measure zero amps.

Because the current throughout the circuit is zero, the digital circuit (board) will not carry any current. With no current, there will be no voltage drop across the 10-kΩ resistor. A voltmeter attached to its ends would read zero. Also, a voltmeter

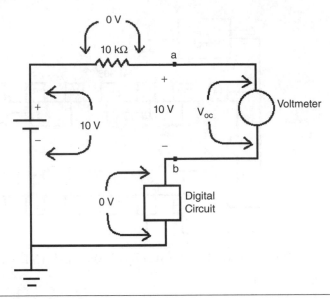

FIGURE 7.37   Open-circuit diagram.

measuring any components across the digital circuit board would read zero, due to no current flow. The open wire, which has the highest resistance or impedance in the circuit, will cause all 10 V of the power source to appear across a-b.

Digital systems are generally not simple series circuits like the example above, but they all have one thing in common. If there is a broken or loose connection (high resistance), you will measure a voltage across the high resistance. Normally there would not be any voltage across a low-resistance conductor like a wire, cable, or board connector.

Note if there are no breaks in this circuit and there is a connection between a and b, it would be a closed loop and current would flow throughout the circuit. As a result, because the wire has very small resistance, current through the wire should drop a very small voltage V_{ab} on that wire. A voltmeter connected across the ends of a good connector or cable should show a value near 0 V.

Open Inputs

In TTL circuits, if an open or disconnected input occurs, the assumption might be that the input would be regarded as logic 0 or OFF state, but this is not the case. For most types of TTL logic, the disconnected input reacts like a logic 1 (TRUE or high) input. For example, if a TTL OR gate has a broken or disconnected input, or a bent pin that is not making contact, it may "lock" the output of the gate high (logic 1). An open input will pick up noise; however, it may also be switching at times that has no connection to the ordinary operation of the circuit.

Compared to TTL logic, CMOS logic, will often react to an open or disconnected input by continuing to hold the state that was at the input just before the disconnect occurred. This is temporary and will become unreliable after a short time based on the capacitance of the gate inputs of the CMOS devices in the chip.

Open Outputs

There are also a number of logic devices with outputs that can be high (logic 1), low (logic 0), or in a high-impedance (high-Z) state very similar to an open circuit. In the high-Z state, for instance, a 7401 open-collector NAND gate can be placed in a state where other gates connected to the same output line can "take over" the line. The line would be driven then with 1s and 0s without interference from the gate in the high-Z state. Digital devices with tristate logic outputs can be found in many forms and are often used in digital designs to allow multiple devices to take turns controlling the signals on a conductor. When the device is signaled to put its outputs in the high-Z state, its gates are said to be disabled. Tristate logic is covered in Chap. 8 in detail.

Short Circuit

An undesirable low resistance path between two points in a circuit is called a *short circuit*. For example, when a solder bridge occurs between two solder joints (a and b), it will cause a short circuit between a and b and the current flowing through the least resistant path will be called short-circuit current, I_{SC}. Figure 7.38 shows an example of a digital circuit with a short. Two parts of the digital system are each 10 kΩ, and should divide the voltage up 50-50, with 5 V across the resistor and 5 V across the circuit board. However, there is a short circuit across points a and b. In this case, it is an ammeter, which is near 0 Ω, that is used to both short out the circuit board and to measure the short-circuit current I_{SC}. I_{SC} is diverted around the digital circuit board. If this were a real troubleshooting situation, I_{SC} would be the current.

I_{SC} is calculated according to Ohm's law, from which 10 V/10 kΩ, which will be 1 milliampere (mA) read by the ammeter. Because the voltage across the low-resistance ammeter is near zero, V_{ab} will be near zero. Note that, if there were no

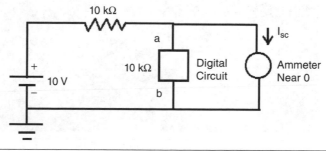

FIGURE 7.38   Short-circuit diagram.

shorts in this circuit, current through the 10-kΩ resistor would have been 0.5 mA as follows:

$$10 \text{ V}/20 \text{ k}\Omega = 0.5 \text{ mA}$$

As we observe, the short caused the overall current to increase to twice as much as the normal current. In practical applications, the short current can sometimes be extremely high; overheat, and even burn other component(s) in the loop, and cause other damages to the circuit. Therefore, it is a good idea to use fuses, which would protect the circuit from being damaged due to an excessive amount of current.

Installation and Replacement of an IC Chip

Figure 7.39 demonstrates soldering on a circuit board. Soldering is done by heating up the surface of the board with a hot soldering iron and melting the solder at the joint where an electrical connection can be made between the board circuit lines and the electrical component such as an IC chip. An IC should be placed on a circuit board with an insertion tool, as shown in Figs. 7.40 and 7.41, rather than by hand.

In order to extract an IC chip, it first needs to be de-soldered by heating up the solder joints. Then a vacuum pump can suck the melted solder out of the holes. Once the holes are clear of solder, the IC chip can be removed by carefully using an IC chip remover. Another option is to use the solder WIC, which is pre-fluxed copper braid that is used to remove solder. Once the joints are melted through the soldering iron, WIC can be used to remove all the solder out of the holes. Figure 7.42 shows

FIGURE 7.39 Soldering on a circuit board.

FIGURE 7.40 Chip insertion tool.

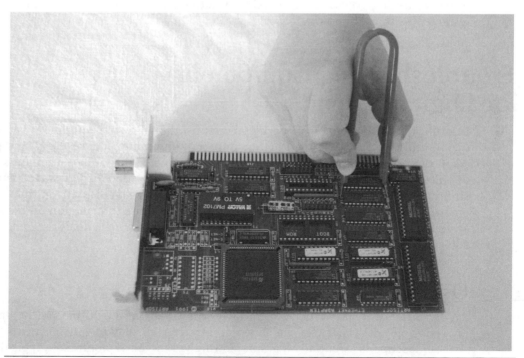

FIGURE 7.41 Placing an IC chip with the insertion tool.

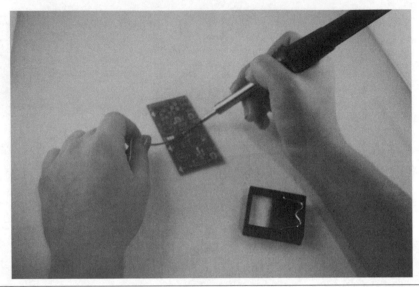

FIGURE 7.42   De-soldering Process with a Solder WIC.

the de-soldering process. After making sure that all solder is off, the IC chip can be removed using a chip extractor.

Troubleshooting Equipment for Digital Circuits

In order to test and troubleshoot digital circuits, the following equipment is used: digital multimeter (DMM), voltmeter, ammeter, ohmmeter, logic pulser, logic probe, oscilloscope, and logic analyzer. DMMs are used to take digital measurements of voltage, current, and resistance. Oscilloscopes are used to display waveforms simultaneously one, two, three, or four. Logic analyzers can display multiple channels such as 8, 16, and 24. Chapter 12 provides more information on recurrent-sweep, triggered-sweep, and digital storage oscilloscopes; logic analyzers; and signature analyzers.

Oscilloscope

An oscilloscope displays the voltage waveforms at any of the outputs of the digital or analog circuit. The waveform shows an electrical display of voltage versus time. An oscilloscope is essentially a graph-plotting voltmeter. A picture of a typical oscilloscope is shown in Fig. 2.3. Also, Fig. 7.43 shows a voltage measurement across a component placed in a circuit on a breadboard. On an oscilloscope, up to

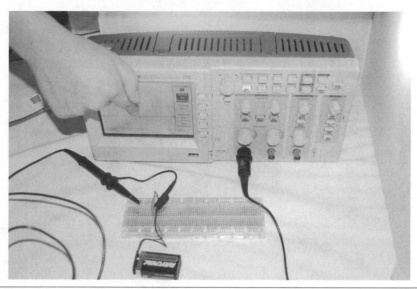

FIGURE 7.43   Digital oscilloscope.

four waveforms can be displayed at the same time so that they can be compared for troubleshooting purposes. In order to observe waveforms with high frequencies clearly, the scope has to be rated to operate at high enough frequencies.

A digital storage scope (DSS) uses digital memory. With DSS technology, more stability and clarity of waveform displays are achieved. It digitizes the analog signal and stores it in memory. As a result, the waveform can be analyzed using advanced digital signal processing techniques. Samples of the waveform are taken at equal intervals and processed through ADCs. According to the Nyquist theorem, in order to display the signals clearly, the sampling frequency should be greater than or equal to twice the highest-frequency component of the input signal. If the sampling frequency is less than twice the highest frequency of the input, distortion occurs, which is called *aliasing*.

Logic Analyzers

Figure 7.44 shows a picture of a logic analyzer and the waveforms it displays. Logic analyzers are used to capture and display graphs of voltage from many different inputs. Similar to oscilloscopes, logic analyzers are used for troubleshooting, where more analyses of signals can be performed. When high-voltage resolution and high-time-interval accuracy are needed, oscilloscopes are used. For example, rise time and fall time can be measured best by an oscilloscope. Because they do not require high-voltage resolution; data, control, and address bus information on a microprocessor can be observed best on logic analyzers. The timing relationships among signals and triggering on the patterns of logic 1s and 0s can be observed using logic analyzers.

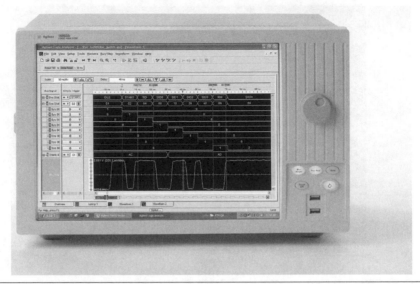

FIGURE 7.44 16803A logic analyzer. (*Courtesy of Agilent Corp.*)

Self-Examination

1. A logic output from a NAND gate is stuck high (logic 1) when it should be switching. What kind of fault could cause this result?
   A. An input shorted to ground
   B. An open or disconnected input
   C. An input shorted to a positive or high (logic 1) state

2. A circuit board in a system with multiple boards is not functioning. Checking voltages on the board reveals that all components are receiving 0 V when they should be showing voltages between 0 and 5 V. What kind of problem could cause this?
   A. An open circuit in the connectors carrying power to the board.
   B. A short circuit bypassing all current around this board.
   C. Could be either problem; more information is needed.
   D. Neither of these problems would cause this; the power must be off.

3. In the previous example, another indication is that there is a burnt smell inside the enclosure and there may be some smoke coming from another board. What kind of problem could cause this?
   A. An open could do this because current is blocked.
   B. In a short circuit currents are usually larger and overheating could occur.
   C. It is a software problem; reboot the system and start it up again.

4. In the example of item 1, all the indicator lights on the outside of the system cabinet are off and the fan in the enclosure is not running. What do you do first?
   A. Check if the plug was pulled out; if not, look for a blown fuse.
   B. It is a software problem. Shut down the system and restart it.
   C. Buy a replacement unit; this one is dead.

5. A recently repaired digital system has been working fine and suddenly "locks up." When its owner loses his/her temper and kicks the cabinet, it starts working again. This may indicate
   A. Intimidation works.
   B. A loosened connector (the kick tightened it up).
   C. A short, like a loose solder-ball (the kick jolted it loose).
   D. This never works; do not even think of it.

6. As a troubleshooter it is not advisable to learn the theory behind the circuit or component.
   A. True
   B. False

7. The first thing to do in troubleshooting is to
   A. Check for solder bridges, cracks, burned components.
   B. Make sure the power is on while testing the circuit and the power is off when replacing a component.
   C. Learn the manufacturers' names of the IC chips you are working on.

8. Overheating of electronic components can be minimized by installing _____ around the devices.
   A. Copper
   B. Silver
   C. Heat sinks
   D. Insulators

9. The tool that injects logic 0 or 1 pulses into the digital circuits is called a
   A. Logic pulser
   B. Logic gate
   C. Logic chip
   D. Logic probe

10. To troubleshoot an IC chip, the first step is
   A. Apply certain conditions at the inputs of one of the gates.
   B. Disconnect the IC chip from the rest of the circuit.
   C. Observe the logic probe lighting on and off at the same time as the pulser.
   D. All of the above.

11. The voltage across a short circuit is
   A. Very high
   B. Equal to the source voltage
   C. 0 V or near 0 V
   D. 1 V

12. The current through an open circuit is
   A. Very high
   B. 0 A
   C. 1 A

13. If a TTL or gate has a broken or disconnected input, it may "lock" the output of the gate
   A. Undecided
   B. High (logic 1)

C. Low (logic 0)

D. B or C.

14. In most digital systems, _____ values are used to represent voltage or current.
    A. 2
    B. 1
    C. 4
    D. 3

15. TTL logic has _____ switching speed than CMOS logic.
    A. Faster
    B. Equal
    C. Slower

16. For a two-input NAND gate, when either A or B inputs are logic 0, output C is logic
    A. 1
    B. 0
    C. Undecided

17. For a two-input XNOR gate, when A and B inputs are both logic 0, output C is logic
    A. 0
    B. 1
    C. Undecided

18. The last two digits on an IC chip designate
    A. Commercial or military product
    B. CMOS or TTL chip
    C. Propagation delay
    D. Type of gate

19. Other senses such as smelling and listening can be used to troubleshoot a circuit.
    A. True
    B. False

20. Static charges can be discharged through the body by the process called
    A. DES
    B. ESD
    C. SED
    D. None of the above

Questions and Problems

1. A closed-loop circuit is constructed from a voltage source in series with a resistor and a digital circuit. If the wire is broken in the loop, what is the open circuit voltage across the open wire equal to?
2. How does CMOS logic react to an open or disconnected input?
3. Describe tristate logic output.
4. Where are tristate logic outputs used?

5. Describe an open circuit.
6. What is a short circuit?
7. Give an example of a short circuit.
8. What damages can a short circuit cause and how can the circuit be protected?
9. Describe the scenarios that could occur by contacting the point of test on a circuit with the tip of logic probe.
10. Where are logic analyzers used?
11. What is the difference between digital and analog signals?
12. Give examples of applications for analog-to-digital converters (ADCs).
13. Give examples of applications for digital-to-analog converters (DACs).
14. When an 8-bit ADC is used, how many discrete levels is the analog signal quantized into?
15. What is logic 0 represented by?
16. What is logic 1 represented by?
17. Give a practical example for an AND gate.
18. Give a practical example for an OR gate.
19. Describe the difference between TTL and CMOS technologies.
20. For a 74LS04 chip, what are the voltage specifications for logic 1?

Chapter 8

Troubleshooting Combinational and Sequential Digital Circuits

All of the gates covered in Chap. 7 are independent such as AND, NOT, and OR gates. The input entered in a gate affects only the output of that particular gate. This chapter covers combinational and sequential circuits where one gate's output affects the next gate's input. The previous logical states of the components such as flip-flops (FF) affect their present and next state of the logic outputs.

Combinational and sequential digital circuits such as counters, flip-flops, and registers are used in advanced devices such as microprocessors. Industrial controls and machinery use complex logic with microprocessors, microcontrollers, or other embedded systems.

Flip-Flops

Flip-flops (FF) are memory storage devices that remember the previous set of inputs entered into a system. For example, in a digital lock of a safety box as you enter the lock combination 2546 sequentially, a group of flip-flops remembers all the numbers 2, 5, 4, 6 in sequence, and when 6 is entered last, it will unlock the safety box. The most popular kind of flip-flop is edge-triggered, which means that the output of the flip-flop changes state only on the rising or falling edge of a clock pulse. D and JK are the most common flip-flops.

D Flip-Flops

A D flip-flop has two synchronous inputs: D and CLK (Clock); two asynchronous inputs: PRN (Preset) and CLRN (Clear); and one output: Q. Figure 8.1 shows the symbol of a D flip-flop.

Asynchronous inputs override the synchronous inputs. An active low pulse on PRN will set the flip-flop output Q to 1 and active low pulse on CLRN will reset the flip-flop output Q to 0 regardless of the synchronous inputs D and CLK.

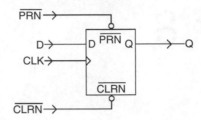

FIGURE 8.1 D flip-flop.

PRN	CLRN	Mode	CLK	D	Q
0	1	Asynchronous	X	X	1
1	0	Asynchronous	X	X	0
1	1	Synchronous	↑	0	0
1	1	Synchronous	↑	1	1

X = Don't care = Doesn't matter what the input is
↑ = Rising edge of the clock pulse

FIGURE 8.2 D flip-flop truth table.

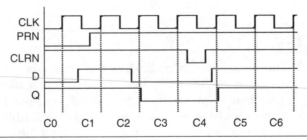

FIGURE 8.3 D flip-flop timing diagram.

The truth table shown in Fig. 8.2 demonstrates the operation of a D flip-flop.

Figure 8.3 shows a timing diagram for a D flip-flop in operation:

At C0 and C1, asynchronous input PRN overrides the synchronous inputs D, CLK. Therefore, the flip-flop sets Q to 1 until C2.

At C2, PRN = 1 and CLRN = 1; therefore no changes occur due to asynchronous inputs. Q follows D, which is 1.

At C3, PRN = 1 and CLRN = 1. Q follows D, which is 0.

At C4, CLRN = 0. Therefore the flip-flop resets Q to 0.

At C5, PRN =1 and CLRN = 1. Therefore Q follows D, which is 1.

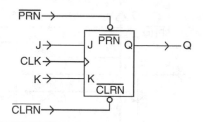

FIGURE 8.4   JK flip-flop.

JK Flip-Flops

A JK flip-flop has three synchronous inputs: J, K, and CLK; two asynchronous inputs: PRN and CLRN; and one output: Q. Figure 8.4 shows the symbol of a JK flip-flop.

Just like the D flip-flop, asynchronous inputs override the synchronous inputs in JK flip-flops. Active low pulse on PRN will set the flip-flop output Q to 1, and active low pulse on CLRN will reset the flip-flop output Q to 0 regardless of the synchronous inputs J, K, and CLK. Figure 8.5 shows the truth table of a JK flip-flop.

Figures 8.6 and 8.7 show an example of a JK flip-flop in operation and its timing diagram.

At C0, the asynchronous input CLRN is 0, which sets Q to 0.

At C1, both asynchronous inputs PRN and CLRN are 1. Therefore, synchronous

PRNS	CLRN	Mode	CLK	J	K	Q	State
0	1	Asynchronous	X	X	X	1	
1	0	Asynchronous	X	X	X	0	
1	1	Synchronous	↑	0	0	Previous Q	Hold
1	1	Synchronous	↑	0	1	0	Reset
1	1	Synchronous	↑	1	0	1	Set
1	1	Synchronous	↑	1	1	Previous Q	Toggle

X = Don't care = Doesn't matter what the input is
↑ = Rising edge of the clock pulse
Hold = the output Q continues to be whatever the last output was
TOGGLE = the output Q becomes the opposite state of last output

FIGURE 8.5   JK flip-flop truth table.

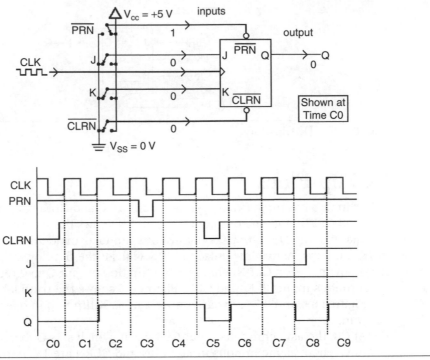

FIGURE 8.6 JK circuit and timing diagram.

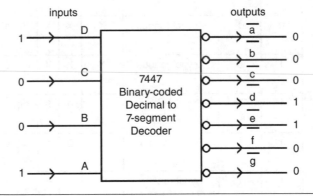

FIGURE 8.7 Binary-coded decimal (BCD) to seven-segment decoder.

inputs J and K make the decision. As J = 0, K = 0, Q will get its previous state, which is 0. Q will remain 0 until C2.

At C2, PRN = CLRN = 1. Therefore, synchronous inputs decide the output Q. As J = 1, K = 0, Q is set to 1.

At C3, asynchronous input PRN goes active (PRN = 0) and it overrides the synchronous inputs. PRN resets the output Q to 1. Q was already 1 and it remains 1.

At C4, PRN = CLRN = 1. Therefore, synchronous inputs decide the output Q. As J = 1, K = 0, Q is set to 1.

At C5, asynchronous input CLRN goes active and it overrides the synchronous inputs. CRLN = 0 resets the output Q to 0.

At C6, both asynchronous inputs are inactive. PRN = CLRN = 1. Synchronous inputs decide the output Q. J = 1, K = 0 will set Q to 1.

At C7, PRN = CLRN = 1. Synchronous inputs make the decision. J = 0, K = 0 cause the output Q to hold at the same state it was before, which is 1.

At C8, both PRN and CLRN are 1. Synchronous inputs make the decision. J = 0, K= 1 will reset Q to 0.

At C9, as PRN and CLRN are 1, synchronous inputs make the decision. J = 1, K = 1 will cause the output to toggle. Therefore, Q toggles from 0 to 1.

Decoders and Encoders

For security purposes, electronic messages sometimes need to be encrypted. Encoding is the process where the known code is converted to an unknown binary code in order to keep the message secure. Once the message is transmitted through a medium such as air, fiber-optic cables, twisted-cables, or trunk lines for computers, the decoder then converts the unknown binary code to a known (binary) code for the receiver. Another reason to encode is to make the signal more usable and easier to transmit through the digital devices such as computers, smart phones, and tablets. When transmitting, encoding can be used in encryption and compression of data, and decoding can facilitate decryption and decompression when receiving data. The most common decoder is the BCD decoder seen in Fig. 8.7, which has four inputs and seven outputs.

Seven-segment display as shown in Fig. 8.8 is used to decode the combination of four binary BCD inputs (A-D) to seven-segment outputs (a-g).

Each segment (a-g) is represented by a light-emitting diode (LED), which turns on or off according to the decimal number. In the following table shown in Fig. 8.9, each decimal number is represented by a set of inputs (a-g).

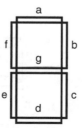

FIGURE 8.8   Seven-segment display light-emitting diode (LED) pattern.

BCD Input				Output							Decimal Digit in the Display
A	B	C	D	a	b	c	d	e	f	g	
0	0	0	0	ON	ON	ON	ON	ON	ON		0
0	0	0	1		ON	ON					1
0	0	1	0	ON	ON		ON	ON		ON	2
0	0	1	1	ON	ON	ON	ON			ON	3
O	1	0	0		ON	ON			ON	ON	4
0	1	0	1	ON		ON	ON		ON	ON	5
0	1	1	0			ON	ON	ON	ON	ON	6
0	1	1	1	ON	ON	ON					7
1	1	1	1	ON	ON	ON	ON	ON	ON	ON	8
1	0	0	1	ON	ON	ON			ON	ON	9

FIGURE 8.9 Decimal number representation.

For example, input combination 1001 displays (decodes) decimal 9 in the output, which is achieved by a, b, c, f, g LED segments being turned on, as shown in Fig. 8.10.

Calculators are prime examples where encoding and decoding are used. An encoder is the sequential digital device that converts decimal code to a binary code.

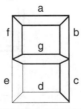

FIGURE 8.10 A seven-segment display showing a "9."

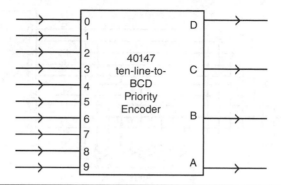

FIGURE 8.11   A ten-line-to-BCD encoder.

The most common encoder is the decimal-to-BCD encoder (10-to-4 encoder), which has 10 inputs and 4 outputs as seen in Fig. 8.11.

In a calculator, each decimal encoder input is converted from a push button to digital signals after passing through a digital circuit constructed from logic gates. When a decimal number in a calculator is pressed, a combination of fictitious A, B, C, D LEDs light up representing a BCD code, as shown in Fig. 8.12. In reality, instead of A, B, C, D LEDs, calculating circuits are connected to the BCD outputs in order to receive the input signals in BCD code.

For example, as shown in Fig. 8.13, decimal input of 7 at the encoder will provide a BCD output of 0111, which will be represented by LED D being Off, and LED C, LED B, LED A being On.

Shift Registers and Troubleshooting

Storage and movement of data bits are achieved by shift registers in digital systems. Shift registers are made up of flip-flops such as D or JK. Figure 8.14 shows four kinds of shift registers as follows:

1. Serial-in, serial-out (SISO)
2. Serial-in, parallel-out (SIPO)
3. Parallel-in, serial-out (PISO)
4. Parallel-in, parallel-out (PIPO)

Figure 8.15 shows a 4-bit SIPO shift register. Each bit is moved from the input of a flip-flop to the output at each rising edge of the clock pulse.

For example, assume that originally 1110 are inputted to D_0, D_1, D_2, D_3 of the shift register before clock pulse 0 (C0). At the rising edge of C1, each input of a flip-flop shifts to its output. Therefore Q_1, Q_2, Q_3, Q_4 become 0111 as the timing diagrams show in Fig. 8.16. By the same reasoning, at the rising edge of clock pulse 2 (C2), Q_1-Q_4 becomes 0011. At the rising edge of clock pulse 3 (C3), 0001 is the output, and at the rising edge of clock pulse 4 (C4), Q_1-Q_4 becomes 0000.

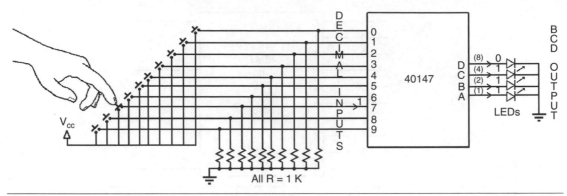

FIGURE 8.12 Decimal keys converted into BCD code.

Decimal Input	LED A	LED B	LED C	LED D
0				
1				ON
2			ON	
3		ON	ON	
4		ON		
5		ON		ON
6		ON	ON	
7		ON	ON	ON
8	ON			
9	ON			ON

FIGURE 8.13 BCD code to LED representation.

Therefore, it takes four clock cycles to shift the data completely out of the 4-bit shift register. If the CLRN input is set to zero, all flip-flop outputs Q_1, Q_2, Q_3, Q_4 change to zero.

Figure 8.17 shows a typical communication between PC1 and PC2. Computers make the calculations in digital and parallel format to process information fast. For example, PC1 transmits 8 parallel bits of information to the shift register 1. Shift register 1 as a PISO shift register converts the data from parallel to serial in order to transmit it through the communication lines such as fiber-optic cables.

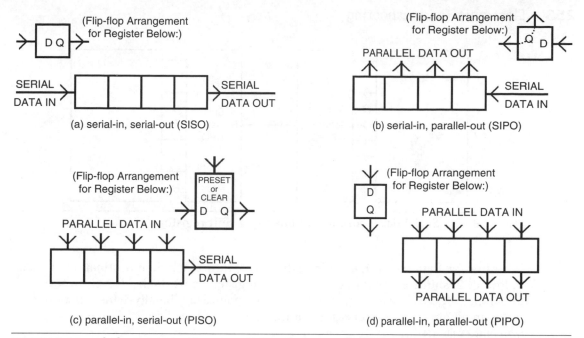

FIGURE 8.14   Shift registers. (Assume all flip-flops are clocked and attached to the same clock-pulse source.)

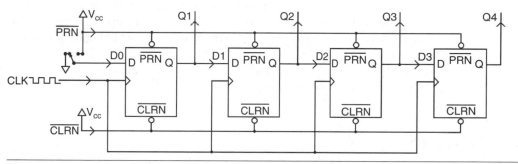

FIGURE 8.15   SIPO shift register.

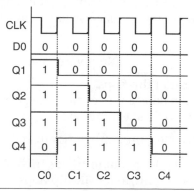

FIGURE 8.16   Shift register example.

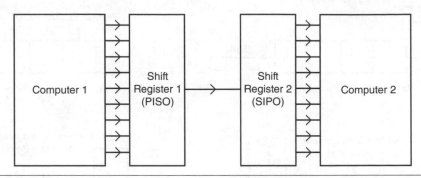

FIGURE 8.17 Serial data communications using shift registers.

After the data travel short or long distances, they reach their destination at PC2. Before PC2, shift register 2 in the SIPO shift register, converts the data from serial to parallel. PC2 receives the parallel data and analyzes it digitally. This function is found in a chip or microprocessor feature called a *universal asynchronous receiver-transmitter* (UART).

Troubleshooting of the shift registers can be performed by sending an 8-bit signal from the PC1 keyboard and receiving it at PC2. For example, typing a character at the PC1 should show up on the PC2's screen. If it doesn't, test the cables between PC1 and shift register 1 with a cable tester. Look up the 8-bit equivalent ASCII (American Standard Code for Information Interchange) code for a keyboard character and check it against the cable tester. If this step checks fine, connect the tester in between shift register 2 and PC2. If the tester doesn't show the expected code, shift register 2 needs to be replaced.

Counters

Counters perform counting up or down operations. Flip-flops are connected together to construct counters. Connecting the output of the flip-flop to the clock input of the next flip-flop creates the cascaded configuration. The number of flip-flops decides the maximum count, or the number of states (modulus), which is calculated by the following formula:

$$\text{Number of states} = 2^n$$

where n is the number of flip-flops.

For example, 4 flip-flops are needed to construct a counter that counts up to 15. $2^4 = 16$ different states of the counter are

0000, 0001, 0010, 0011, 0100, 0101, 0110, 0111, 1000, 1001, 1010, 1011, 1100, 1101, 1110, 1111

Figure 8.18 shows a 4-bit counter made up of 4 JK flip-flops. All J and K inputs are held at logic 1. According to the JK flip-flops truth table, this represents "toggle" mode. A toggle event happens, for instance, when you pull the chain on a light

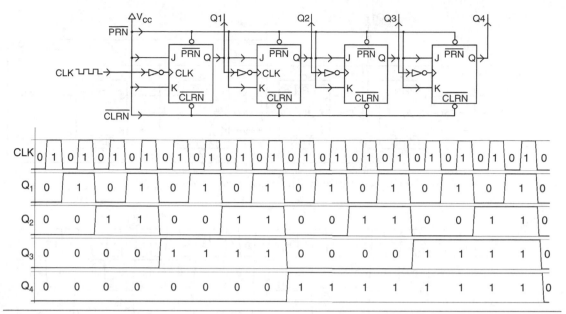

FIGURE 8.18   Modulus-16 ripple counter timing diagram.

socket that turns it on when you pull once, then turns it off when you pull it the next time. The triggering edge of the clock pulse has that effect on a toggling flip-flop.

At each falling edge of the clock pulse, flip-flop changes state (toggles). Figure 8.18 also shows the timing diagram, where the count starts with 0000 during the clock pulse 0, which are the outputs of Q_1, Q_2, Q_3, Q_4. As you observe clock pulses 1, 2, 3, etc., with each falling edge of the clock pulse the counter counts one binary number up.

During clock pulse 15, the counter completes counting with a count of 15 (1111), where outputs Q_1, Q_2, Q_3, Q_4 are all 1s. At the beginning of the clock pulse 16, counter goes back to 0 (0000) and starts counting again up to 15. As the output of each FF is connected to the clock input of the next FF, the signal propagates or ripples through in time from left to right. This is the reason that the counter is called *ripple counter*.

Counters are used as frequency dividers. For example, a radio transmitter uses a counter to divide the incoming frequency to obtain different channel frequencies to broadcast stations. Counters are also used in digital watches. For example, a counter is used to count up to 60 minutes and when it reaches 60, then the clock's hours-counter are incremented by one.

The number of different states a counter can exhibit is its modulus. Any modulus number (MOD#) counter can be designed by connecting a combination of flip-flops and NAND gates. According to the MOD#, inputs of the NAND gate need to be connected to the appropriate outputs of flip-flops. The output of the NAND gate needs to be connected to the CLRN inputs of the flip-flops.

Figures 8.19 and 8.20 show a MOD-10 (decade) counter, where the counter resets to 0 after it goes through 10 states and keeps repeating the same process.

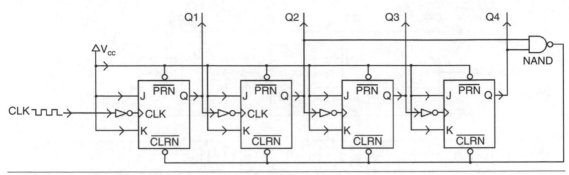

FIGURE 8.19 Modulus-10 ripple counter with forced-reset counting.

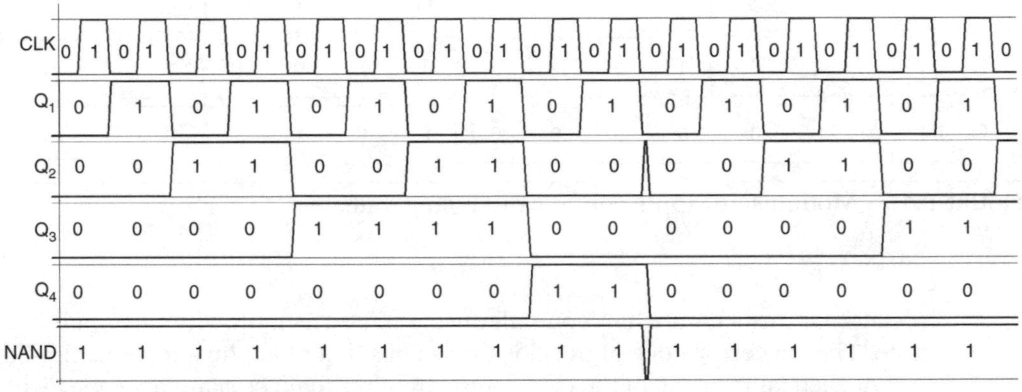

FIGURE 8.20 Timing diagram for modulus-10 ripple counter with forced reset.

In order to construct a MOD-10 counter, a MOD-16 counter needs to be constructed first. MOD-16 is made up of 4 flip-flops ($2^4 = 16$).

Figure 8.19 shows that the NAND gate inputs are connected to Q_2 and Q_4 flip-flop outputs. The first time these inputs are logic 1 simultaneously is when $1010 = 10$ is counted as the counter counts from 0 to 15. At this point the NAND gate outputs a logic 0. Because the NAND gate output is connected to all FF CLRN (Reset) inputs, when the NAND gate output is 0, it resets all flip-flops to 0. The counter starts counting the sequence from 0 to 10 again and the process repeats itself until the MOD-10 counter is turned off.

Digital-to-Analog Converters and Analog-to-Digital Converters

Most of the world is made up of analog signals such as voice, heat, humidity, and pressure. Computers do not understand analog; therefore, analog signals need to be converted first into digital signals. This process is called analog-to-digital conversion (ADC). Digital signals then get processed through the computer and/ or network of computers where they are analyzed. After the signal analysis, an electronic circuit connected to the computer filters unwanted signals which results in smoother waveforms. This process is called *digital signal processing* (DSP).

Further transmission occurs through the twisted-pair cables and air or optical fibers. Once the signals are received by the computer, digital-to-analog converter (DAC) converts the digital data to analog for the analog world to understand. DACs can be constructed from operational-amplifiers (op-amp) or from integrated circuits (ICs). Figure 8.21 shows an inverting-amplifier that is a summing amplifier op-amp configuration representing a 3-bit DAC. V_1, V_2, V_3, inputs are 5 volts (V) each representing "111" 3-bit digital code. The summing amplifier processes the input signal, and the analog output voltage can be calculated as follows:

$$V_{out} = -V_1(R_f/R_1) - V_2(R_f/R_2) - V_3(R_f/R_3)$$
$$V_{out} = -5 \text{ V } (10 \text{ k}\Omega/50 \text{ k}\Omega) - 5 \text{ V}(10 \text{ k}\Omega/25 \text{ k}\Omega) - 5 \text{ V}(10 \text{ k}\Omega/12.5 \text{ k}\Omega)$$
$$V_{out} = -7 \text{ V}$$

The output voltage can be inverted through an analog inverter to produce +7 V, which is analog decimal equivalent of the "111" input code with the proper polarity. Figure 8.22 is the DAC0808 that is a common IC chip version of a DAC. It has eight digital inputs (A1 – A8) and an output current (I_o) that is proportional to the digital input voltages. Analog V_{out} is obtained by multiplying I_o (from pin 4) by R_1 of 10 kilohms (kΩ) using an op-amp.

ADCs can be constructed from op-amps or from ICs. Figure 8.23 shows an ADC with 8 op-amps representing voltage levels 0 to 7 V. Each op-amp is a comparator that compares the analog input V_{in} with its reference input voltage (V_{ref}). If V_{in} is equal to the reference voltage, the op-amp output will be logic 1, and the 74148 encoder will output the 3-bit digital code according to the truth table shown in Fig. 8.24. This circuit is called a *flash converter* because it is faster than other ADC circuits and doesn't require multiple clock pulses to complete its conversion; hence, it is done "in a flash."

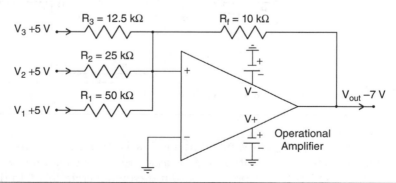

FIGURE 8.21 Simple digital-to-analog converter (DAC).

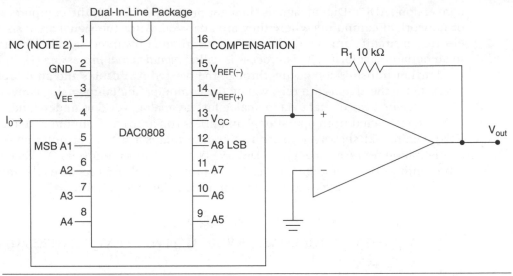

FIGURE 8.22 DAC0808 digital-to-analog converter (DAC) circuit. (*Courtesy of Texas Instruments®.*)

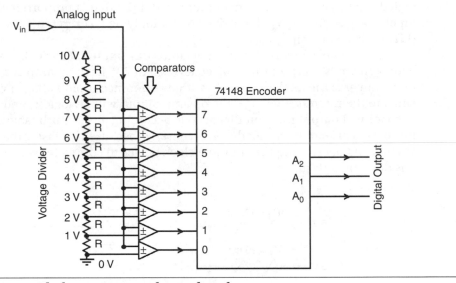

FIGURE 8.23 Flash converter analog to digital.

For example, if $V_{in} = 3$ V, op-amps 4 to 7 will output logic 0, as $V_{in} < V_{ref}$ for each one. Op-amps 0 to 3 will output logic 1, as $V_{in} > V_{ref}$ for each one. According to the truth table, the 74148 encoder will output the digital code "011." In this example ADC converts 3-V analog input voltage to a digital output of "011."

Analog Input	Output A_2	Output A_1	Output A_0
0 V	0	0	0
1 V	0	0	1
2v	0	1	0
3 V	0	1	1
4 V	1	0	0
5 V	1	0	1
6 V	1	1	0
7 V	1	1	1

FIGURE 8.24   Truth table for flash converter.

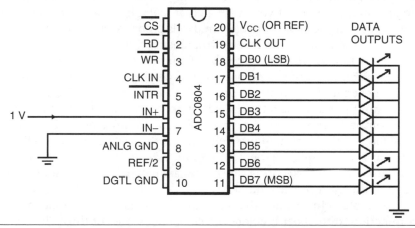

FIGURE 8.25   Analog-to-digital converter ADC0804. (*Courtesy of Texas Instruments®.*)

Figure 8.25 shows the ADC0804 that is the IC chip version of an ADC. It has an analog signal input V_{in}, (IN+), eight digital outputs, supply voltages, and eight LEDs. Digital output can be calculated from the following formula:

$$\text{Digital output} = (255/5 \text{ V}) \times V_{in}$$

For example, in order to calculate the digital output for the ADC with $V_{in} = 1$ V, use the preceding formula as follows:

$$\text{Digital output} = (255/5) \times 1 \text{ V} = 51_{10} = 110011_2$$

The outputs will be: $D_7 = 0$, $D_6 = 0$, $D_5 = 1$, $D_4 = 1$, $D_3 = 0$, $D_2 = 0$, $D_1 = 1$, $D_0 = 1$

The LEDs will light up according to Fig. 8.25. Each LED connected to a logic 1 output will light up and LEDs that are connected to Logic 0 outputs will turn off. Through the ADC, analog input signal of 1 V will convert to a digital output binary signal of 00110011.

Troubleshooting Systems with Synchronous Logic

Troubleshooting digital systems that interface with power equipment often involve a visual inspection for physical damage before any other diagnostics. If the system is "dead" and no outputs at all are apparent, the first thing to check for is power at the board. If a connector is loose or a fuse is blown, there may be power at the plug, but nothing is operational.

If the digital system being checked is a commercial product and you can find another one like it that works, you can compare the sequences of digital signals between the two units. If an output is not working, use an oscilloscope or logic analyzer to compare the signals stage by stage, working backward through the signal path toward the original input. Wherever you reach a point that the signals from the two units match, the next stage that the signals are delivered to is probably where the fault is located.

To do the type of side-by-side troubleshooting mentioned previously, it is necessary to know how the signals flow in the system works. This is generally a matter of finding inputs and outputs on the chips or boards, and tracing the signals to the next set of components or boards that utilize that signal. If a signal-flow path from input to output can be laid out, it becomes easier to trace a fault by the side-by-side comparison method. Especially if there are no schematics or timing diagrams available for the units being tested, this method is a good one to find system faults. Using this technique will be a little more exhaustive than if you have timing diagrams or troubleshooting flowcharts available, but it should help find the faulty stage of the system fairly quickly.

Most often in systems that are exposed to vibration or mechanical shock, a bad connection to the board, where the signal does not get through, is the cause of the failure. Exposure to moisture or corrosive chemicals is also a source of damage that can cause connectors to lose contact.

If the troubleshooting problem relates to a circuit you are building for a prototype or "breadboarding," you may have different sources of failure. The components may be inserted incorrectly, or the polarities or voltage levels on the power connectors are incorrect. These are not normally problems that appear in commercial electronics, but commonly happen when a system is being assembled as a prototype. In this case, there may also be a problem with the original design. It may be necessary to look over the theory of how the circuit is supposed to process the signals, which is not a problem with commercial or industrial digital systems.

When you remove and replace components, it is important to turn off all power to the boards and components before you begin that process. The removal tool can cause a short circuit and damage components or the power supply if power is left on when the component is being changed. Additionally, protection of the components from static electricity involves grounding the components. Also, you do not want to attach components to the electrical ground when the power is on, as it will probably blow a fuse or cause other damage.

Testing components with a logic probe and a pulser is called *static testing*. This test is not done at the speed that the system actually uses. Static testing might not work where timing is critical in a system such as dynamic memory. Dynamic memory must be clocked at a certain speed. Otherwise, the memory cells will start losing bits. Dynamic testing is testing the memory or other digital systems at their normal clock speed with an oscilloscope or logic analyzer. Dynamic testing may reveal problems where static testing does not.

Buffers

Buffers, also called "tristate enable gates," are a type of logic with three states, as the name suggests. In Fig. 8.26, you see a package containing eight of these gates:

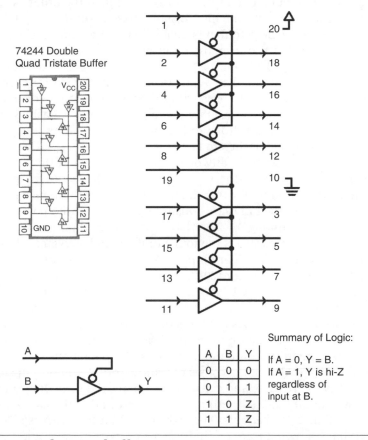

FIGURE 8.26 Quad tristate buffer.

The basic gate has two inputs, shown here as A and B, and an output, Y. Input A is the control input; it allows the input B to be transferred through to the output Y when it is active. The circle where A contacts the symbol indicates that it is active low. According to the truth table, when the A input is logic 0, the 1s and 0s at Y match what is put in at B. When the A input is logic 1, the output is in the high-impedance (high-Z) state. At this state anything entered in B has no effect on Y. The gate is said to be disabled when A is logic 1 and the B input has no effect. The same gate is said to be enabled when the A input is logic 0 and any logic waveform entered at B will be the same at Y.

The 74244 and other buffer packages such as the 74125 contain four gates. These gates are used for attaching signals from different sources onto one set of wires. Issues occur with connecting the outputs of two gates in to the same wire. When one is trying to pull up the wire to a logic 1 and the other is trying to pull the same wire down to logic 0, especially in metal-oxide semiconductor (MOS) logic, the transistors can burn out in one or both of the drivers. In either event, the final state is usually not a legitimate logic 1 or logic 0, but something in between whose effect on the system is unreliable.

When the enable gate is used, many outputs are attached to a line, often called a *bus line.* Only one output at a time is enabled and all the others are disabled or held back at their buffers. With the high-Z state, these gates do not pull up or down on the wire, and the single gate that is enabled is free to control the logic state of the wire without interference from the other outputs.

As shown in Fig. 8.27, one 7448 BCD to seven-segment decoder-driver package is used to run four digits of a display, instead of using a 7448 for each digit.

Calculators and other devices with many numerals on a display actually run all the digits at the same time. Only one digit at a time is flashed on for a short period and then the next in rotation. Hence, at any moment only one LED digit is displayed. The switching goes so fast that to the human eye the digits all appear to be lit, but in this example, any one digit is only lit ¼ of the time. When the light flashes, it is on at full brightness and has not lost that impression by the time the turn comes around again for that digit. The display light appears as if it was at full brightness, and yet the LEDs are consuming only ¼ of the power.

A counter is counting 0,1,2,3,0,1,2 . . . and so forth, known as a *modulus-4 binary sequence.* A decoder identifies the four numbers and activates a line low when its number comes up in the binary inputs. At the bottom of the picture, four registers contain the BCD numbers for "2017." At the count of 0, bits from the 10° register (0111, the number "7" in BCD) are enabled through the rightmost set of buffers. Those 4 bits are placed on the four-wire data bus, which is attached to the other registers through buffers. As only one of the two- to four-line decoder's outputs is active, the other three 4-bit buffers are disabled and have no effect. The number 0111 is what gets transmitted onto the four wires of the data bus.

Now, the 7448 decoder receives the number 0111 and outputs logic 1s to segment-drive wires a, b, and c. At that moment, the same line that is enabling the rightmost quad of buffers is also providing a ground to the cathodes at the rightmost digit of the four LED readouts. The rightmost digit lights up as a "7." The other three digits all have a high-Z state on their cathodes, wires that would normally allow

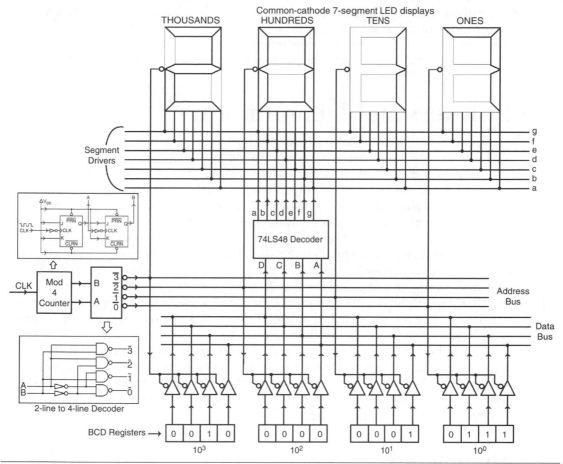

FIGURE 8.27 A "strobed" four-digit display.

current from the anodes to reach ground. Those three digits do not light up, even though the anodes of a, b, and c LEDs are positive.

When the clock completes another cycle, the counter changes from 0 to 1. The buffers at the 10° position go off and the buffers at the 10^1 position go on. Instead of the number 0111 reaching the 7448, the number 0001 arrives. The rest of the buffers remain at high-Z state. At the same time, the low output of the two- to four-line decoder grounds the cathodes of the tens' place digit. As the segment drivers from the 7448 decoder send logic 1s to b and c, the code for the shape of a "1" lights up at the tens' place.

As the counter counts 2 and 3, this process continues for the hundreds' and the thousands' place; then it starts all over again at 0. The process is so fast that your eye does not know that the digits are being "strobed" and they appear to be lit continuously. The buffering action allows the operation of further gates down the line without loading down the gate that supplied the original signal. These drive-current specifications are called *unit loads*, and in this case, each buffer requires only one unit load of power. The protection this provides its input device from overloading is called *buffering*. The buffers are like small power amplifiers.

Multiplexers and Demultiplexers

Multiplexers and demultiplexers are digital logic circuits that perform the job of an old-fashioned rotary switch. When you turn the knob of a rotary switch, it clicks into the next position and allows you to connect any one of several channels to a common conductor by mechanically rotating a contact from point to point around a circle. In the past, this was how television channels were selected, but that has largely been replaced by systems without moving parts.

In Fig. 8.28, rotary switches are shown at the top of the diagram, operating in two different ways. On the left, a switch is being used to select one of eight channels of incoming data to travel out of its output. On the right, a switch is being used to take data coming in and distribute that data to one of its eight outputs. The first switch is a multiplexer, which transmits the data on one of its multiple channels into one output. The second switch is a demultiplexer that receives data from the input and places it onto one of its multiple outputs.

Below the manual switches in Fig. 8.28, there are solid-state circuits that perform the same operation. In both examples, the selection of which channel is connected to the common conductor is done by a decoder. That is similar to what is done in the strobed-LED readouts of Fig. 8.27, and in fact, that is also an example of multiplexing and demultiplexing.

Multiplexers

Let's look at the two multiplexer circuits on the left side of Fig. 8.28. The upper circuit uses tristate enable gates to connect one of its eight inputs to the output. Only one of the decoder's outputs is active (low). For example, if the number on input CBA is 111 (seven), the data from line 7 will enable the lowest buffer in the stack. The outputs of the buffers are tied together, but only the buffer 7 is enabled to provide outputs that are high and low. The rest of the buffers are disabled and their outputs are at high-Z state. As a result, buffer 7 controls the output, which is the same as channel 7's input.

On the left bottom of Fig. 8.28, the circuit is constructed from more conventional AND and OR gates. As any data waveform into an AND gate requires the second input to be a logic 1 before the output can switch, the decoder chosen here by the designer is one with active high outputs. Only one of the decoder's outputs is active at the time so that the AND gate attached to that output will be able to switch on and off. The rest of the AND gates' outputs will be stuck at logic 0 because one of the two inputs of each AND gate receives logic 0 from the decoder.

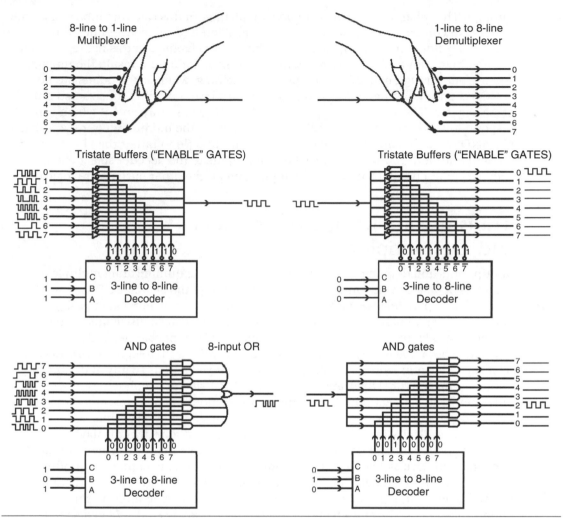

FIGURE 8.28   Multiplexer and demultiplexer.

If the CBA code selecting a channel is 101, decoder line 5 will be high and all the others will be low. AND gate 5 will be able to switch on and off as its input-5 signal varies, whereas all the other gates output a logic 0. The eight-input OR gate receives these eight signals and outputs a logic 1 wherever an AND gate outputs a logic 1. In this case, AND gate 5 is switching 1s and 0s, so the logic waveform at the output of the OR will be the same as the signal counting in channel 5.

Demultiplexers

On the right side of Fig. 8.28, the first demultiplexer circuit is made with tristate enable (buffer) gates and only one of the eight buffers is conducting the signal through. The other seven buffers are nonconducting. In this case, the inputs of all the buffers are connected to the same input signal. Therefore, whichever one of the seven buffers is conducting, that signal goes out the output channel. In this figure, code 000 is going in to CBA at the decoder, and that is the code for 0. As a result, the input is coming out through buffer 0 and output 0 is switching, whereas the others are not.

On the right bottom of Fig. 8.28, the demultiplexer circuit is constructed out of AND gates. The inputs are applied to eight AND gates, only one of which can respond to the input. It has a similar performance to the buffer circuit, except that the AND gates that are switched off all have output 0 (low) rather than being in a high-Z condition. In this case, the CBA = 010 inputs to the decoder select channel 2. The output at channel 2 matches the input, and all the other outputs are logic 0.

Transmission Using Multiplexers and Demultiplexers

Multiplexers and demultiplexers are used together by communicating multiple channels of data or voice on a single line, by "strobing" the channels. This is similar to the idea of parallel-to-serial and serial-to-parallel data transmission on a single path. However, in this case, the many channels of data can be distributed as though multiple parallel lines connect with each having a different channel of data. An example where this is used is to transmit multiple (digitized) telephone conversations on a single current loop. By "rotating" through many channels swiftly and returning to the original channel thousands of times a second, all the telephone users can hear only their own conversation. Many other conversations are being carried on the same line, and yet any given user is connected only for a short period of time.

This complexity saves money just like the "strobed four-digit display" example given before. It costs a lot less to connect one circuit carrying a hundred conversations (trunk line) across the country than it does to connect a hundred circuits between telephone switching centers. The circuit with the four digits of strobed readout in Fig. 8.27 is an example transmission using multiplexers and demultiplexers for the four channels. Each digit display used the 7448 decoder-driver chip for ¼ of the time, yet each one is being kept in synchronization with what is in the appropriate register for its decimal place.

Self-Examination

1. A momentary-contact switch is like a pushbutton used for a doorbell; once you let go of it, it forgets the state it was in when you pushed it. A latching switch is one that remembers the last state you left it in, like a light switch that you switch on, walk away, and the light stays on. Which type of switch is most like a flip-flop?
   A. The momentary-contact switch
   B. The latching switch
   C. Neither of these

2. A D flip-flop can be preset (PRN) or cleared (CLRN) through two inputs that are called its asynchronous inputs. What happens at output Q when one of these inputs becomes active?
   A. The output changes immediately.
   B. The output changes at the start of the next clock cycle.
   C. The output follows D at the next clock cycle and does not respond to the asynchronous input.

3. A D flip-flop without any active asynchronous inputs will have a state at its output Q that is controlled by input D. When does the input at D affect the output at Q?
   A. Immediately
   B. At the beginning of the next clock cycle
   C. At the beginning of each alternation of the clock (at every rising or falling edge)

4. A JK flip-flop can be PRN or CLRN through two inputs that are called its asynchronous inputs, just like a D flip-flop. With its two synchronous inputs, it can do more when it is clocked than a D flip-flop can do. J = 1 is used to set Q high, K = 1 is used to reset Q low, and when both J and K = 1, what happens to Q?
   A. Q stays as whatever it was until another clock cycle.
   B. Q becomes 0.
   C. Q becomes 1.
   D. Q changes from whatever it was to its opposite logic state.

5. What will be a JK flip-flop with J = 0 and K = 0? Assume that both of the asynchronous inputs are inactive.
   A. Q stays as whatever it was until another clock cycle.
   B. Q becomes 0.
   C. Q becomes 1.
   D. Q changes from whatever it was to its opposite logic state.

6. One kind of logic circuit takes some binary pattern like BCD code and converts it into a different pattern that is in some way easier for an ordinary person to read. For example, like the pattern of seven LED or LCD segments looks like an 8 when all are active. This kind of circuit is called
   A. An encoder
   B. A decoder
   C. A diverter
   D. An inverter

7. Another kind of circuit converts a simple input like an alphabetic *W* key pressed on a keyboard into a code that is useful in a digital system. For example, an ASCII code binary equivalent 01010111 is the pattern used for the letter *W*. This kind of circuit is called
   A. An encoder
   B. A decoder
   C. A diverter
   D. An inverter

8. A shift register circuit is made of four D flip-flops connected to transfer data between flip-flops. This circuit is set up so that bits shifted into the first flip-flop of the chain at D_1 are shifted-right out the Q_4 output of the last flip-flop in the chain 1 bit at a time, until after four clock pulses of all the data bits put in at D_1 have come out at Q_4. Basically, the data sequence at D_1 comes out Q_4 delayed by four clock pulses. What kind of register/data transfer system is this?
   A. SISO (serial-in, serial-out)
   B. SIPO (serial-in, parallel-out)
   C. PISO (parallel-in, serial-out)
   D. PIPO (parallel-in, parallel-out)

9. A shift register circuit is made of four D flip-flops connected to transfer data between flip-flops. This circuit is set up so that four bits are shifted into all four flip-flops from their asynchronous inputs, then shifted-right out the Q_4 output of the last flip-flop in the chain one bit at a time, until after four clock pulses of all the data bits put in at the start have come out at Q_4. Basically, the data sequence at D_1 to D_4 comes out Q_4 during four clock pulses. What kind of register/data transfer system is this?
   A. SISO (serial-in, serial-out)
   B. SIPO (serial-in, parallel-out)
   C. PISO (parallel-in, serial-out)
   D. PIPO (parallel-in, parallel-out)

10. A counter made of four JK flip-flops can produce the following outputs: 0001, 0010, 0011, 0100, 0101, 0110, 0111, 1000, 1001, 1010, 1011, 1100, 0001, 0010 and continuing from the binary code for 1 to the binary code for 12. What is the modulus of this counter?
    A. 16
    B. 13
    C. 12
    D. 10

11. A ripple counter is made of four flip-flops connected to count up in binary without any premature resets. If it has a natural binary count, what is its modulus?
    A. 4
    B. 8
    C. 16
    D. 10

12. A circuit is constructed to take in one of 16 different 4-bit binary code inputs and output 16 evenly spaced levels of voltage proportional to the code processed from the inputs. This kind of circuit is called
    A. A decoder
    B. An encoder
    C. A digital-to-analog converter (DAC)
    D. An analog-to-digital converter (ADC)

13. Using an operational amplifier (op-amp), the circuit in Fig. 8.21 allows a number of 0- and 5-V inputs to produce an output from 0- to 7-V. The abbreviation for this type of circuit is
    A. ADC
    B. DAC

C. ENC (encoder)

D. DEC (decoder)

14. The DAC0808 has eight binary inputs. How many levels of output voltage should this circuit be able to produce?

   A. 8

   B. 64

   C. 256

   D. 4096

15. Comparators are used in a flash converter with an encoder. This circuit is used for conversion from

   A. Digital to analog

   B. Parallel to serial

   C. Analog to digital

   D. Serial to parallel

16. The ADC0804 has eight binary outputs. It is used for conversion from

   A. Digital to analog

   B. Parallel to serial

   C. Analog to digital

   D. Serial to parallel

17. The most common source of open-circuit faults in a system with circuit cards plugged into a backplane is

   A. Loose connection

   B. Defective chip

   C. Low power supply voltage

   D. Solder bridge on the circuit board

18. If a system is "dead" and has no activity at all, the first place to check is

   A. Loose boards in backplane

   B. Power plug, line cord, blown fuse, or bad on/off switch

   C. Blown chip resistor or capacitor on board

   D. Overheating

19. A badly repaired circuit board has been soldered with bridged pins or foils on the board. This will cause

   A. Open-circuit fault

   B. Short-circuit faults

   C. Power failure

   D. Slower operation

20. Before removing or replacing components, it is important to

   A. Ground all pins

   B. Put on gloves

   C. Cool down the circuits with ice

   D. Disconnect all power

Questions and Problems

1. What kind of waveform is used for an edge-triggered flip-flop's clock?
2. There are other types of flip-flops besides D and JK types. What are some of these, and how do they differ from the two types in this chapter?
3. Do all latching circuits have to have a clock pulse?
4. If a JK flip-flop is operated with J and K both logic 1, its output will switch (toggle) once every cycle of the clock. If the frequency at the clock is 60 Hertz (Hz), what is the frequency at the Q output?
5. The most common code for sending alphanumeric characters between computers is ASCII code. ASCII uses 7 bits per character and 8 bits for special purpose and special-font characters. What was the first code used commercially between stations to transmit alphanumeric messages electrically, and what were its characteristics?
6. Teletype machines encoded characters from a typewriter-like keyboard into Baudot Code. How many bits were used for each character in that code, and how did it work?
7. Assume that Fig. 8.6 is faulty. At C2, PRN = CLRN = 1 and Q remains at logic 0. What could be the reason?
8. For a commercial product, troubleshooting for a faulty digital circuit can be performed by comparing it side by side with a circuit that works. Describe briefly how this kind of troubleshooting is done.
9. You want to light one of eight lights from a circuit, according to whether the input is 000, 001, 010, 011, 100, 101, 110, or 111. Is the circuit an encoder or decoder?
10. You want to take an 8-bit binary number that is received 1 bit at a time into a shift register, and when all the bits arrive, decode the register's outputs and light a nine-segment readout with that character on it. What kind of shift register is needed for this application?
11. An ASCII keyboard generates 7-bit ASCII code every time a key is pressed. It also generates a "keypressed" signal that identifies the code coming out of the board is from pressing a legitimate key. If that keypress is used to load the code into a 7-bit shift register and then it is clocked out of the register 1 bit at a time, what kind of register is being used here?
12. Eight flip-flops are connected in a shift register whose end is connected to its beginning. With a set of lights attached, whatever number is preset into the ring of flip-flops will circulate around the lights. What kind of a register does this ring of lights display?
13. What is the number of flip-flops required in a digital clock to count the 60 numbers, from 0 to 59 and to display the minutes on the readout?
14. A decimal counter that is also called a decade counter with four outputs counts 0 to 9 and repeats itself. If the frequency of the clock input to this counter is 60 Hz, what is the frequency of the wave coming out the most significant bit (MSB) of this counter? Is this a symmetrical wave having its output high and low in equal amounts of time during a cycle?

15. If a binary counter with 10 flip-flops counts up from 0000000000 until the count reaches 1111100111, and it is reset at the next count, what is its modulus? If used as a frequency divider, what is the frequency at the MSB output compared to the frequency of its clock at the input?

16. A transducer is a device that converts other forms of energy into an electrical signal, or vice versa. A thermistor is a transducer that is used in a temperature sensor, providing a voltage that varies with the temperature. What sort of device would you use to couple that voltage into a display system that shows the temperature on LED readouts?

17. MP3 files are music stored in digital form and played back at 16 bits per channel. To play the music onto a speaker, what kind of converter is needed to drive the amplifiers and speakers?

18. Given a DAC with 12 bits of output operating from a 5-V source, what is the smallest step by which its output voltage could be raised or lowered?

19. If you cannot locate a service manual for an unfamiliar piece of equipment you are troubleshooting, what is the next best thing?

20. A motor controller is not working. The controls and display panel appear normal, but the motor is getting only two of the three phases of power it is supposed to get and is working poorly. Where do you start troubleshooting?

21. The engineering department has built a prototype computer numerically controlled (CNC) mill that worked for a while and then "quit." The lights that are normally lit fail to turn on, the cooling fans are not spinning, and the components that should normally be warm are cold. Where would you look first?

22. A controller for a punch press has failed. It is a system with multiple boards plugged into slots in a backplane. Where would you look first for a problem?

23. Testing for overheating can involve specialized test equipment. It is one of the few places where testing might have to be done with the power on. Suggest a safe way to test for hot spots in a working system?

24. Testing encoders and decoders is a matter of matching inputs and outputs to a truth table. What kind of chart is needed to test a counter or register circuit?

25. Preventive maintenance for digital systems is generally like preventive maintenance for any electrical device including a television. What special risk is found in systems with high-voltage sections inside them?

Chapter 9

Troubleshooting Microprocessor-Based Systems

Introduction to Microprocessors and Microprocessor-Based Systems

By the early 1970s, progress on integrated circuits (ICs) had become advanced enough that all the circuits of a simple computer central processing unit (CPU) could be placed on one chip. Because of their small size, such chips were dubbed "microprocessors." Small portable computing machines such as calculators could now be made using these devices, and they were developed around the same time.

Most computer designers envisioned using these small chips in the development of more products. Such computing equipment is flexible because its operation depends on software. *Software* is a list of instructions in binary code stored in a memory device that operates parts of the microprocessor, one by one, in a sequence called a *program*. Sometimes, the binary-coded instructions loaded into memory are called *code*. By using such programs, a microprocessor can perform the same tasks as any other digital circuit or system. In order to accomplish these functions, a microprocessor should have as many input connections and output connections as the circuit boards it would replace.

The advantage of this approach is that changes and improvements in a design can be done more simply with a microprocessor than with a circuit designed from logic gates and flip-flops. To redesign or update a "hard-wired" logic board would require adding or removing devices and changing the wiring connections. Instead in a microprocessor, the program is replaced by loading in a new code that is not hard-wired. A new code with a different list of instructions is the only step required. You may need some changes in what is connected to the chip, but it is generally a lot less than what you would need to do with a circuit designed using gates and flip-flops.

The vision of the microprocessor designers was to have digital systems that could change their operation easily without modifying any of the circuit hardware. John von Neumann designed EDVAC (electronic discrete variable automatic computer), which was the first computer programmed by instructions stored in an electronic memory device. The reason to program a memory was to enable software to learn and relearn in order to perform different tasks. Reprogramming is not desirable in a microwave oven or a washing machine, but in more advanced robotic systems, it is extremely useful to have equipment that could "learn from its mistakes." For the purposes of these examples, we will stay with machines that can only "learn" new tricks when we load in a program with new instructions. As a result, each model encountered will have the same predictable behavior until it is upgraded.

Parts of a Microprocessor System

Microprocessors have three main components: memory, CPU, and input/output (I/O).

Memory

A microprocessor will not operate without a memory storage device to hold the program that controls it. Some chips have memory embedded on the same chip as the CPU that carries out the instructions. In other designs, there are separate memory chips that are connected to the microprocessor externally.

In some systems the program instructions are located in one memory system (program memory), and information the computer is working with as it does its operations is in a different location (working memory). That prevents an operating program from storing results of an operation in a place that accidentally erases the program that keeps it running. Both these types of memory storage may be on the chip with the processor circuits, but the space available is limited.

The advantage of having a separate memory is an ability to expand or add more memory, while the system-on-a chip can never hold more than the capacity it was designed for. The memory operates something like a shift register where a number of binary digits are transferred into the device and then shifted out in a sequence that controls the CPU.

Unlike a shift register, what is taken out of the memory can be numbers that start in any of a large number of possible locations and the processor can start "playing back" instructions from that location, wherever it may be. A system like that is called *random access memory* (RAM) because the location where a program's list of instructions starts may be anywhere in a large array of storage flip-flops, not necessarily starting in the same place every time.

Whether the memory is included on the same chip as the CPU or elsewhere, we now have a microprocessor system shown in Fig. 9.1. If we want this system to replace a logic circuit made of gates and flip-flops, there must be more to Fig. 9.1. There has to be a place where the inputs from the outside world arrive at the board, and a place where outputs from the board go to control whatever the circuit operates in that same outside world.

FIGURE 9.1 Incomplete microprocessor system.

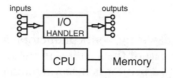

FIGURE 9.2   Basic microprocessor system with I/O handler.

CPU and Input/Output

As with the memory, special connections on the processor chip may be used if the number of input and output signals is small. In larger systems, external chips may be used to select which of several inputs is reaching the CPU and which of several outputs it is receiving signals from. This type of system allows expansion more easily than one where all the I/O connections are on a single chip. Now our system has developed as shown in Fig. 9.2 and it has an I/O handling section to control when signals are picked up and delivered.

The CPU is an electric device and it has both gate logic and synchronous logic. There are other CPU components to complete its operation. There are parts of the system that are not shown in Fig. 9.2. Although these parts are not shown in the block diagram, we assume they must be part of the real computer system. These parts include power supply, clock, and control signals. There are many other control signals that are required to coordinate the actions of memory, inputs, outputs, and other systems that vary with the style and model of the microprocessor system.

In the example of Fig. 9.2, the input and output ports are physical ports also called *hardware ports*, which are for devices plugged into specific connectors in the system. Examples of these hardware ports are universal serial bus (USB) sockets for a mouse or a printer, and red, green, blue (RGB) or video graphics array (VGA) ports for a video or display monitor.

Hardware ports are used to connect a computer to external devices such as printers or monitors. These should not be confused with virtual ports, which are virtual connections created by software programs to allow one computer to communicate with another computer. For example, when you use a browser to connect to an Internet site, your computer usually uses virtual port 80 to connect to the server that hosts that site. There are over 65,000 virtual ports on personal computers (PCs).

Troubleshooting a Microprocessor-Based System

Digital system that uses a microprocessor rather than gates and sequential circuits can fail in two completely different ways. If the system is not operating properly, it can have a hardware problem or a software problem. The resulting failure system may look the same.

Hardware

Hardware is the actual wiring, semiconductors, connectors, and other electric components that are part of the system. If one of these fails, it is a hardware failure. Suppose that, on a PC, a printer is disconnected due to a loose plug or connector. The printer will fail to operate, while the rest of the system may operate normally. This is a hardware problem. No amount of fiddling with the code in the computer's memory will restore the operation. The plug has to be put together physically for that printer to operate again.

Another example of a hardware problem—suppose the printer has its power switch turned off or it is out of paper. As a result the printer will not work. If the operating system in the computer is fairly "smart," you may see a message on the screen of the PC, telling you the printer is out of paper or appears not to be connected.

Software

Software is the series of binary numbers stored in memory, which is the program that controls the actions of the processor. Each number is brought into a part of the CPU that activates one or more circuits in that processor. Each circuit or combination of circuit actions has a different coded number in binary and each instruction executed in the processor causes an action.

For example, the printer is not working because the software driver in the memory has been damaged. The rest of the system operates normally, but when the user tries to print, a single bit in an output instruction has been erased and the data intended for the printer is directed to a different socket with no output device on it at all. Because this is a software problem, fiddling with the connection cables and plugs of the printer will not have any effect. Instead, the driver for the printer has to be reloaded with all the correct codes so that the printer will work properly again.

For another example, the computer may have software for several types of printers such as an Epson and a Hewlett-Packard. These printers have been used with that computer in the past, but only one of them is presently attached to the system. Assume that the user has accidentally left the computer "pointed" at the wrong printer. In that case, there is nothing wrong with the code, but the operating system is directing the print data, using the wrong set of instructions. The driver program does not have to be reloaded, but the control panel has to be used to change the default printer. The printer data will go to the default printer automatically.

Hardware and Software Troubleshooting Scenarios

The first step in troubleshooting is separating hardware from software problems. Some hardware problems are easy to check visually. For example, check whether the power light on the printer is lit. If there is no light, check whether its power switch is turned off or take the power plug out of the socket. Check to see whether there is smoke coming out of the printer cabinet. If you can see or smell smoke, very quickly unplug the printer from the power source. Check the wall for any fire damage. Unfortunately once the printer is burnt, it will be more cost-effective to replace the printer than to have it repaired. If you have checked all the hardware connections and the printer appears to be on, load the paper and try to make a copy or print. Failure to copy or print indicates a software problem.

For troubleshooting software, the first course of action is to restart the computer, turn it off and on again, and to reboot. In most cases, when a computer restarts, its programmed instructions are loaded into RAM from a storage device such as a hard-disk drive. The computer's operating system, drivers, and other components are coded to allow the computer to operate and are transferred into the computer's electronic memory from the boot drive. This is the hard-disk drive, which is essentially an input device where the programming that actually runs the computer resides. The names "boot" and "boot drive" come from the fact that the system starts up with no instructions in its memory, except those that transfer the code into RAM. As the program is loaded, the computer "pulls itself out of the mud by its bootstraps" to become the system you see when it is fully "awake."

The PC rebooting or reloading process can fix software problems because of a faulty instruction in RAM. RAM that has been damaged or erased while it was running will be replaced by the original code in the boot drive. As long as the code stored in the boot drive storage device is operating properly, the reloaded software will run normally.

When the data in the hard drive is faulty and/or the wrong default printer is selected, rebooting the system will not fix the problem. Then when the computer is turned off, it will store the wrong default printer. The computer will be restored to its former, incorrect state when it wakes up.

If the boot drive has a faulty code, the solution is to install the operating system and other programs into the hard drive again from the original source. This way you will treat the computer as though it is a new computer being set up for the first time. This is a more drastic step and you may lose valuable information placed on that drive from the time the computer was first operated. It is like fixing a hardware problem by replacing an entire circuit board instead of replacing a single chip in a socket.

In an embedded system, such as a microprocessor in a computer numerically controlled (CNC) milling machine on an industrial production line, the system may be expected to run continuously, for years without an error. To prevent an error from occurring, unreliable components with moving parts such as disk drives, switches, or buttons may be avoided. In these cases the system might be controlled through a touch-screen panel and the hard drive might be replaced by flash memory or read-only memory (ROM) with an operating system program "burned-in" permanently.

The motors and power equipment in that environment might put higher levels of noise on the electric power lines and radiate into the air as electromagnetic interference. The system then would need to be more robust, or "hardened" against power surges and radio-frequency noise.

Identifying Hardware and Software Problems

Though a CNC milling machine was originally perceived to have a software problem, it had a hardware problem instead. This new milling machine on a factory floor was equipped with a magnetoresistive memory instead of a disk drive. That system stored bits magnetically on a disk, but they could be read without moving parts; thus the system was expected to be much more reliable for long-term continuous operation. Instead, the system kept failing, and each time it was found that the code in memory was either wiped out completely or large parts of the stored code were corrupted.

Troubleshooters tried everything to diagnose the problem, but the system operated normally whenever they were there. Throughout the troubleshooting sessions they could not duplicate the fault that was happening. One day, the engineers came in and found the massive lifting electromagnet of a nearby overhead crane had been lowered to the factory floor and was resting propped up against the cabinet of the milling machine.

Although the lifting magnet was not activated at the time it was set down, the iron core of the magnet had a fairly strong residual field. Resting the magnet head against the side of the CNC machine cabinet brought it within a short distance of the memory board inside. The engineers found the data and code in the memory board closest to that side of the cabinet had been "wiped." The CNC machine was stored in a heavy iron case so that it would be shielded from strong magnetic fields. A simpler solution was to tell the crane operator to rest the magnet head somewhere else rather than against the side of the milling machine.

Troubleshooting Hardware Problems

Hardware problems usually happen suddenly or are evident when you are turning on the machine that contains the microprocessor. Observation with your senses can help you identify certain hardware problems. Listening carefully can be just as important as looking carefully when something is wrong. The following are examples of hardware troubleshooting:

1. Not enough noise: There are certain sounds you normally hear when the machine is turned on. You may hear a fan, the hard drive starting up, maybe the monitor powering-up if it is a cathode-ray tube (CRT) display. Systems such as Windows will generate a sound signature that tells you that the computer is booting up. If you do not hear some or all of the noises you expect, there may be a hardware problem. The power supply and/or any on-off switch may be bad if you do not hear any noise and see nothing at all. If only some of the sounds you expect are present, you may be able to narrow down what to look for based on the sound you do not hear.

2. Too much noise: The example of too much noise is when you hear the disk drive running continuously instead of operating normally for only a short period of time, while searching hardware/software components. This can

happen while you are waiting for something to happen on power-up and it never seems to get started. The drive sound might be loud or you may hear sounds that you normally do not hear such as grinding or buzzing. This may indicate that the disk drive may be damaged and needs replacement. If you hear the cooling fan running loudly and/or unusual noises, it means the fan is about to fail. Failure of the cooling fan can cause overheating and damage to the CPU and to the other components. The cost of replacing the fan is much less than replacing the CPU chip.

3. Unexpected turn off of the machine: If the machine turns off unexpectedly a few minutes after starting, it may be overheating. That could be from a bad fan, but look first to see whether there are any obstacles blocking the computer's cooling vents and limiting the airflow through the cabinet. The cooling fans need proper air circulation in order to avoid overheating.

4. Unexpected stoppage of an I/O device: When a specific input or output device suddenly stops working, although it had been working normally, a hardware problem is probably the cause.

   A. Keyboard failure: Keyboard keys are switches and their failures are basically due to accumulation of dust, other particles, and liquid spills. Keeping keyboards clean at all times will ensure the proper operations.

   B. Mouse failure: Failure of a mouse is due to a hardware or software problem. The mouse is plugged into a USB port. The mouse cable can get flexed and eventually it will fail. Plug it into another unit to ensure that the issue is the mouse and not the USB connection.

   An optical mouse with a hardware problem is determined if you fail to see the light. First check the batteries, and if the mouse is still not operational, there may be a software issue. Reload the software for the optical mouse, and if it still does not work, check the mouse on another computer. If the mouse works with another computer, that may be an indication of a faulty component.

   Wireless controllers do not have the wires that eventually break from metal fatigue and are more reliable. In a factory line, though, electrical noise and interference may cause a wireless mouse to be less reliable than a directly wired mouse or control unit.

   C. Display failures: A bad or jumbled-up display may indicate a faulty video card, or a loose connector. Check all the connections including the power source and VGA or RGB cables at the monitor. Reload the software if all the connections are correct, and if the monitor still fails, try on another unit. If the monitor operates properly on the other unit, it indicates that you have a port issue.

Troubleshooting Software Problems

When an operational microprocessor-based system stops working after a download or update, the cause is almost certain to be a software problem. New versions of an operating system may not be compatible with older versions of software that are still in the system. If the new drivers for a printer are not synchronized properly with existing hardware, the printer may fail to operate. New drivers for the printer may have to be located and installed, which is a relatively minor fix.

If the drivers for a memory controller do not synchronize properly, the CPU cannot fetch instructions from memory, and the entire operation of the system will collapse. Both these examples are software problems, but the latter problem will require reinstalling the older operating system all over again. This is much more serious and will require a great amount of time.

A sudden and excessively slow operation is a symptom of malware, which is malicious software that has invaded your system. An example of malware is software that takes over your e-mail account and starts mailing out unwanted advertising messages (spam), using up unreasonable amounts of computer time and slowing your ability to use the computer. If your computer suddenly seems to be too busy to pay attention to your mouse and keyboard, you may need to download a new Internet security system.

Bad software can seldom damage hardware. An example of a damaged hardware is when a software error causes a motor to run continuously. A program that is only supposed to be operating 50 percent of the time but instead runs 100 percent of the time can cause the motor to overheat. Or, if a controller starts one motor pulling up on something and another motor pulling down on the same part at the same time, those motors can be burned-out by the error. A correct hardware design can be damaged by faulty software instructions. Software errors can be especially dangerous in health informatics. For example, there have been cases where a software failure caused loss of human lives.

Replacement of Hardware by Software

For an example, suppose we want to build a tester that can compare one 8-bit binary number to another and indicate if there is any difference between the two. This could be used to compare the output of a working parallel data cable from one machine with another cable from a machine that is not working well.

Figure 9.3 shows a design that lights the output light Z any time a bit at one line on the cable does not match its counterpart on the other cable. Even if data is flying down the cables at millions of bytes per second, if 1 bit fails 20 percent of the time, the light at the output will be illuminated to 20 percent of its maximum brightness. This logic system uses two-input XOR gates, whose output is high (logic 1) only when its pair of inputs do not match. An eight-input OR outputs a 1 which turns on lamp Z if any bit does not match. Figure 9.3 shows the hardware for this tester where A_0 through A_7 are the output data of one machine and B_0 through B_7 are the output data of another machine.

For comparison, the same circuit could be done using a microcontroller, which is a microprocessor with its own memory and I/O controllers built in. Assume, for example, a hypothetical 8-bit microcontroller is already

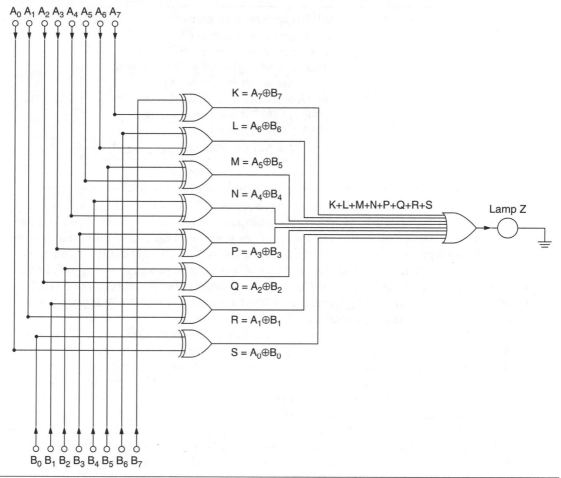

FIGURE 9.3   Hardware tester for logic.

supplied with all its power, clock, and control signals. Figure 9.4 shows the microcontroller in terms of the inputs and outputs attached to it. In this figure the same two 8-bit cables are attached to ports A and B of the processor, instead of the inputs of XOR gates.

The instructions to perform the comparison would actually be binary codes stored in consecutive locations in the chip's memory. Assume that there are two registers to be used in order to execute the instructions. These two registers are A and B and are used for storing 8-bit numbers. The instructions can be written in English instead of binary numbers. This is called *pseudocode* and it is similar to the instructions found in actual microprocessor instruction sets. The following textbox shows the pseudocode.

```
START       OUTPUT register A to port C
            INPUT port A into register A
            INPUT port B into register B
            EXCLUSIVE OR register B with register A:
                (put result in register A)
            COMPARE register A with 00000000
            JUMP to START if register A was zero
            STORE 1 into register A
            JUMP to START (always)
```

The software goes down step by step until it reaches one of the instructions that says to JUMP back to the beginning. Note that if the test whether port A contains the number zero is FALSE, the computer continues to move down the list of instructions. A logic 1 entered into register A will end up being delivered to lamp Z (lighting it) when control of the program goes back to START. On the other hand, if the test is TRUE and register A contains 00000000, the program never gets to the step that puts a "1" in register A. Control of the program will arrive at the START with register A still holding the number 00000000, so that OUTPUT at port C will send a logic 0 to lamp Z, turning it off.

If the program in the preceding textbox runs in the processor shown in Fig. 9.4, it performs the same operations as the hardware shown in Fig. 9.3. The only difference is that the microcontroller may be a little slower at doing the comparison than the array of logic gates. If the binary on the two 8-bit cables changes between lines 2 and 3 of the program, the comparison could show an error even if the two cables were matching perfectly, moment by moment. Therefore, it is necessary to clock this processor through its instructions faster than the data on the A and B cables.

Instead if a 16-bit processor were used, both 8-bit numbers could be INPUT into one port at the same time, then broken apart and compared by the software. That would be better in terms of synchronization, but the software would be harder to understand. In this slightly less reliable example, we hope for data on cables A and B to change more slowly than the processor is clocked. Figure 9.5 is the flowchart representation of this software.

Microprocessor Fundamentals

The first microprocessor was the Intel® 4004 brought out by Intel® Corporation in 1971. It handled data in 4-bit chunks called *nibbles*. A byte is 8 bits and half a byte is only a nibble. Forty years have passed, and microprocessors have been elaborated into a menagerie of different styles. According to Moore's law, the number of devices and degree of complexity has doubled every 18 months. The original 4004 microprocessor looked like the chip shown in Fig. 9.6.

The processor was developed for a special-purpose application, which was not very good for general purpose computing. A few years later, Intel® brought out the

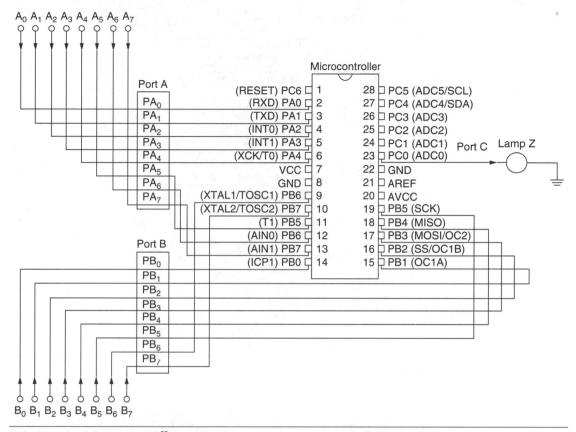

FIGURE 9.4 Microcontroller.

8080 that was the first microprocessor with enough capability to use as a general-purpose computing machine. In 1973, Forrest Mims built a computer based on the 8080 called the Altair 8800, and sold it to hobbyists as a kit. These kits are generally considered the first personal computers.

The 4004 model and all later designs still had a data bus, an address bus, control signals, power, and a clock. The main design had not changed, except for signals for parallel and serial data had been added. Digital-to-analog and analog-to-digital conversions on chip were also added. Word sizes had grown from 4 to 8 to 16 to 32 to 64 bits, address spaces have doubled according to the Moore law. CPUs even have more than one central processor; dual-core designs have succeeded single CPU designs. Quad-cores double the number in a dual core; eight processors may be found in current designs, and again microprocessors are expected to get more and more elaborate.

The 4004 has grown into an Intel® LGA2011 CPU. The LGA2011 chip with its 2011 pins is shown in Figs. 9.7 and 9.8.

Flowchart:

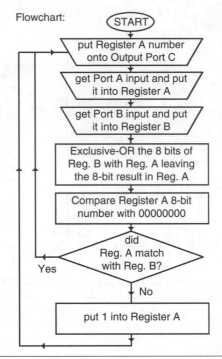

FIGURE 9.5 Flowchart for testing software.

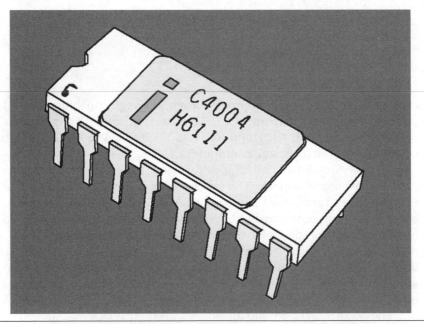

FIGURE 9.6 Original 4004 microprocessor chip. (*Courtesy of Intel®.*)

FIGURE 9.7   LGA2011 chip with 2011 pins. (*Courtesy of Intel®.*)

FIGURE 9.8   LGA2011 socket. (*Courtesy of Intel®.*)

The LGA2011 processor may be close to the current state of the art, but if you look at the fundamentals of how a microprocessor works, an LGA2011 is highly advanced and complicated. To illustrate the fundamental operation of a microprocessor, we will look at an example from an earlier and simpler generation of microprocessor instead.

Zilog Z-80180, an 8-Bit Microprocessor

The Zilog Z-80180 is an 8-bit processor with the ability of addressing ½ megabyte (MB) of memory, with clock speeds up to 10 megahertz (MHz). Ten MHz is 10 million clock cycles per second. It is backward-compatible with Zilog processors going all the way back to its ancestor the Z80 from 1976. Compared to the 4004 and the 8080, the first 4- and 8-bit processors, the Z-80180 microprocessor had already begun to accumulate the extra features offered by its competitors. Each of the microprocessor competitors raced to add new features to get in front of the competition in the 1970s and 1980s.

Serial input and output have been added as new features. Also, internal circuits that allow a crystal to be attached to provide clock pulses for synchronization have been added. The use of the internal crystal did not require a whole oscillator circuit to be built outside the chip. Special interrupt and memory management circuits have been added as well. In Fig. 9.9, data and address buses are identified with arrows showing signal flow.

Data Bus

The data bus in Fig. 9.9 also shows the eight pins labeled D_0 through D_7 and double-ended arrows. These arrows identify the fact that an 8-bit number can be inputted to the microprocessor through the eight pins. Or, conversely the processor can output an 8-bit number through the same pins. As these pins take turns sending or receiving signals, other pins, such as Read (RD) and Write (WR), identify the direction data is sent and received.

Address Bus

The address bus has 19 outgoing pins. These 19 pins send outgoing signals through the address bus wires A_0 to A_{18}. These bus wires are used to select where in a memory device the processor's next instruction is coming from. This means that most of the time the address bus is sending out 19-bit binary numbers that count up as the processor rolls through a program. On occasion, a JUMP takes place, and the next number on those pins is completely unrelated to the one that was there previously.

Another use for these 19 pins is placing data into a different location than where the program instructions are located in the memory. The same 19 pins are used for retrieving data out of the memory as well. As each pin has a state of either 0 or 1, if all 19 of them are used, there are $2^{19} = 524,288$ different numbers possible. This means ½ MB of memory can be addressed by this chip. It is a small amount for a

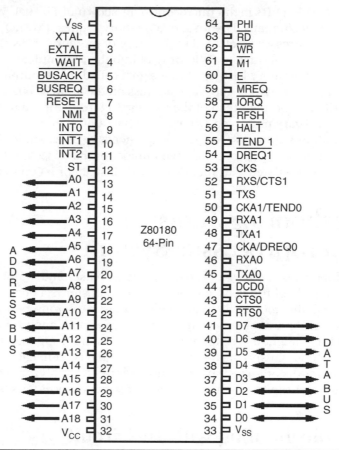

FIGURE 9.9   Zilog Z-80180, 8-bit microprocessor. (*Courtesy of Zilog Corp.*)

computer today, but 128 times more than the 4096 locations a 4004 could select. By contrast, the LGA2011 processor in Fig. 9.8 can address 34,359,738,368 bytes which is 32 gigabytes (GB) of memory. It is 65,536 times what the Z-80180 can handle and over 8 million times as much as the 4004.

In addition to memory, the microprocessor is connected to hardware devices called *input ports* and *output ports*. These are built-in parts of the microcontroller in Fig. 9.4 but are added externally in a microprocessor system. The same address pins that select one among thousands of memory locations for reading or writing can also select one among thousands of input ports and thousands of output ports.

Input and Output

It is hard to imagine any microprocessor system with more than a few dozen hardware input or output ports. The Z-80180 could certainly operate more than the three available ports in the microcontroller shown of Fig. 9.4. When the RD or WR signals are accompanied by an active memory request (MREQ) signal, one of the ½ million addresses is selected in the array of memory chips. This array of memory chips has a byte written to (WR) or read into the accumulator (RD) of the CPU from that specific memory location. If the signal called input/output request (IORQ) is active instead of MREQ, the same address on the address bus pins is used to select one of the I/O ports instead of the memory address. Although a computer may use several hardware ports, you will not find half a million input devices and another half a million output devices. If a design needs to be expanded beyond the limits of the address bus pin count, a large number of the 524,288 I/O addresses could actually be populated with memory boards instead of I/O ports.

Troubleshooting Tools for a Microprocessor System

Individual gates can be checked using a logic pulser and a probe. For counters or registers made of flip-flops, a logic analyzer that shows a dozen or so signals would be the desirable tool for testing. When you test a microprocessor chip with 64 pins, each one of the waveforms would be displayed on an analyzer screen 4 in (inches) high, with a height of a sixteenth of an inch. In order for each trace to show clearly and allow enough spacing between traces on the screen, each trace would end up a few hundredths of an inch high. Consequently the analyzer is not the right piece of test equipment to test the performance of a microprocessor system.

Troubleshooting with In-Circuit or Hardware Emulator

All digital systems in a microprocessor are a series of pins with a set of rules about what happens on one set of pins when certain signals are applied to other pins. An in-circuit emulator (ICE) or hardware emulator is a device that can be attached to a computer.

If the memory or an I/O port has a fault, the hardware emulator attached in place of the CPU will detect the problem(s). ICE can diagnose hundreds of different currents entering in and out of the CPU through the pins of the emulator. ICE is a sophisticated microprocessor chip that tests the system by pretending to be the microprocessor. If the emulator is built with the right diagnostic capabilities, it can look for error readings better than any analyzer with thousands of different traces and timing diagrams.

ICE matches the pin array on the microprocessor chip and it can be placed into the socket of the original microprocessor. While it has the ability to run the same software, that software does not have to run at full speed. The software can be stepped forward and backward, stopped at any step to check the behavior of all the circuitry outside the CPU socket. The diagnostic display of the emulator can also register activity at each point of the socket which is connected to circuit parts outside the CPU.

Once this type of hardware emulator is inserted, circuitry around the CPU is tested by running the appropriate diagnostic software. If the system works normally but does not work with the original CPU, it is time to replace the old CPU with a new processor. In the absence of an ICE, replacing the CPU chip with a new one will accomplish the same task as with an ICE. With the new CPU chip, run the diagnostic software to test to see whether the problem is the CPU or the outside circuitry. This, however, is not as useful as testing the system with a simulated CPU controlled by a diagnostic program with ICE.

The in-circuit or hardware emulators are also useful in debugging software, by allowing the software to run and stop. As the software runs, it looks at each location in the memory and I/O map, trying to see whether each step of the program performs as it was designed. When the program goes astray or "blows up" by creating an array of numbers, the memory runs out of space. In order to debug the software, the tester can walk through each step of the program.

Examining Software in RAM and Memory Dump

Debugging software is an art similar to computer programming. In fact, good programmers make the best software troubleshooters. This section is intended to introduce the simplest principles of the microprocessor operation. It is important to know the software inside the RAM of a computer and how the software is represented as well as a simple example of how to debug.

A listing of what is stored in the RAM of the microprocessor system is called a *memory dump*. If you would like to know what addresses in the RAM hold the part of the code you want to investigate, you want to perform a memory dump. The most common appearance of such a display is shown in the following textbox:

```
Address Data----------------------------------------------------------------Data

06E50:  2C 20 70 72 6F 67 72 65   73 73 20 6F 6E 20 69 6E
06E60:  74 65 67 72 61 74 65 64   2D 63 69 72 63 75 69 74
06E70:  73 20 68 61 64 20 62 65   63 6F 6D 65 20 61 64 76
06E80:  61 6E 63 65 64 20 65 6E   6F 75 67 68 20 74 68 61
06E90:  74 20 61 6C 6C 20 74 68   65 20 63 69 72 63 75 69
06EA0:  74 73 20 6F 66 20 61 20   73 69 6D 70 6C 65 20 63
06EB0:  6F 6D 70 75 74 65 72 20   63 65 6E 74 72 61 6C 20
06EC0:  70 72 6F 63 65 73 73 69   6E 67 20 75 6E 69 74 20
06ED0:  28 43 50 55 29 20 63 6F   75 6C 64 20 62 65 20 70
06EE0:  6C 61 63 65 64 20 6F 6E   20 6F 6E 65 20 63 68 69
06EF0:  70 2E 20 53 75 63 68 20   63 68 69 70 73 20 77 65
```

Contrary to appearance, this is a list of numbers. The number 06E50 to the left of the colon (:) represents the 20-bit address of the first byte of data displayed. In the memory, the binary address where the first memory byte is located is 00000110111001010000. Reading all these numbers in binary code is awkward and hard to follow for an address bus that has 20 bits. Instead, this address has been condensed into its hexadecimal representation as 06E50. Each digit represents a 4-bit binary group, using the decimal digits 0 through 9 for the first ten 4-bit numbers (0000 to 1001) and using letters A through F for the next six numbers (1010 to 1111).

Each of the 16 two-letter groups to the right of the colon (:) symbol is a hexadecimal code for the 8 bits of data at a memory location. In the first row, for instance, address 06E50 contains 2C, which is 00101100 in binary. The next address, 06E51, holds 20, which is 00100000 in binary. Moving to the right, address 06E52 holds the number 70, which is 01110000 in binary. The remaining 13 hexadecimal bytes, 72, 6F, 67, 72 … 6E, occupy locations 06E53 to 06E5F.

According to the way the hexadecimal system of counting numbers is organized, each line shows 16 bytes of data, and the starting address of each line down is 16 more than the line above. The same method shown above for reading data from a RAM can also be used to examine data stored in addresses on a hard drive or other mass storage devices.

The memory dump example shown in the textbox earlier happens to be a listing of a text file that holds the first few lines of text in this chapter. Each byte is actually a letter of the alphabet encoded in American Standard Code for Information Interchange (ASCII) code. A listing of machine codes in a program would look exactly like the memory dump example. The only difference is that bytes would be selected for which circuits they turn on in a CPU instead of which letters they print on a printer. The binary numbers can have a variety of meanings, depending on the kind of equipment where the transmission and reception occur.

Disassembled Memory Dump

In the case of binary instructions for a CPU, different processors may do different things with the same numbers. To see the results of what a series of numbers in a memory mean to a specific processor, there is a program written called a "dis-assembler." The dis-assembler matches the binary or hexadecimal codes up with the specific machine instructions. The following textbox shows a disassembled code.

Address:	Data:	Disassembled Code:	
0x6E50:	5F	pop	edi
0x6E51:	33C0	xor	eax, eax
0x6E53:	5E	pop	esi
0x6E54:	8BE5	mov	esp, ebp
0x6E56:	5D	pop	ebp
0x6E57:	C20400	ret	4
0x6E5A:	0090	nop	

Notice that the addresses do not count up by 16s. Instead, depending on the instruction, they may be counted by one, two, or three bytes long. The next line of the listing counts down to the address where the next instruction starts. To the right of the addresses, the bytes of each instruction are displayed in hexadecimal, each two characters being a consecutively stored byte of binary. Finally, on the right-hand side of each line is the assembly language representing what the instruction does. This is not as clear as the pseudocode shown earlier in the textbox, but if you know the names of the registers in the CPU this program is running on, you can figure out where the numbers being processed are going to and coming from.

To find a list of all the instructions that mean something to a microprocessor, you can look for them online as PDF files from the manufacturer. These documents are called *data sheets* and in early days they were only a few sheets long. By the time of the Zilog Z80180, the data sheets had been 80 pages long. For the Intel LGA 2011, a sheet is 564 pages long, which can be downloaded in two parts because it is 1128 pages long.

Decompiled Memory Dump

Most emulators will show what is in memory using a disassembled listing. Disassembled code shown in the previous textbox shows the "assembly language" for the specific processor it is designed to replace. Programs for computers, however, are not usually written in the assembly language of the microprocessor chip. It is more likely they will be written in a high-level language such as C or Java. Programs written in these languages are then translated into one of the many machine codes that exist for different processors. Regardless of what high-level language the source code of a program is written in, the computer runs the program from its memory. The source code ends up being a list of binary machine code instructions.

Like the dis-assembler, which simplifies analysis of binary programs found in memory, there are also specialized analysis programs that will take the bytes in RAM or mass memory storage and decode them into higher-level languages such as C or Java. These are called *decompilers* and will decipher binary into a close match to the original source program.

If the binary code was created by a C-language compiler that developed the right lines of machine language from a program in C, the matching C-language decompiler will convert the binary back to a good approximation of the source code in C language. Using a decompiling memory dump depends on knowing the language of the original source code, and where that code is stored in the memory. Then comparing the source code for the program with the decompiled listing from RAM or mass storage is how a decompiler can help a troubleshooter find where software is failing.

Memory Upgrade

Sometimes, the problem with an older PC is not software or hardware, but the issue is the amount of memory available. One reason for slow operation is insufficient memory and the solution is to expand the memory.

When a computer program does not have enough room in a RAM, pieces of the program code are fetched into the RAM that is available. The PC runs the program as much as it can until it has to stop and get the next piece of the program into RAM.

If more RAM can be added to the computer, less time will be spent waiting for a new piece of the program to load from a slow mass-memory device into the RAM. If the memory is big enough for the entire program, once it is loaded, it can run continuously without waiting to load another piece of code.

Laptops, for instance, are not always sold with the maximum amount of RAM that is possible to install in the machine. There is always a slot or two for memory expansion available. With the advancement of the technology, it is possible to replace an old RAM cards with larger and newer memory chips. The following procedure lists the main approaches to upgrading a PC to a larger amount of memory:

1. Find out what upgrade cards exist for your model PC.
2. Buy the right card for your model. The card usually comes with instructions.
3. Make sure the PC is completely powered-down before you flip the PC over and open the case to reach the expansion slot(s).
4. Unscrew and open the panel that covers the memory which may involve removing the battery pack. Usually, the only tool you will need is a screwdriver for small Phillips-type screws.
5. Locate where the new memory card fits. There is usually a notch in the card that matches up with a ridge on the socket. This notch will help you insert the card into the slot with the correct end down.
6. Insert the card into the socket. It usually fits in at a shallow angle rather than completely flat down.
7. Press the card down flat. There are usually holders on either side which will snap into place when the card is secured.
8. Replace the cover and screw it into place.

The main dangers in handling the memory cards are static electricity that damages the card when you touch it, and inserting the card the wrong way. For the "static" problem, make sure you remove static charge from your hands by using a grounding strap. Grounding straps are often used to handle PC components when you are doing this in a professional environment. They are like bracelets that connect to an earth ground and are worn around the wrist. It is now safe to take the memory card out of the package and open the PC.

The "insertion" problem will require that you really pay attention to the specific diagrams packed with the expansion card, in order to get things facing exactly the way they are in the manufacturer's diagrams. In the event you do not have instructions, go online and find instructions before you start or use the slot-and-notch rule in step 5 as a guide.

The example in Fig. 9.10 shows a laptop opened up for a memory upgrade. It shows the 4-GB RAM card used for the expansion, the slot in the opened-up laptop, and steps 5, 6, and 7 of the installation process. When it is completed and the computer is closed, replace the battery pack and turn on the computer power. If the computer starts up normally, the operating system should detect the additional memory and begin to run more quickly than before for programs that had "insufficient memory" problems.

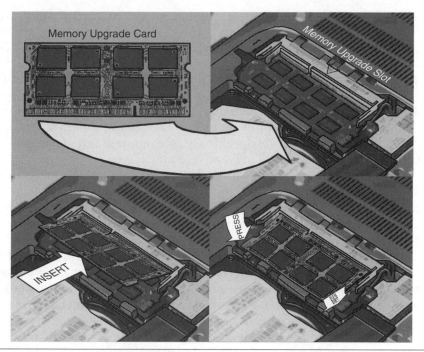

FIGURE 9.10   Memory upgrade card.

Preventive Maintenance

As the most common failures in microprocessor equipment are in places where there are moving parts, the preventive maintenance is of mechanical nature. For example, if there is a keyboard used for input and control instead of a touch screen, it would be prudent to have a regular schedule for air-cleaning with compressed air. Caution is recommended in choosing the amount of pressure and what size nozzle to use. Too much pressure might blow the key-tops off the keyboard. A properly conducted pressurized air-cleaning could keep the keyboards from clogging up and becoming unusable.

At the cost of a little downtime, a microprocessor-based system exposed to dust or corrosive spray should be maintained by unplugging plugs, cleaning contacts, and reconnecting them on a regular schedule. In particularly noxious environments, a fine screen or filter might be added to block particles or droplets from getting into the cooling vents. Make sure a reasonable flow of air for cooling can be maintained.

For less extreme environments, it would be useful to use a vacuum cleaner every few months to clean dust buildup from cooling holes or vents on the sides and top of the microprocessor equipment cabinets or racks. Any blockage or clogging of cooling vents is a risk for overheating the electronics. The vacuum cleaner can also clean dust buildup off boards inside the cabinets. Always remember to turn off power before opening up the case. There are special vacuum attachments made just for this purpose. Periodic vacuum cleaning reduces the chances of an electrical "short" especially in high-voltage sections such as motor controllers. Care should be taken with the vacuum nozzle, not to push, pull, or bump into components on circuit boards that might contact other components and cause short circuits.

It is wise to periodically plug and unplug USB-connected peripheral devices. This action cleans the contacts and prevents buildup of oxidation on the metal surfaces. Inspection is necessary to help identify if buildup of corrosion on the contacts has taken place, which might require more extensive cleaning. If the system has mechanical components that need periodic lubrication, rebalancing or other attention that are not specifically microprocessor related, they should still be part of regularly scheduled preventive maintenance.

Infrared monitoring and vibration analysis equipment can be used to inspect for overheated machines or equipment that is not sufficiently secured such as microprocessor-based equipment. Readings from these analysis equipment can identify potential overheating problems, or whether a piece of mechanical equipment is working improperly.

Microcomputer-based equipment should be inspected more often if it is very new, or very old. When equipment is new, it is still in its "break-in" period. If it has a manufacturer's defect, the problems will generally show up early. Spotting a "hot spot" on a new microprocessor board may indicate a short that can be repaired or corrected before the equipment fails catastrophically. Undue vibration or "wobbling" may indicate equipment that was improperly installed before shipping. Catching the problem early can prevent much more costly repairs later on when the machine breaks down.

When equipment is very old, its mechanical or electrical components will simply wear out. Electrolytic capacitors dry out and stop storing charge effectively. Atoms migrate in nanometer-size microstructures, and their conducting and non-conducting portions begin to overlap. Mechanical parts wear down to dust. Lubrication helps forestall the latter, but not forever. Technical obsolescence will probably require replacement of circuit boards long before the capacitor and other nanostructures run down. Vacuum tubes have been replaced by transistors, and transistors may eventually be replaced by memristors, optical switches, or some technology yet to be discovered.

At some point, increasing frequency of maintenance for old equipment may become more expensive than replacing that equipment with newer machines. Replacement parts may become unavailable. Vendors may stop supporting the software generation in use. All of these are factors in deciding when preventive or predictive maintenance is less cost-effective than replacement. One thing that can be predicted with certainty is that nothing lasts forever.

No system works well if the equipment operator does not understand the functions. For any machine, but especially for a microprocessor system that is constantly being updated, it is critical that the operator be trained adequately about what has changed and how that affects the performance of the machine.

A failure to adequately educate the workers or users can result in operator errors. The operator might have trouble even if there is no trouble in the system to troubleshoot. The only preventive maintenance possible for this kind of fault is education. A plan should be in place for every worker to be periodically recertified on the machine. Periodic lubrication or software updates for the equipment should also be a part of the maintenance plan.

Self-Examination

1. Why do you need a microprocessor for a design where logic gates could do the same thing?
   A. Easier to redesign
   B. Lower cost
   C. Fewer chips
   D. Less power use
2. Microprocessors and other types of central processors must fetch a series of binary codes from what part of the microprocessor system?
   A. Input
   B. Output
   C. Memory
   D. Ports
3. A printer is an example of a(n)
   A. Input device
   B. RAM
   C. EPROM
   D. Output device
4. A microprocessor can detect conditions in the outside world through sensors, which measure conditions in the exterior and convert them to binary numbers. When the computer picks up information from sensors, this is an example of
   A. Input
   B. Output
   C. Memory read
   D. Disassembly
5. USB sockets on a computer are an example of
   A. Virtual ports
   B. Physical ports
   C. None of these
6. A problem that affects a machine with a microprocessor due to a bad connection or failure of electric power is a
   A. Hardware problem
   B. Software problem
   C. None of these
7. A problem caused by faulty instructions in a code or an inappropriate program is
   A. Hardware problem
   B. Software problem
   C. None of these
8. A computer program is running and "hangs," although the computer appears to be operating normally otherwise. What is a good first step toward straightening out the problem?
   A. Replace the microprocessor.
   B. Replace the monitor.
   C. Restart the processor.

9. You turn on a machine with a microprocessor and nothing happens. You do not hear the fan whirring, you do not see any lights come on at the keyboard, and the start-up "beep" you normally hear is not heard. Where would you look first?
   A. Is disk in drive?
   B. Is power on or off?
   C. Turn system off and on again.

10. The message "new updates available" appears on the screen of your machine and you click "download now." After the new version is installed, the screen appearance is different and you cannot read the screen. Which of the following is a better cause of action?
    A. Replace the driver chip for the screen display.
    B. Delete the newly installed software and reinstall the older version.

11. Figures 9.3 and 9.4 show two versions of a simple diagnostic machine to compare data on two 8-wire cables. Each design solves the same problem. Figure 9.3 is a
    A. Hardware solution to the problem
    B. Software solution to the problem

12. Figures 9.3 and 9.4 show two versions of a simple diagnostic machine to compare data on two 8-wire cables. Each design solves the same problem. Figure 9.4 is a
    A. Hardware solution to the problem
    B. Software solution to the problem

13. The pseudocode example provided in a textbox and Fig. 9.5 shows how the microprocessor is used to do the same task in the circuit shown in Fig. 9.4 as the XOR and OR gates in the circuit shown in Fig. 9.3. The pseudocode example and Fig. 9.5 are
    A. Hardware
    B. Software
    C. None of these

14. Microcomputer terminology calls a group of conductors with a common purpose a bus. One of these buses carries signals that select which location in memory the next instruction comes from. Which of these buses does that?
    A. Control bus
    B. Data bus
    C. Address bus

15. In the course of running a program, the microprocessor in Fig. 9.9 fetches each group of binary machine-code bits into itself through a group of wires called a bus. Which bus is used for this purpose?
    A. Control bus
    B. Data bus
    C. Address bus

16. What kind of problem is a memory dump used to diagnose?
    A. Hardware problem
    B. Software problem
    C. None of these

17. A troubleshooter suspects a defect in the microprocessor chip of a machine. When the chip is removed and an in-circuit emulator (ICE) is put in its place, everything works right. What does this troubleshooting session indicate?
    A. Bus fault
    B. Software problem
    C. Bad CPU chip
    D. Faulty power supply

18. What sort of problem would a disassembled memory dump be most useful for?
    A. A disconnected bus connector.
    B. A faulty memory chip
    C. A mistake in an assembly language program
    D. A mistake in a high-level language program

19. What sort of problem would a decompiled memory dump be most useful for?
    A. A disconnected bus connector.
    B. A faulty memory chip
    C. A mistake in an assembly language program
    D. A mistake in a high-level language program

20. Preventive maintenance and predictive maintenance are done to prevent serious problems from appearing in microprocessor-based machines. What kind of problems are anticipated by most maintenance procedures?
    A. Hardware problems
    B. Software problems
    C. Operator errors

Questions and Problems

1. Since the 1970s microchip complexity has doubled every 18 months. This is called Moore's law. In 2013, the scale of components on a microchip is 28 nanometers (nm). A single silicon atom is about 0.2 nm in diameter, but a single atom cannot be made into a transistor. Can Moore's law go on forever? If not, speculate about how much longer electronics can continue to be further miniaturized at this rate.

2. Imagine a computer or robot is running a program that can "learn from its mistakes" by rewriting its own code. What could go wrong?

3. Suggest an example of a machine that would require updating (reprogramming) in order to keep it useful.

4. Name an example of a household appliance that contains a microprocessor. What makes you suspect that it contains a microprocessor or microcontroller?

5. When you connect your computer to an Internet e-mail provider, do you think you are using a hardware I/O port or a software I/O port?

6. A digital clock is basically a fancy counter counting pulses from an oscillator and showing the results on a display. Could a digital clock be made using a microcontroller?

7. What should you do if your computer's performance suddenly slows down after a download of some pictures?

8. When you transfer data into your system from the Internet, it is called a download. When you put data such as pictures or files onto an Internet site, it is called an upload. "Warehousing" of data on the Internet is called putting that data into the cloud. Why is it that you are "down" and the Internet is "up"?

9. In what way is binary code in memory "softer" than the hardware of the electric circuits?

10. If you were designing a tester for high-speed digital communications, would you prefer to build it in gate logic or using a microprocessor? Why?

11. As the scale of semiconductor devices has gotten smaller, the speed of these devices has gotten greater. Speculate as to the reason why that has happened.

12. If a disk or other mass storage device contains software that is all right, but it does not run properly after being loaded into RAM, where would you look first for a problem?

13. Look up the difference between standard and mil-spec (military specification) semiconductors and ICs. If you were repairing a microprocessor-based outdoor weather station's CPU and some memory chips that failed, would you replace them with mil-spec or standard parts? Why or why not?

14. Even using the same clock speed, an 8-bit processor (having an 8-bit data bus) should not be able to run programs as fast as a 16-bit processor, and a 32-bit processor should be even faster. Why?

15. Every data byte from a temperature sensor at an 8-bit input port appears to be "stuck" at the value of "FF," which is the decimal number 255. The troubleshooters know the temperature is not 255° at that location, so they suspect a connector to the sensor has become unplugged and is open. If the input port socket contains transistor-to-transistor logic (TTL), why might the troubleshooters think it is disconnected?

16. Equipment is more likely to fail early in its useful life than after it has been running a few years. Why?

17. Microprocessors using a 64-bit data bus are currently in use. If the size increases to 128 or 256 bits, how would that change the performance of the computer?

18. If the size of a microprocessor's address bus increases by 2 bits, how does it affect the amount of memory that can be used in the system?

19. A user updated the computer's Internet browser to the next generation, and suddenly an application that displays animated graphics has stopped working. What would you recommend the user do to get the computer fully functional again?

20. A mouse stopped working and the cursor on the screen was "stuck." The user unplugged the mouse and plugged it back into the USB socket again. Suddenly it was working again. What was the probable cause of the problem?

21. A flash-drive music player has a microcontroller containing a digital-to-analog converter. Why is that a useful feature in this microprocessor application?

22. Teachers of online courses at a certain college have to take a mini-course and recertify on the Internet teaching software they use every 2 years. Why?

23. Is it reasonable to look for a truth table of all possible states in a LGA2011 microprocessor? Would it be "logical" to test the processor by running it through every possible input signal combination and comparing the outputs to that table?

24. Telephone numbers in the North American Dialing Plan consist of a decimal "1," followed by 10 other decimal digits. In ASCII code, each number, letter of the alphabet or punctuation mark is represented by an 8-bit binary code. If telephone numbers are stored in a binary memory using ASCII code, how many bits of memory would be needed to store a phone number? Supposing that packed BCD code (4 bits for each decimal number) is used instead of ASCII, how many bits of memory would be needed in that case?

25. Based on the example above, can you think of any reasons why it might be better to store telephone information in ASCII instead of BCD?

Chapter 10

Troubleshooting Biomedical Equipment

All biomedical engineering technology (BMET) equipment has its individual protocols, unique features, and usages. Most BMET equipment has personal computers (PCs) and peripheral devices that require cable(s) or wireless connections. All BMET equipment should have maintenance logs so that you can efficiently troubleshoot the machine.

Troubleshooting of BMET equipment is also performed by consulting with the technicians using the machines as well as the manufacturers' service manuals. Proper preventive maintenance procedures specified by the manufacturer must be followed in order to avoid downtime due to equipment failure. To prevent loss of life or limb to patients and to yourself, follow all safety procedures set by the equipment manufacturers and the institutions.

Electrical Safety and Safety Equipment

Electrical safety relates to hazardous electric shock, explosion, fire, or damage to equipment and buildings. Electric shock refers to both macro-shock (high-value) and micro-shock (low-value) shocks that result from improperly wired or maintained electrical equipment or power systems.

Explosive gases such as ether can be ignited and cause explosion by contacting electrical sparks. Heat produced by overloading, incorrectly wiring, or improperly maintaining equipment and power systems could cause fire. Explosions, fires, or electrical overloads can damage equipment and buildings. Another definition of safety is the condition of being safe from pain, injury, or loss. Although no situation is totally safe, safety can be achieved mainly from harmless situations. In the hospital, electrical safety is one of the primary concerns requiring continuous involvement, including detection of potential problems.

There are three primary effects to the human body when electric current passes through: injury to tissue, uncontrollable muscle contractions, or unconsciousness and fibrillation of the heart. Cells and muscles in the body act as tiny batteries called *polarized units*. Effects of controlled electric shock include electrical stimulation of tissue, tingling sensation, extreme violent reactions of muscles, ventricular fibrillation, or death. Skin contact with a voltage source will cause macro-shock that is defined as a high-value current level [microampere (μA)] passing from arm to arm through the body. Even though skin is an excellent insulator, current passing through the heart could cause ventricular fibrillation resulting in death.

The definition of micro-shock or cardiac shock is a low-value current (μA) that passes through the heart by a catheter or needle in a vein or artery. Leakage current is a low-value current (μA) that passes from electrical parts of the appliance to the metal chassis. This undesirable current is the natural result of electrical wiring. Ground-fault interrupter (GFI) is an automatic switch that disconnects power if excessive leakage current is present in a circuit. It uses a toroidal coil where hot and neutral conductors are wrapped as seen in Fig. 10.1. When the currents in the hot and neutral wires are equal, no leakage current is present in the circuit. The relay stays closed and the current flow will not be interrupted.

When the currents in hot and neutral wires are not equal, magnetic flux develops and causes a leakage current in the circuit. Sensing coil will send a signal to the relay coil, relay contacts open, and current is interrupted in the circuit. Up to 6 μA of leakage current can be detected through the GFI. GFI is usually used in the wet areas of the hospital such as hemodialysis sections. It is dangerous to use GFIs in any room where biomedical equipment and patients are located.

Electric power wiring, distribution, and grounding are very important in electrical safety. Distributing power from a central junction box and keeping wires to all outlets at approximately equal lengths are major guidelines to follow. *National Electrical Code* (NEC) describes the wiring standards, including the ground wires required to be 15 feet (ft) long or less between outlets. BMET technicians are responsible for performing preventive maintenance on the equipment and documenting the results of the electrical safety inspection and tests. The purpose of these tests is to uncover hazards such as faulty lamp sockets, frayed power cords, and

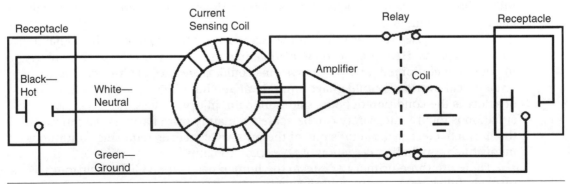

FIGURE 10.1   Ground-fault interrupter (GFI).

broken plugs. The following are the test equipment needed to perform proper tests and take measurements for safety:

- Tension tester: This tester tests the tension of the neutral, hot, and ground lugs of the wall receptacle. Good physical contact of the plug is provided by at least 8 ounces (oz) of pull tension.
- Ground wire loop resistance tester: This tester measures resistance between the ground (green) and white (neutral) wires.
- Receptacle polarity tester: This tester tests for correct wiring.
- Resistance tester: This tester measures resistance between the hot (black) wire to chassis and the neutral (white) wire to the chassis insulation.
- Leakage current tester: This tester tests leakage current for the chassis to ground and electrocardiographic (ECG) leads to ground.

Defibrillators

A defibrillator sends electric shock to the heart muscle experiencing a fatal condition. It stores a direct-current (dc) charge that is delivered to the patient. Figure 10.2 shows the voltage and current applied to the patient's chest with the common delivery system called the *Lown defibrillator*. With the application of about 3 kilovolts (kV), current reaches about 20 amperes (A) very fast. The waveform comes down to zero in 5 milliseconds (ms), and a small negative pulse is produced within another 5 ms. The dc voltage is supplied by the high-voltage power supply.

The dc voltage charges the capacitor. The stored energy in the capacitor is calculated by

$$E = \tfrac{1}{2}\, CV^2$$

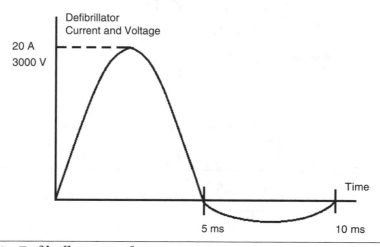

FIGURE 10.2   Defibrillator waveform.

where E is the energy in joules (J); C is the capacitance in farads (F); and V is the voltage across C in volts (V).

As shown in Fig. 10.3, the Lown defibrillator consists typically of a 16-microfarad (μF) capacitor (C_1), 100-mH inductor (L_1), resistance of the inductor (R_1), and the patient's resistance (R_2). After the capacitor is charged fully to about 400 J, energy will be stored and 7 kV will develop across the capacitor. During discharge the inductor's (L_1) magnetic field will produce a negative waveform during the last 5 ms. The following are the events that occur sequentially:

1. User turns the set-energy control level to the desired level and presses the charge button.
2. Capacitor C_1 starts charging and will fully charge up to the voltage reading of the power supply.
3. User places the defibrillator's electrodes on the patient's chest and hits the discharge button.
4. Relay K_1 switches to the other position where capacitor connects to the patient.
5. Capacitor C_1 discharges through L_1, R_1, and the patient R_2. During the first 5 ms, high positive voltage is delivered to the patient.
6. During the last 5 ms the magnetic field across L_1 collapses, producing a negative current and a voltage shown in Fig. 10.3.

Figure 10.4 shows a modernized defibrillator that has an operational-amplifier (op-amp) comparator to turn the power supply on and off. The user presses the charge button in order to start the charge process. The comparator compares the

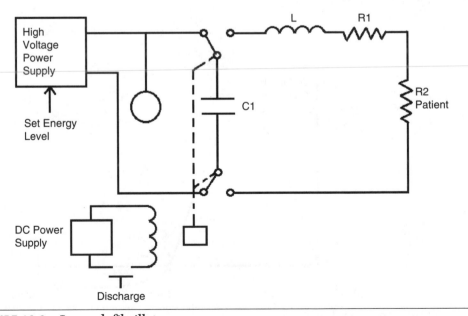

FIGURE 10.3   Lown defibrillator.

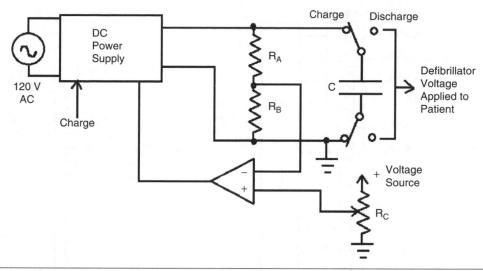

FIGURE 10.4   Modern defibrillator with an op-amp.

set-level voltage positive input with the 0 V at the negative input. As the positive input voltage of the op-amp is higher than the negative input, comparator output becomes high. The high op-amp output turns on the power supply.

The capacitor keeps charging and the voltage at the comparator's negative input rises. When the negative input voltage reaches the set-level voltage at the positive input of the op-amp, the comparator outputs 0 V. As a result, the power supply turns off and the charging cycle stops. A discharge button is pressed through the paddles applied to the patient's chest. The energy in the capacitor gets discharged through the patient's body.

Defibrillators are tested by defibrillator testers. A typical defibrillator tester is a voltmeter that measures energy in Watt-seconds. The tester is connected to a 50-ohm (Ω) dummy load simulating a patient. The load is also connected between two electrodes. The capacitor gets discharged through the load and through the paddles that are located against the electrodes. If the meter reading does not meet the specification of the defibrillator, the tester needs to be repaired or replaced.

Preferably the tester would have an oscilloscope output jack that can be connected to the oscilloscope. The output waveform of the defibrillator can be observed on the scope. The defibrillator performance is determined through the scope and the defibrillator tester.

ECG Systems

The electrocardiogram (ECG) machine records the electrical activity of the heart. Figure 10.5 shows a typical block diagram of an ECG machine. The ECG records data

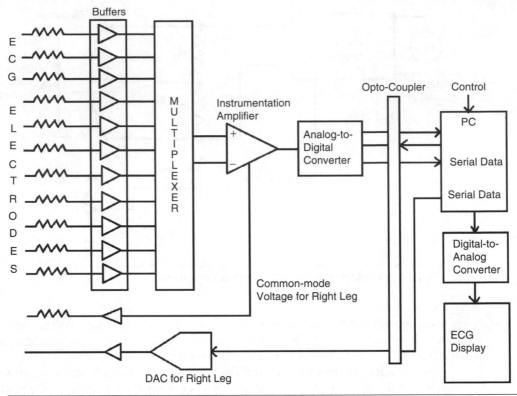

FIGURE 10.5   ECG machine block diagram.

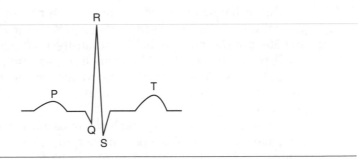

FIGURE 10.6   ECG waveform.

from different parts of the heart in the form of waves seen in Fig. 10.6. The P wave is the depolarization of the arterial muscle tissue. The electrical changes caused by the depolarization result in arterial contractions. The contractions related to the ventricles produce the Q, R, S, and T waveform.

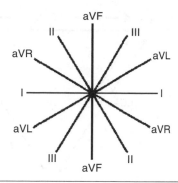

FIGURE 10.7   The six leads: I, II, III, aVR, aVF, and aVL.

ECG Lead Systems

Standard lead system ECG recording requires the connection of five electrodes to the patient: right arm (RA), left arm (LA), left leg (LL), right leg (RL), and chest (C). Through a lead selector switch, inputs of differential amplifiers are connected to these five electrodes. Different waveforms are displayed across different pairs of electrodes. Each one of these waveforms called *lead* displays unique information that cannot be obtained by any other leads. As shown in Fig. 10.7, these six leads are I, II, III, aVR, aVF, and aVL. Each one of the leads could provide information about a different heart disease.

The patient's right leg is used as the common electrode in ECG machines, and the lead selector switch makes the connection between a limb or chest electrode and the input of the differential amplifier. The bipolar limb leads I, II, and III form the Einthoven triangle as shown in Fig. 10.8. Figure 10.9 shows the connection for the bipolar leads that are described as follows:

1. Lead I: RA is connected to the inverting input of the amplifier (buffer), and the LA is connected to the non-inverting input.
2. Lead II RA is connected to the inverting input of the amplifier (buffer), and the LL is connected to the non-inverting input.
3. Lead III: LA is connected to the inverting input of the amplifier (buffer), LL is connected to the inverting input, and RA is connected to RL.

Figure 10.10 also shows the connection for unipolar or augmented leads in order to observe the composite effect from all three limbs at the same time. The voltages from two limbs are summed and applied to the inverting input of the amplifier, and the voltage from the third limb is applied to the non-inverting input.

The unipolar leads are the following:

1. Lead aVR: LA and LL are summed at the inverting input of the buffer, and RA is connected to the non-inverting input.
2. Lead aVL: RA and LL are summed at the inverting input of the buffer, and LA is connected to the non-inverting input.

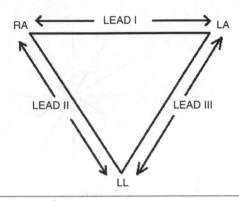

FIGURE 10.8 Einthoven triangle.

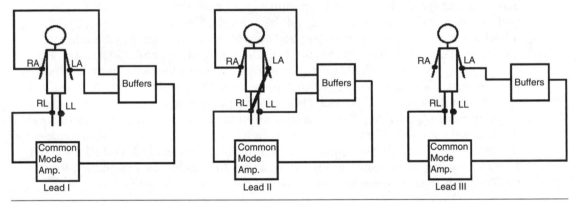

FIGURE 10.9 Bipolar leads.

3. Lead aVF: RA and LA are summed at the inverting input of the buffer, and LL is connected to the non-inverting input.

ECG Machine Troubleshooting

Figure 10.11 shows an ECG machine. ECG machines are not immune to breakdown just like any other machine. Malfunctions will occur and it is the technician's experience and log entries that will help lessen the amount of time spent on diagnosing the fault. It should be noted that most issues arise from operator error and simple adjustments or repairs will correct the situation.

The following examples are common occurrences and solutions to problems related to ECG machines:

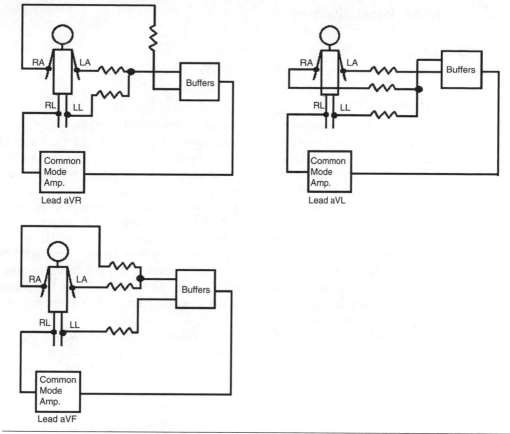

FIGURE 10.10 Unipolar leads.

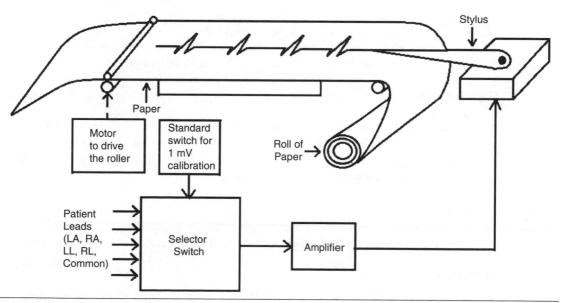

FIGURE 10.11 ECG machine.

Problem

Stylus tip is not writing or writing too lightly.

Solutions

1. A lack or decrease of heat on the stylus tip would indicate the voltage is incorrect and there would not be any black lines appearing. It is important to know that the tip under normal circumstances is hot enough to write on paper. As a result the tip can cause second-degree burns when touched. It is highly recommended to use an insulated probe with the tip. As voltage differs from one manufacturer to the other, it is important to refer to the service manual for the correct voltage. After finding the correct voltage, make a note in the log for future reference and check the voltage according to the instruction for that specific ECG model.

2. Stylus pressure is insufficient to write on the paper. When very dark lines appear, the pressure on the stylus is not enough. Find the correct pressure setting from the manual and mark it in the log also. The ranges of setting can be between 2 and 20 grams (g). Make any adjustments to the correct setting by applying the following steps:

    A. To calibrate, hold stylus pressure vertically at eye level with a 2-g calibration cap on a hook, then slide knurled ring to read 2 g.

    B. To measure the stylus pressure, hook the tester under the writing arm tip near where it makes contact with the paper. Read the gauge just as the stylus contacts the paper.

Problem

Smeared traces are due to incorrectly loaded paper and/or a worn stylus.

Solutions

1. The paper might not have been loaded correctly. The error results from bypassing the paper brake or tension bars. Check the paper to see whether it is loaded correctly, and if it is not, refer to the manufacturer's procedures and reload. Occasionally regardless of appearance, reloading the paper will correct the problem.

2. Styluses that are worn can also affect the print quality. Check the stylus for wear or any other irregularities. If it needs to be replaced, replace it with the manufacturer's part for best results. There might be equivalent parts available, but check to make sure the print quality is up to par.

Problem

Poor readings can be the result of a multitude of issues that include bad input connector or lead switch, faulty patient cables, or improper connections to the patient.

Solutions

1. To check mechanical issues, set the lead selector switch to STD, short all electrodes together, and then press the 1-millivolt (mV) cal button. If the

normal calibration pulse appears, that means the issue of the poor reading is due to the connections made to the patient.

2. Repeat the first step, using a functioning patient cable or a deadhead plug. If the poor reading has improved, replace the bad patient cable. If, however, the ECG still has a poor reading, consult the service manual for further instructions.

Interference is caused by many factors, including the 60-hertz (Hz) wall plug inducing signals into the electrode wires from a common-mode (CM) signal. As all electrodes and wires are affected equally, these signals do not interfere with the preamplifier of the ECG. Any electrode defects, open patient cable wires, or poor contact to the patient will result in an unbalanced input. This causes the preamplifier to create an unwanted differential signal and provide false ECG readings. Lack of electrode jelly, a loose contact to the patient, and especially moist or sweaty skin are other causes for interference. Any broken or loose power ground on the ECG machine along with other instruments connected to the patient can cause interference. Interference isolation can be achieved by the following procedures:

1. Short all electrodes.
2. Check each position on the lead selector switch:
   A. If the interference issue was a lack of electrolytic jelly, poor skin preparation, or a bad electrode, check each one until the interference ends.
   B. If the interference exists only in a certain lead position of the selector switch, this means there is an open wire in the patient cable. Use a conductance checker or ohmmeter to find the defective electrode.
   C. When the interference exists in all the positions of the lead selector switch, this means the issue is internal. Refer to the manual for further specification on the particular model.

ECG Machine Maintenance

Routine inspection and maintenance should be performed and logged for ECG machines to ensure accuracy. Checking the recommendations of the manufacturer's manual for routine care and maintenance is essential. Also, follow all recommended safety measures. Most guidelines include the following steps:

1. Make sure the machine is allowed to warm up after being turned on.
2. Test to see whether there is a trace present. Turn the function switch to RUN then the lead selector switch to STD.
3. While pressing the 1-mV button several times, be sure to take note of the following:
   A. Is there visibility of the vertical edge of the pulse?
   B. Does the pulse look square?
   C. Can the sensitivity control be adjusted to provide less than 10 millimeters (mm) of deflection?

4. Test whether the stylus will travel to the top and bottom margins of the paper. Adjust the position control to its complete range.
5. Make sure there is no open wire in the cable by:
   A. Connecting electrodes together.
   B. Then moving the lead selector switch to all positions. Each position should show a quiet and stable baseline on the paper.
   C. Observing that any activity on a position indicates an open cable which in turn will identify the affected leads and electrodes.
6. Set the sensitivity to exactly 10-mm deflection and turn the lead selector to the STD position.
7. Check the low-frequency response of the ECG machine. Press the 1-mV cal button; the stylus will deflect 10 mm and will drop slowly back to its original position.

EEG Systems

The paper representation or the CRT display of the brain's electrical activity is called *electroencephalogram* (EEG). The cerebral tissue's ionic currents are transformed by the EEG electrodes into electric currents used in EEG preamplifiers. Characteristics of the currents are due to the type of metals used in the electrode discs. The most common type of electrodes used can be categorized as follows:

- Scalp electrodes are made of discs, caps, silver pads, and stainless steel rods. The reusable electrodes are placed on the head with the use of a conductive cream. To ensure an accurate EEG signal, using conductive cream keeps the resistance below 10 kilohms (kΩ). It is important to test the resistance, and the best method is to use an alternate-current (ac) ohmmeter and test between two electrodes. A dc ohmmeter can also be used to test the polarization of the electrodes.
- Sphenoidal electrodes are made of alternating bare wire and insulated silver. Its chloride tip is inserted through muscle tissue by a needle.
- Nasopharyngeal electrodes are made with silver rod and with a silver ball at the tip which is inserted through the nostrils.
- Electrocorticographic electrodes are made of cotton soaked in saline solution and they rest on the surface of the head. They are used to remove the cerebrum-generated artifacts.

EEG signals are dependent on the placement of electrodes on the patient's frontal, occipital, temporal, and cranial areas. The head is measured, the points are mapped, and 19 electrodes are placed. The twentieth electrode is used as a ground and can be placed on an earlobe. As shown in Fig. 10.12, either bipolar or unipolar arrangement may be used as follows:

- Interconnecting scalp electrodes create the bipolar arrangement. The voltages between the electrodes can be measured.
- Unipolar arrangements are made up of leads and the electrodes are common to all channels. A lead is connected to a common point like the earlobe.

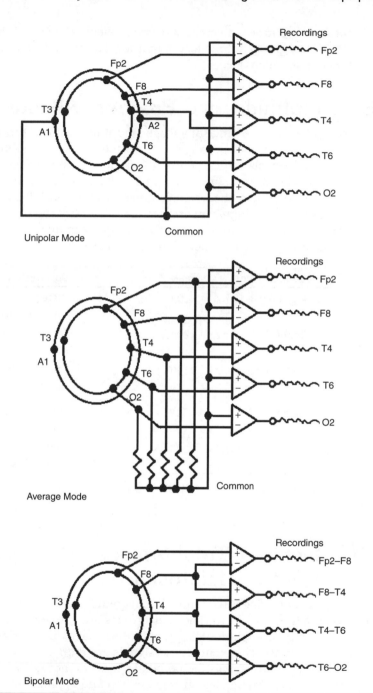

FIGURE 10.12 Uni- and bipolar arrangements for EEG.

- Patterns or montages are created between electrodes. All of the combinations of inputs are sent to a three-lead amplifier. The three-lead amplifier uses a third connection as reference.

EEG Amplitudes and Frequency Bands

At the cranial surface, the range of voltage amplitudes is from 1 to 100 microvolts (μV) peak to peak at the frequency range of 0.5 to 100 Hz. At the cerebrum surface, the amplitude of the signals can range up to 1 mV at the same frequency range. At the brain stem, the signals measure less than 0.25 μV peak to peak at the 100- to 3000-Hz frequency range. Chest-to-chest signals range from 500 to 100,000 μV peak to peak. Preamplifiers that have high gain are needed to amplify the weak EEG signals.

The frequency bands are listed in five groups with different frequency ranges, peak-to-peak amplitudes, and their descriptions are as follows:

Frequency Bands	Peak-Peak	Description
1. Alpha (α) 8–13 Hz	<10 μV	Posterior brain awake with eyes closed*
2. Beta (β) 13–22 Hz	<20 μV	Central region, patient at rest
3. Delta (δ) 0.5–4 Hz	<100 μV	Central region, adult patient is sleeping
4. Gamma (Υ) 22–30 Hz	<2μV	Sensory stimulation and attention
5. Theta (θ) 4–8 Hz	<100 μV	Central region, adult patient is sleeping

*Open eyes reduce alpha waves.

EEG Machines, Troubleshooting and Preventive Maintenance

The process of obtaining EEG output signals requires 20 electrodes attached to the patient as shown in Fig. 10.13. These electrodes are connected to eight different switch-selected preamplifiers. The preamplifiers have differential inputs and single-ended outputs. The eight output signals are then amplified more to create a current which in turn drives the pen deflectors. To monitor whether the system is in operation, a calibration signal is generated and applied to the differential-amplifier inputs. A calibration signal waveform is generated by the calibration signal generator. The correct sensitivity settings depend on the calibration signal amplitude. Adjustment of the amplifier should be made if the readings are not within the calibration settings.

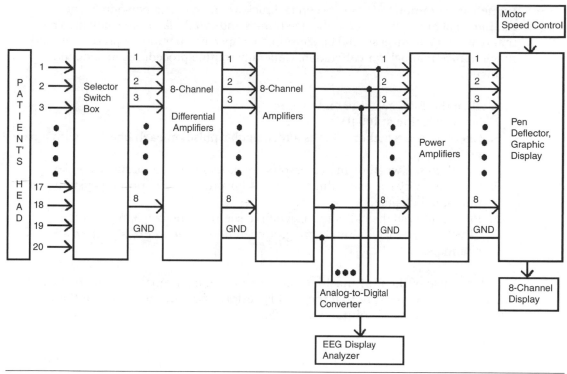

FIGURE 10.13   EEG machine.

EEG analog output signals are processed through an analog-to-digital converter that is connected to an EEG signal analyzer. The results are stored digitally in the analyzer. EEG signals and readings can be easily interfered with, making it impossible to diagnose the patient's condition. The most common interferences are eye blinking which results in spikes and muscle activity, to name a few. Removal of higher frequencies by adjusting the external EEG machine control to 30 percent high-frequency attenuation at 25 Hz can reduce the interference. Common errors can occur due to the following problems:

- High-impedance connections to the scalp.
- Broken cables, indicator knobs, switches, or wires.
- Bent connector pins.
- Switches positioned incorrectly.
- Faulty switch contacts.
- Electrical problems due to faulty circuits in the individual channels, power supply, or system control.
- Malfunction of the graphic recorder can be due to clogged ink pens, the roller slipping, and ink pens that are not seated well.

Regular inspection should be conducted on EEG machines, and a log should be kept of what was discovered and repaired during the maintenance and repairs.

Immediate communications between technicians and medical personnel who use the EEG machine will decrease the downtimes and avoid major breakdowns. The manufacturer's manual should be consulted for specific operational procedures and maintenance manuals for calibration. Generally routine procedures include the following:

1. Turn the EEG machine on to warm up.
2. Set calibration at 100 μV.
3. Check to see whether there is a rectangular pulse on each channel with a pen recorder.
4. Check to see whether ink reservoirs are at manufacturer's suggested levels.
5. Set sensitivity corresponding to 100 μV for proper deflection for system and individual channels.
6. Press and hold the calibration switch to see the time constant decay.
7. Check to see that all inputs are grounded to observe a zero signal on all channels.

Heavy use of the EEG machine causes faults, and all repairs should be logged and checked to make sure that the EEG is performing properly. Some of the most common problems and solutions are as follows:

Problem
Recording either too light or too dark is a common problem.

Solutions

1. The paper might not have been loaded correctly; check to see whether it is loaded correctly, and if it is not, refer to the manufacturer's procedures and reload. Occasionally regardless of appearance, reloading the paper will correct the problem.
2. Pen tips that are worn can also affect the print quality. Check the tips and if they need to be replaced, replace them with the manufacturer's part for best results. There are equivalent parts available, but check to ensure that there is no leakage and the print quality is up to par.

Problem
Interference of 60 Hz is the most common problem in EEG machines.

Solution
Signals on EEG are very low with amplitudes of only 5–100 μV_{p-p}. It is very important to connect and shield all the leads. A shorted voltage regulator and/or open power supply filter capacitors appear to be lead problems, but they are not. Check the manufacturer's guidelines for proper settings.

Problem
Irregular and wandering baselines.

Solution

Baseline irregularities and wandering are caused by extended connection times and/or bad connections. It is important to wipe clean and check that all electrodes are securely connected. They need to have a resistance of less than 10 kΩ. Ensure the dc drift of the amplifier is not excessive. Check also to see whether all junction box leads are correctly connected.

Problem

Muscle jitters occur because the patient is not relaxed.

Solution

Even filtering of the 60-Hz frequency signal will not correct the issue of muscle jitters until the patient relaxes. Contacting the medical personnel in charge is the only solution.

Problem

Noisy and/or poor recording of EEG reports.

Solution

Reset switches to standard calibration positions, then check for noise and any improper operation. Once normal operation is obtained, check the patient connection by physically inspecting all electrodes and the connectors. Look for physical defects such as cracks, rips, or faulty connectors, replace and repair as needed, and then check. If noise is still present, ground all EEG leads and check for straight line tracing. Check tracing by using an EEG simulator. If there is still noise on the trace, the problem is internal. For internal machine problems refer to the service manual for instructions.

Problem

Tracing is missing on one or more channels.

Solutions

- Clogged ink tubes: Gently remove the clogged tube with the pen and soak it in warm water. Carefully push the clog through using a very fine wire without puncturing the tube.
- Dry ink: Check the ink reserves if they are low. Fill the ink a little below the suggested manufacturer's limit. Too much ink can damage circuitry and drip into the machine. If the problem is dry ink, check for clogged ink tubes after the ink is filled.
- Pen(s) are not touching the paper: Check to see whether any pens are bent upward. Using an instrument or even a finger, carefully push the pen to the paper. Check to see whether tracing appears. If there is no tracing, remove the faulty pen and carefully bend downward. These pens can crack; therefore they do not bend enough to create a right angle.

Hemodialysis Machines and Troubleshooting

Hemodialysis is the process that removes the chemical substances from the blood and passes them through the tubes covered by semipermeable membranes. The tubes (dialyzer coil) are inserted in a tank filled with a dialysate solution composed of salts. Toxic material such as uric acid passes through the plastic tube into the dialysate solution. Salts and small molecules pass through the semipermeable material and blood cells continue circulating in the tube.

The hemodialysis machine carries the following systems: blood pump, control panel, dialysis bath delivery, dialysate drain, dialysate recirculation, positive/negative pressure monitor, power, and temperature shown in Fig. 10.14.

The following are the typical faults and troubleshooting solutions in hemodialysis machines:

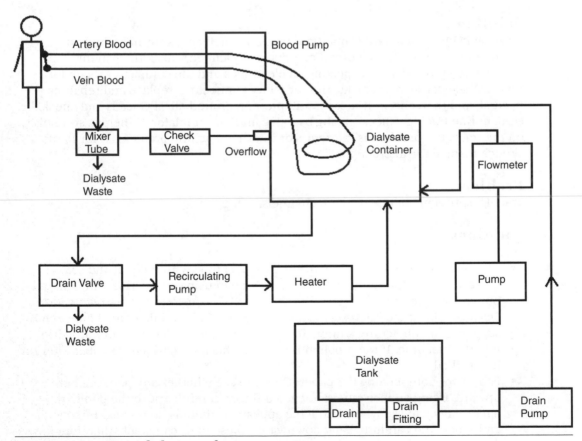

FIGURE 10.14 Hemodialysis machine.

Problem
All lights are off.

Solution
If the power switch is on and all lights are off, it most likely is due to a leakage of dialysate solution. A short has occurred in the system. Check also for any loose wires that could cause an ac fuse to blow.

Problem
Control panel without lights.

Solution
When a light or all lights on the control panel are off, it indicates that the circuit board has been damaged and/or a light bulb might have blown.

Problem
Dialysis bath solution does not run.

Solution
The dialysis bath solution that does not run through the dialyzer indicates one of the pumps is damaged. Maybe a foreign object is stuck in the recirculating pump or a switch is damaged.

Problem
Leaking pumping system.

Solution
Pumping system leaks are due to fittings or the seal of the pump(s) being worn, and hoses having holes and/or loose clamps.

Problem
Leaking dialysate tank.

Solution
Dialysate tank leaks are due to cracks occurring on the tank drain cap when it is not in place or the seals being worn.

Problem
Inaccurate temperature.

Solution
Inaccurate temperature display or the dialysate bath not heating up are common problems when the heater switch, heater, or heat sensor is not working.

Problem
Blood roller pump motor is not working.

Solution

When the blood roller pump is inoperable, it is due to a faulty motor control circuit and or a blown fuse.

Problem

Blood roller pumping at the wrong speed.

Solution

When the blood roller pump does not indicate the correct speed, this problem is due to the short in the speed control circuitry or the speed control potentiometer is not operating properly.

MRI Machines and Troubleshooting

Magnetic resonance imaging (MRI) was created by Dr. Raymond Damadian in 1977 along with his staff and students. The first image took 5 hours (h) and the original concept was to use magnets to take images of the body in the most noninvasive manner. A great amount of technological advances have taken place since the 70s such as the development of more accurate, open and faster MRI machines.

Magnets are the most important and largest component of the MRI machine. Gauss (1 tesla = 10,000 gauss) is the measurement of the magnetic field, and MRI machines create magnetic fields of 5000 to 20,000 gauss. Earth's magnetic field is only 0.5 gauss. The three gradient magnets (180–270 gauss) along with the main magnet create an intense, sustained large magnetic field.

The following are the typical problems and troubleshooting solutions in an MRI machine:

Problem

Computer console issues including lack of image acquisitions.

Solution

When the computer console does not react to commands or does not respond to standard commands, do a temporary protected status (TPS) reset that should only take 5 to 10 minutes (min). It is much faster than resetting the main processor. To perform a TPS reset:

1. Click on the utilities button on the scanner desktop on the workstation.
2. Then choose TPS reset and check system. It is important to check the manufacturer's manual for model specific procedures.
3. If the problem is still unresolved, initiate a main power cycle.

Problem

Pressing the emergency stop due to power outage.

Solution

When anyone presses the emergency stop or needs to initiate a main power cycle, restart the entire system as follows:

1. Shut down the computer console.
2. Locate the main central processing unit (CPU) cabinet in the server room and press the red power off button.
3. Wait 30 seconds (s) before moving the lever to the three o'clock position and then back to the six o'clock position.
4. Press the green reset button.
5. Turn the computer console on. Test to make sure the system has been corrected. Always check the manufacturer's specifications for appropriate restart procedures. If restart has not fixed the issues, contact the manufacturer.

Problem

Poor image quality and scanner is not scanning.

Solution

1. Check to make sure that the coil is connected. Pressing against the inner wall can upset the RF transmit coil. The coil is very sensitive and any pressure can cause a malfunction.
2. Replace the coil after checking the connection and that there is no pressure on the inner wall.
3. Ship replaced coils for manufacturer's repair.

Problem

Helium is low.

Solution

1. Check the helium percentage and then check the maintenance log to see when the last refill was done. If it was refilled recently, there can be a helium leakage; call for service immediately.
2. After calling for a refill and checking the manufacturer's manual, see whether the level is safe to operate the MRI.

Problem

Overheating.

Solution

The water pump is not adequately cooling. It is important to determine where the problem originated from in order to call the correct service agent. Take the following steps to solve the problem:

1. If the heat exchange is lit, trip the breaker. If tripping the breaker does not solve the issue, contact the MRI service.
2. Check the incoming water line that should feel like a cold bottle of water. If the water line is warm, contact the HVAC (heating, ventilation, and air conditioning) vendor.

Problem
Pre-scan failure or a weak signal.

Solution
When pre-scan fails or the signal is too weak, check the calibration.

Ultrasound Machines

Ultrasound is a technology that utilizes the echolocation technique similar to what dolphins use in navigating themselves. With ultrasound machines physicians can view the patient's interior, using noninvasive procedures with immediate responses. Ultrasound waves travel at frequencies above human hearing range of 20,000 kilohertz (kHz).

Figure 10.15 shows an ultrasound machine that transmits ultrasound waves into the human body. When the waves contact the body tissue or bone, they reflect signals back to the machine. The speed of the signal is calculated and displayed

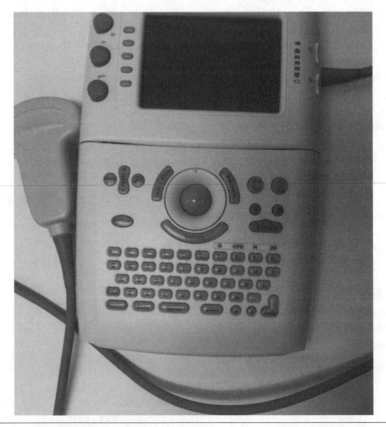

FIGURE 10.15   Ultrasound machine.

by the ultrasound CPU. The greater the distance to the organ, tissue, or bone, the dimmer the reflection of the image. Closer images are reflected much clearer in a two-dimensional (2D) image format.

There are two types of ultrasound technologies: Doppler and 3D ultrasound imaging. The Doppler ultrasound is used mainly to measure the rate of blood flow through the heart and major arteries. Frequency changes of echoes are measured in order to calculate how fast an object is moving. The frequencies get lower as the probe moves away from an object and higher as it moves closer. In 3D ultrasound imaging technology, 3D images are produced by the use of a combination of 2D images.

As a separate unit, a transducer probe is attached to the ultrasound machine. There are different shapes and sizes of probes. Ultrasound gel is applied on the skin above the location of the organ to be scanned. The gel is needed to create conduction between the skin surface and the transducer. The ultrasound waves are transmitted and received by the transducer probe.

The components inside all ultrasound transducer probes include an oscillator to create electrical pulses through the piezoelectric materials. Electrical pulses produced by an oscillator inside the ultrasound machine reach the piezoelectric material, which is made up of quartz. When an electric current is applied, the quartz rapidly changes shapes and vibrates which in turn produces outward moving sound waves. The shape and size of the probe determines the view field. Some transistor probes contain more than one quartz crystal.

These signals created by the quartz crystals arc thcn scnt to the organ or tissue that is located directly under the skin. Once the signal hits an organ, for example, the ultrasound wave returns and hits the piezoelectric material, resulting in vibrations that produce the electrical signals. These electrical pulses are processed in the computer and developed into pictures. Transducer pulse control changes the amplitude, duration, and frequencies of the pulses emitted from the probe. Keyboard/cursors are used to enter patient data, etc.

Mathematical calculations are performed by the computer to include distance from the top of the skin to the organ under the transducer. The velocity of the returning wave is calculated. The ultrasound picture develops through the distance and velocity information of traveling waves.

Troubleshooting of Ultrasound Machines

Typical symptoms and solutions for ultrasound machines include the following:

Problem
Lack of power.

Solution
Lack of power to the system indicates a power connection problem. Check the battery; first make sure the battery is connected properly, and then test the charger. In some models a low charge will not be enough to power the system. If the issue is not the connections or the battery power, remove the battery for at least 10 s and then reconnect.

Problem
Unclear image.

Solution
A poor system image requires the adjustment of the liquid crystal display (LCD) screen until a better view is achieved. Remember to adjust the angle and the brightness to obtain the best contrast to improve the image quality.

Problem
Printer is not working.

Solution
When the printer is not operating, there might be several issues:

1. Check the connections and restart the printer.
2. Check the printer cartridge and replace.
3. The paper might not have been loaded correctly. Check to see if it was loaded correctly, and if it was not, refer to the manufacturer's procedures and reload. Occasionally, regardless of appearance, reloading the paper will correct the problem.
4. Worn pen tips can also affect the print quality. Check the tips and if they need to be replaced, replace them with the manufacturer's parts for best results. There are equivalent parts available, but check to ensure there is no leakage and the print quality is up to par.
5. If the printer is still not operating, refer to the manufacturer's model specifications.

Problem
Age (fetal) calculation error.

Solutions
Inaccurate fetal age calculation errors are results of the information that is entered into the system. Reenter all the information with the patient to ensure no other errors exist.

Problem
Transducer is not operating.

Solution
The transducer is not recognized by the system. This can be just a connection issue:

1. Even if the connection appears correct, disconnect and restart the system after reconnecting. If the transducer still does not recognize the system, check the manufacturer's manual for the specific model.
2. Some models might require transducer codes to be entered. Check that the codes identifying the transducer match.

Problem
Character recognition issues.

Solution
The digital imaging and communications in medicine (DICOM) system cannot read English characters for the patient name, date, etc. Ensure optical character recognition (OCR) features are correct in the system setup.

Problem
Incorrect calculations.

Solution
The CPU of the ultrasound machine is where all the calculations are performed. For troubleshooting the CPU, refer to Chap. 9.

X-Ray Machines and Troubleshooting

X-ray machines generate high-energy and high-frequency electromagnetic waves. The X-ray waves pass through the bones that are dense and they penetrate more through the tissues, muscles, and organs that are not dense. As a result of the different amounts of X-ray penetration, different light patterns are formed on the X-ray images. These images can diagnose abnormalities in the human body. The major sections of X-ray machines are outlined in Fig. 10.16.

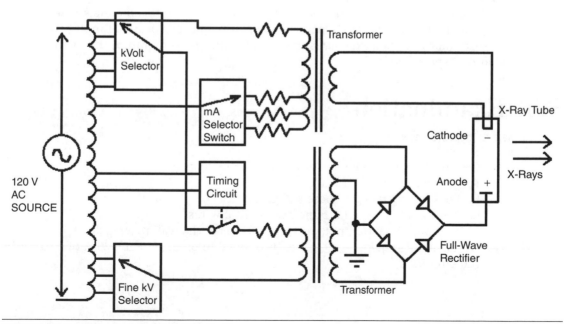

FIGURE 10.16   X-ray machine block diagram.

The following explains the parts of an X-ray machine:

- Multitap ac transformers depend on the application. Through the transformer, the operator can choose different voltages to apply to the X-ray machine.
- The X-ray tube filament circuit and transformer receive the power from the power supply and heat the cathode filament.
- The X-ray tube high-voltage circuit and bridge rectifier convert the ac line power supply to dc voltage to accelerate electrons from cathode to anode. The total energy delivered to the patient can be monitored by selecting different taps of the multitap auto transformer.
- The timing circuits control the turn-on and turn-off times of the X-ray exposure. There are three main controls to control the radiation dose in X-ray machines:
- The filament heat control adjusts the current level (mA) for exposure strength.
- The kilovolt control adjusts the voltage level (kV) for depth.
- The timing control adjusts the length of the X-ray radiation exposure.

Most problems in the X-ray machines occur in electromechanical devices. As an example, the servomechanisms that position the patient tables might not work properly. Relays being stuck on open or closed and broken wires are some of the faults that can occur in X-ray machines.

When X-ray tube filaments open, rotating anodes stop working and high-voltage rectifiers open. Factory-trained employees are the only ones qualified to replace the X-ray tubes. They have special instrumentation that can get rid of the radiation leakages. Refer to the manufacturer's manual and follow all the safety precautions for a possible radiation leak.

Electronic components in modern X-ray machines are reliable. As a result, electronic faults rarely occur. Operational errors can cause deaths. Employees measure their radiation exposures periodically and try to minimize the amount of radiation exposures.

Self-Examination

Select the best answer:

1. Ground loop resistance tester tests the following:
   A. Correct wiring
   B. Leakage current for chassis to ground
   C. Resistance between ground and neutral wires
   D. Tension of neutral, hot, and ground lugs of the wall receptacle
2. Defibrillator requires _____kV to be applied in order to reach 20 A of current in the circuit.
   A. 3
   B. 1
   C. 4
   D. 2

3. Defibrillators are tested by the
   A. Ohmmeter and defibrillator tester
   B. Voltmeter and defibrillator tester
   C. Ammeter and defibrillator tester
   D. Oscilloscope and defibrillator tester

4. During testing of the defibrillator, the tester needs to be connected to _____ Ω dummy load in order to simulate a patient.
   A. 100
   B. 50
   C. 75
   D. 200

5. While using the ECG machine, the stylus stops writing or writes too lightly. The reason can be
   A. Stylus pressure
   B. Heat decrease on the stylus tip
   C. Incorrectly loaded paper
   D. A and B

6. Stylus of the ECG machine is not writing because of decrease in heat on the stylus tip. What is the reason?
   A. The pressure of the stylus is not enough.
   B. The voltage applied to the stylus is not correct.
   C. The paper is not loaded correctly.
   D. The preamplifier in the ECG machine is not providing any output voltage.

7. Interference is caused in an ECG machine by the following:
   A. Electrode defects
   B. 60-Hz wall plug signal
   C. Open patient wires or poor contact to the patient
   D. B and C

8. Interference isolation in an ECG machine can be achieved by the following procedure(s):
   A. Short all electrodes.
   B. Check each position on the lead selector switch.
   C. If the interference exists in a certain lead position, there is an open wire in the patient cable.
   D. A, B, and C.

9. In EEG systems, scalp electrodes are made of
   A. Cotton soaked in saline solution
   B. Silver rod with a silver ball
   C. Disks, cups, stainless steel rods
   D. Teflon-coated gold or platinum wires

10. Some of the common errors that can occur in EEG machine readings are
    A. Sensitivity control is not adjusted to provide less than 10 mm of stylus deflection
    B. High-impedance connections to the scalp
    C. LCD screen is not adjusted well
    D. Transducer is not recognized by the system

11. In EEG machines, interferences of 60 Hz is the most common problem. What is the solution?
    A. Open power supply filter capacitor needs to be connected.
    B. Check all the pen tips and replace them if they need to be replaced.
    C. Connect and shield all the leads.
    D. None of the above.

12. Tracing is missing on one or more channels in an EEG machine. What could be the reason(s)?
    A. Dry ink
    B. Pen(s) are not touching the paper
    C. Clogged ink tubes
    D. All of the above.

13. In a hemodialysis machine, dialysis bath solution does not run. What is the problem and what is the solution?
    A. Turn on the control panel's lights.
    B. Maybe a foreign object is stuck in the recirculating pump and the pump needs to be fixed.
    C. Helium percentage is not correct. Fix the helium percentage.
    D. All of the above.

14. In an MRI machine, the image is poor and the scanner is not scanning, what would be the solution?
    A. Check to ensure that the RF transmit coil is connected. If the coil cannot be connected, send it back to the manufacturer for repair.
    B. Contact the HVAC vendor.
    C. Replace the defective piezoelectric material.
    D. Whip and untangle the cable until it starts to work.

15. In order to process ultrasound signals, what needs to be applied on the skin above the location of the organ to be scanned?
    A. Paste
    B. Gel
    C. Electric current
    D. Plastic sheet

16. If the transducer connected to the ultrasound machine is not operating, what would be the solution?
    A. Reenter the patient's information into the system.
    B. Correct the OCR features in the transducer system setup.
    C. Reconnect the printer.
    D. Reconnect the transducer and restart the system. If the transducer still does not recognize the system, check the manufacturer's manual for the specific model.

17. X-ray waves cannot penetrate through the tissues, muscles, and organs that are not dense.
    A. True
    B. False

18. Electronic components in modern X-ray machines are not reliable. Therefore, electronic faults occur frequently.
    A. True
    B. False
19. In an ECG machine, open wire in the patient cable due to the interference can be tested by
    A. Ohmmeter
    B. OTDR
    C. Logic analyzer
    D. None of the above
20. In testing with EEG systems, interconnecting scalp electrodes creates the following arrangement:
    A. Unipolar.
    B. Multipolar.
    C. Bipolar.
    D. It is not correct to interconnect scalp electrodes.

Questions and Problems

1. What are the reasons for poor readings in an ECG machine?
2. Explain briefly unipolar and bipolar arrangement of ECG electrodes.
3. Explain briefly some of the preventive maintenance procedures for the EEG machines.
4. Explain the solutions if noisy or poor reading is observed in an EEG machine.
5. What is hemodialysis?
6. What are micro-shock, macro-shock, and defibrillation?
7. What does "ground wire loop resistance" measure?
8. Explain briefly the graph in Fig. 10.2.
9. State the names of five electrodes required to connect in an ECG standard lead system.
10. How would you troubleshoot an ultrasound system with character recognition issues?
11. In an ultrasound machine, what would you do in order to fix an "unclear image" problem?
12. In an X-ray machine, what does the "X-ray tube high-voltage circuit and bridge rectifier" do?
13. Most of the problems in the X-ray machines are electromechanical by nature. Give examples.
14. How would you troubleshoot an ultrasound machine that has an "unclear image" problem?
15. Describe the solution to the MRI machine's "lack of image acquisition" problem.
16. Name the systems that a hemodialysis machine carries.

17. In a hemodialysis machine, if the power switch is on and all lights are off, what could be the most likely reason?
18. In an EEG machine, how would you troubleshoot the problem of "irregular and wandering baseline"?
19. State four of the routine preventive maintenance procedures for an ECG machine.
20. The stylus tip is not writing or writing too lightly in an ECG machine. How would you troubleshoot the problem?

Chapter 11

Computer Networking and Network Devices

Introduction to Networking

A *computer network* is a set of personal computers (PCs) and the peripheral devices connected to each other. All equipment shares the same data. A *server* is the computer that stores the shared data in the software in the network operating system (NOS). A *client* is the computer that requires and receives information from the server. There are three types of networks:

- Peer-to-peer network: Each station acts like a server and a client. There is no centralized station to manage the network operation.
- Server-based network: A server stores all the network's shared files and applications. Each client can access the files on the server and the files of other clients. Examples of server-based networks are mail and print servers.
- Client/server network: The server keeps all the files. When a client requests a file, the server sends the file to the client. There is no communication between clients.

Network Topologies

Network topology describes the way the computers are connected to each other. The major factors in design are cost and reliability of different network topologies. The major network topologies include the following:

- Star topology: The computers are connected to a central controller or hub as shown in Fig. 11.1. The hub is the network device that repeats the information it receives and passes it to the other computers. The several advantages of

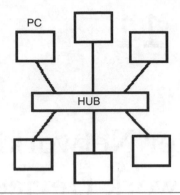

FIGURE 11.1   Star topology.

star topology are that it is easy to set up and expand the network, and that if a computer in the star breaks, the rest of the computers is not affected or disturbed. Disadvantage of the star topology is that if the hub breaks down, the whole network crashes.

- Ring topology: There is a source computer in the ring that checks the address of the received information and passes it to the next station in the ring shown in Fig. 11.2. The destination computer that has the matching address receives the information. The disadvantages of the ring topology are that if a computer in the ring breaks, the rest of the computers in the ring are affected, and that it is hard to add a new computer to the ring.

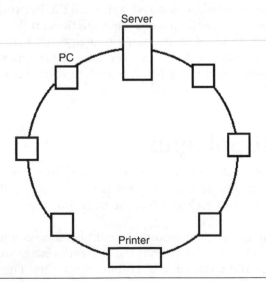

FIGURE 11.2   Ring topology.

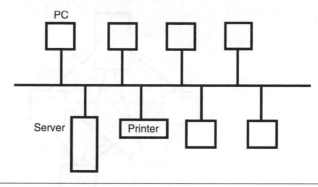

FIGURE 11.3   Bus topology.

- Bus topology: All computers in this topology are connected to a single cable as shown in Fig. 11.3. Ethernet is the most popular bus topology. The advantage of the bus topology is the easiness of adding new PCs with minimum cost to the network. The disadvantage of bus topology is that if the bus cable breaks down, all the PCs on the bus will be affected.
- Tree topology: All of the PCs are connected in a tree fashion. There are hubs throughout the tree. Each hub connects the PCs in all of the branches that are at the same level shown in Fig. 11.4. The advantage of tree topology is that if one hub breaks down, it only affects the PCs in that level of branches.
- Mesh topology: In mesh topology all the PCs are interconnected without a hub or any other device as shown in Fig. 11.5. The advantage of mesh topology is the high reliability it provides. The disadvantage is the high cost of wiring used in this topology.
- Hybrid topology: A mixture of other topologies connected together as shown in Fig. 11.6. The cable, which connects all topologies together, is called the *backbone*. Each network topology is connected to the backbone through a device called the *bridge*.

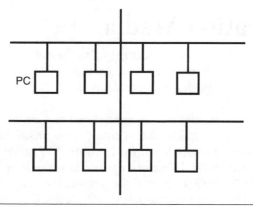

FIGURE 11.4   Tree topology.

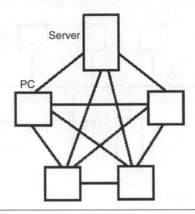

FIGURE 11.5 Mesh topology.

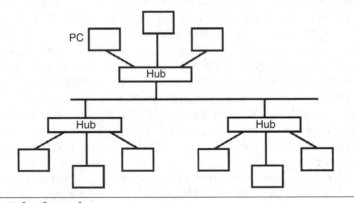

FIGURE 11.6 Hybrid topology.

Communication Media

Digital/data communication media refers to the cabling system that the information travels between the transmitter and a receiver. Copper cables, optical fibers, and air are the three major media of communication:

1. Two types of copper cables are twisted and coaxial.
   A. Twisted-pair cables can be shielded or unshielded:
      1. Shielded twisted-pair (STP) cable: STP cable has a metal shielding. It is expensive and less susceptible to noise than the unshielded twisted-pair (UTP) cable. It is used for long-distance transmission (Fig. 11.7).
      2. Unshielded twisted-pair (UTP) cable: UTP cable uses RJ45 and RJ11 connectors. It does not have any metal shielding; hence it is least

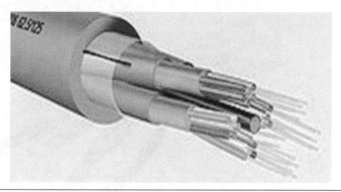

FIGURE 11.7   Shielded twisted-pair (STP) cable.

expensive and most susceptible to noise. Used for short distances, it is shown in Fig. 11.8. Categories of UTP cables are listed as follows:

CAT1:	For voice communications; it does not support data communication.
CAT2:	4 Mbps speed. Rarely used.
CAT3:	10 Mbps speed for data transmission.
CAT4:	16 Mbps speed for data transmission.
CAT5:	100 Mbps data transmission speed. It can handle up to 1 Gbps.
CAT5e:	10 Mbps, 100 Mbps, and gigabit range Ethernet data transmission. Rated at 350 megahertz (MHz). Most widely used network cable.

B. Coaxial cables: They use Bayonet Neill–Concelman (BNC) connectors. They are used for high-speed and long-distance transmission. They have copper conductors in the center and outer shielding under the cover, and in the middle there is a dielectric material shown in Fig. 11.9.

2. Optical fibers: They have the highest transmission rates, and are least susceptible to noise and most expensive. Mode describes the way that light propagates through the fiber-optic cable. Two types of fibers include step index and multimode graded index. Figures 11.10 through 11.12 show different types of fibers. The description of these fibers is as follows:

A. Step-index (single-mode and multimode) fibers: In a single-mode fiber, light propagates in one direction only. The light direction is alongside the center of the axis of the fiber. In a multimode fiber, light propagates in variety of paths.

B. Multimode graded-index fibers: In a multimode graded-index fiber, the index of refraction of the light is the highest at the center and it gets smaller as the light travels toward the cladding sides of the fiber. The disadvantages

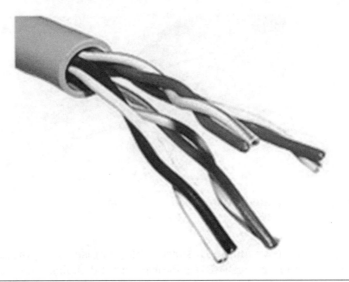

FIGURE 11.8 Unshielded twisted-pair (UTP) cable.

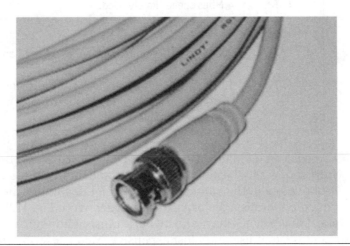

FIGURE 11.9 Coaxial cable.

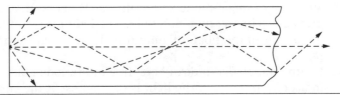

FIGURE 11.10 Multimode step-index fiber.

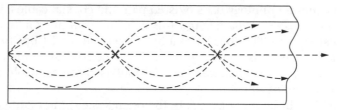

FIGURE 11.11   Multimode graded-index fiber.

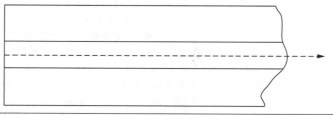

FIGURE 11.12   Single-mode step-index fiber.

are that pulse spreading and loss of light occur as a result of the different modes traveling in a multimode step-index fiber.

C. Single-mode step-index fibers: These fibers are the best to use but are very expensive. Data travels great distances because the light does not separate. Multimode step-index fibers are less expensive but are not used in long distances due to the dispersion of the light signal.

3. Air: Radio frequencies, infrared light, and lasers are the other major ways of transmission of data.

Network Architecture and Ethernet Technology

Network architecture includes physical and logical topology. Physical topology is the layout of the network. Logical topology defines the network access method including how the PC accesses the media, and describes how the data travels through the media. The most important network architectures are Ethernet, token ring, fiber distributed data interface (FDDI), and wireless.

The following list shows the four specifications used for these network architectures:

IEEE802.2:	Logical link control (LLC)
IEEE802.3:	CSMA/CD
IEEE802.5:	Token ring
IEEE802.11:	Wireless

Ethernet technology uses twisted-pair and coaxial cables, including 1000BaseT, 100BaseT, 10BaseT, 10Base2, and 10Base5. For example, for 1000BaseT cable, 1000 designates 1000 Mbps (megabits per second) or 1 Gbps (gigabits per second) and T stands for twisted pair. 10Base2 is the coaxial cable rated at 10 Mbps and the signal travels up to 185 meters (m). In 100Base5 coaxial cable, signal travels up to 500 m at a rate of 100 Mbps.

The data link layer (layer2) in the open system interconnect (OSI) model has two sublayers: logical link control (LLC) and media access control (MAC). LLC provides connection between source and destination PCs. MAC provides access to the network. It uses carrier sense multiple access/collision detect (CSMA/CD) protocol.

The protocol for CSMA is described in the following procedure: If a PC wants to transmit, it listens to the bus. If it does not detect any transmissions or collisions, it transmits its frames. If the bus is busy, PC waits until the bus is free and then it transmits. When a PC transmits a frame, the PCs on the same bus as the transmitting PC will check the frame, and if the frame matches their NIC addresses, they will keep the frame; otherwise, they will discard it.

The following procedure explains CD protocol: If two or more PCs transmit frames at the same time on a bus, collision will occur and the frames will be discarded. The PC that detects the collision first sends the information about the collision to the other transmitting PCs. Then the colliding PCs transmit according to a waiting period or back-off algorithm, which ensures that another collision does not occur.

Network Devices

Network devices consist of computer hardware that connects PCs, local area networks (LANs), and wide area networks (WANs). LANs are networks that are local to a room or building of an organization. WANs span areas beyond buildings next to each other within a geographical area that covers cities, metropolitan areas, and the world. The following are categories of devices: repeaters, hubs, bridges, switches, and routers.

Repeaters

Owing to noise and long distances, degradation or weakening and attenuation of signals occur as they travel through cables. In order for the signals to reach their destinations clearly, they need to be cleaned from noise. *Repeaters* clean the digital signals from noise, regenerate the signals, and retransmit them. As a result, clean signals reach their destinations. Amplifiers should not be used instead of repeaters because they amplify signals and noise as well.

Hubs

A *hub* is a device that connects several computers. Figure 11.13 shows a typical hub. Active hubs are the hubs that are connected to electric power. They also repeat and regenerate signals. Passive hubs are not connected to electric power and they do not

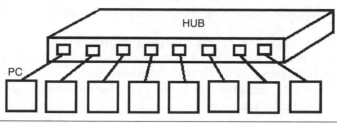

FIGURE 11.13   A hub.

repeat signals. A repeater has one incoming and one outgoing port. A hub has one incoming and multiple outgoing ports.

Repeaters and hubs work at the physical layer of the OSI model that is described in detail later in this chapter. The advantages of hubs and repeaters are that they extend the distance signals travel and reduce the network traffic problems.

Bridges

A network device can transmit packets only if the media is free. Too many devices in the network will cause data collisions. Segmenting the network is the solution to minimize data collisions. A *bridge* is a device that connects two segments of a network together. It has one input and one output port. A bridge works at the data link layer of the OSI model.

MAC address is the physical address that is electronically burned-in and configured in the network interface card (NIC) of the PC or device. Every network device or a computer has a MAC address, which is permanently developed during the manufacturing process. It is composed of 12 hexadecimal digits; the first 6 digits designate the manufacturer and it is called the *organizational unit identifier* (OUI) and the last 6 digits identify the vendor and it is called the *serial number* or the *device ID*.

Once it is connected, a bridge knows the destination MAC addresses of the PCs in the network and builds a MAC address table. According to the table, a bridge forwards the data to the appropriate PC. A bridge performs error detection as well. The advantages of the bridges include the following:

- Bridges reduce network traffic by segmentation.
- Bridges increase bandwidth.
- Bridges create separate collision domains.
- Bridges connect networks with different media.

The disadvantages of the bridges include the following:

- Bridges are slower than repeaters.
- Bridges occur when one device transmits information to all the rest of the devices in the network.
- Bridges are more expensive than repeaters.

Switches

As shown in Fig. 11.14, a switch is like a multiport bridge and it connects the LANs together. It is a network device that operates at the data link layer of the OSI model. The switch creates a virtual circuit between the source and destination PCs. The bandwidth of the network will be the same for the two or more PCs connected through a virtual circuit. For example, if it is a 100 Mbps switch, each one of its ports that are connected to different PCs will carry signals with 100 Mbps bandwidth separately. With the utilization of switches, network traffic will not be congested.

A hub will divide the bandwidth into the number of PCs it is connected to. Contention occurs when PCs sharing the same network experience show data rates. The hub causes contention, whereas the switch minimizes contention. Latency is the delay on a network caused by different factors such as adding more network devices. The advantages of switches include the following:

- Switches increase bandwidth.
- Switches reduce the number of collisions by segmenting the network and using virtual circuits. Each port of the switch will create a collision domain.
- Switches have multiports.

The disadvantages of switches include the following:

- Switches are more expensive than bridges.
- Broadcast could cause traffic through switches because switches do not create separate broadcast domains.

Routers

Routers are network devices that operate in the network layer of the OSI model. They decide the most efficient routing or data path to transmit data between LANs and WANs. They can convert data between different topologies of networks. In order to route the data packets efficiently to their destinations, routing protocols need to be programmed in the configurations of the routers.

Routers are multiport devices. They use logical or Internet protocol (IP) addresses to route the data. IP addressing provides logical grouping of networks so that the routing of packets can be accomplished in large networks. Each IP address is assigned to a specific network segment. Each port or connection of the router has its own IP address.

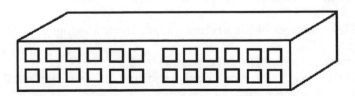

FIGURE 11.14   Back view of a switch.

There are two basic kinds of routing protocols: static and dynamic. In static routing, the routing table is updated manually, whereas in dynamic routing the routing table is updated automatically. Each router has a routing table that has the IP addresses and the routes of the devices it is connected to in the network. Each time a PC or a network device is added to or taken out of the network, the routing table of each router gets updated within the network. The advantages of routers include the following:

- Routers facilitate different network topologies to communicate with each other.
- Routers create collision domains by segmentation that minimizes traffic.
- Routers create different broadcast domains. If a broadcast occurs in one port, it only reduces the traffic connected to that port. The rest of the ports do not get affected by the broadcast.

The disadvantages of routers include the following:

- Routers are more expensive than repeaters, hubs, and bridges.
- Because dynamic routing requires updates of routing tables, the bandwidth is reduced due to the updated traffic.
- Routers are slower than bridges and switches.

OSI Model

The open system interconnect (OSI) model describes how data flows from the source to the destination. The source computer forms the message at the application layer and passes it through the OSI layers to the physical layer. Each layer adds control and address information that is needed to process the data. This process is called *encapsulation*. The OSI model will also help us troubleshoot network problems that are explained at the end of this chapter.

After reaching the physical layer of the source computer, data is put out on the media. The destination computer receives the data at its physical layer and passes it to the application layer by passing through the layers of the OSI model. Each layer strips off the control and address information through the decapsulation process. As a result, the message arrives at the application layer of the destination computer.

The OSI model has seven layers. The first letters of each word in the phrase "**A**ll **P**eople **S**eem **T**o **N**eed **D**ata **P**rocessing" can help us remember the names of the OSI layers as shown:

Layer 7:	**A**pplication
Layer 6:	**P**resentation
Layer 5:	**S**ession
Layer 4:	**T**ransport
Layer 3:	**N**etwork
Layer 2:	**D**ata link
Layer 1:	**P**hysical

Physical Layer

The physical layer breaks down the frames into bits. It also places and takes off binary data on and off the media. Signals through the media can be light, radio waves, or voltages. The physical layer defines the type of media, including smart-serial, copper, coaxial, fiber-optic cables, and wireless. The type of interfaces and connectors are high-speed serial interface (HSSI), DB-25, DB-9, and RJ-45. The physical layer is identified as layer 1. Repeaters and hubs are the network devices at this layer. The unit of encapsulation is bits.

Data Link Layer

The data link layer is composed of two sublayers: LLC and MAC. The MAC layer defines methods (CSMA/CD, Ethernet, token passing, and FDDI) that a computer accesses a line to transmit information. The LLC layer establishes device-to-device connections within the LAN and between WANs.

For transmission, the data link layer receives the packets from the network layer and breaks them into frames. Each frame is composed of fields, including destination and source addresses, frame checking sequence (FCS), data, and length of field. The data link passes the frames to the physical layer.

For reception, the data link layer receives bits from the physical layer and forms them into frames. It performs error detection. If there are no errors, the data link layer passes the frames to the network layer. It does synchronization at the beginning and at the end of the frame by adding opening and closing bytes to each frame. Through link management it manages transmission between source and destination computers. The main protocols for the data link layer are synchronous data link control (SDLC) that is developed by IBM and high-level data link control (HDLC).

The data link is identified as layer 2. Switches and bridges are the network devices that operate at this layer. The units of encapsulation are frames.

Network Layer

The network layer provides connectivity between the computers located on different networks geographically separated from each other. It also performs IP or logical addressing, IP address to MAC address translation, and router-to-router communication. Network layer protocols include IP, Internet control message protocol (ICMP), address resolution protocol (ARP), and routing information protocol version 2 (RIPv2).

ICMP provides message control and error reporting with replies such as packets received, echo request, congestion (source quench), time exceeded, and destination unreachable. ARP performs IP address to MAC address translation. RIPv2 checks the logical address of the frames and forwards them to the correct network according to the routing table of the router. The units of encapsulation are packets.

The network layer is identified as layer 3. Routers are the network devices at this layer. Packets are the units of encapsulation.

Transport Layer

For transmission, the transport layer receives the messages from the session layer, breaks them into segments, and sends them to the network layer. For reception, the transport layer receives the packets from the network layer, forms them into segments, and sends them to the session layer.

The transport layer provides data segmentation, which is chopping off the messages into maximum transmission units (MTUs). It does data sequencing that puts the data together and handles error. Host-to-host (end-to-end) connection, acknowledgment of data, reliability check, and flow (speed) control are also among the transport layer's responsibilities.

There are two protocols for the transport layer: transmission control protocol (TCP) and user datagram protocol (UDP). TCP is connection-oriented, reliable, slow, and it sends an acknowledgment (ACK) for the data received, which guarantees the packet delivery. The second protocol (UDP) is connectionless, unreliable, fast, and it does not guarantee that the packets are received. It does not send an ACK for the data received.

The transport layer is identified as layer 4. Routers are the network devices at this layer. Segments are the units of encapsulation.

Session Layer

The session layer establishes application-to-application connection between two computers. It does session setup/exchange and disconnection. When the server requests user ID and password, it is the session layer's responsibility to provide the "Login" process. An example of a session layer protocol is SQL, which is a database access language.

The session layer is identified as layer 5. Routers are the network devices at this layer. Messages or data are the units of encapsulation.

Presentation Layer

The presentation layer does data conversion, encryption, compression, and formatting. It receives the information from the application layer and translates it into a code that computers understand. Before transmitting the data to the lower layers, the presentation layer does encryption and compression. The opposite process is decryption and decompression, through which the data is passed to the application layer. Compression and decompression can help reduce the data size, which improves network traffic and performance.

Examples of presentation layer protocols are the American Standard Code for Informational Interchange (ASCII), Bitmap (BMP), Moving Picture Experts Group (MPEG), and Waveform Audio File (WAV).

The presentation layer is identified as layer 6. Routers are the network devices at this layer. Messages or data are the units of encapsulation.

Application Layer

The application layer provides interface and services for applications used by the host computers. Examples of application layer protocols are hypertext transfer protocol (HTTP), file transfer protocol (FTP), simple mail transfer protocol (SMTP), post office protocol (POP), domain name service (DNS) and terminal emulation protocol (TELNET). HTTP is the application to access the Internet and it defines the message types that client requests and server responds to. Even though it is slower than HTTP, FTP is the reliable connection to send files securely to a client. SMTP and POP help make connections for e-mail transfer between clients and mail servers.

DNS resolves an Internet web address to an IP address so that the data would reach its destination correctly. For example, when you type www.chi.devry.edu in the web address, it needs to be resolved. DNS provides the IP address corresponding to the web address and sends it to your computer. Then you are able to connect to the DeVry web server. TELNET allows the client to access a switch or a router remotely in order to troubleshoot it.

The application layer is identified as layer 7. Routers are the network devices at this layer. Messages or data are the units of encapsulation.

TCP/IP Protocol Suite

The TCP/IP protocol suite is another model similar to the OSI model that was developed by the Department of Defense (DOD). Layers of the TCP/IP protocol suite and OSI models are compared as shown in the following table:

TCP/IP Protocol Suite Model	OSI Model
Application	Application
	Presentation
	Session
Transport	Transport
Internetwork	Network
Network interface	Data link
	Physical

All the protocols are the same for both TCP/IP protocol suite and OSI models. In addition to the concepts written earlier under the OSI model, transport layer TCP and UDP protocols use port numbers to communicate. For example, if a PC wants to TELNET another PC, it uses port number 23. The following are the most important port numbers:

TCP, FTP:	21, 20
DNS:	53
Telnet:	23
http:	80

Another example of Internet work layer protocol is ARP and reverse ARP (RARP). ARP provides IP address to MAC address resolution and RARP provides MAC address to IP address resolution. During the application of ARP, the source PC broadcasts in the local subnet requesting the MAC address corresponding to the IP address of the destination PC. The destination host is the only one that responds to the ARP request. ARP tables exist in the random access memory (RAM) of the PCs and devices.

The ARP tables are constantly updated dynamically. When a source PC wants to transmit a packet, it checks its ARP table for the MAC address corresponding to the IP address of the destination PC. If it finds the MAC address, it sends the data. If it does not find it, it broadcasts an ARP request first, the destination PC responds, the source PC updates its ARP table with the new MAC address, and then the source PC sends the data as follows:

```
ARP request frame MAC header
Destination:        FF-FF-FF-FF-FF-FF
Source:             00-00-BC-12-34-56
IP header
Destination:        192.168.1.100
Source:             192.168.1.205
ARP request message
What is your MAC address?
```

IP Addressing

The network layer uses Internet protocol (IP) as the protocol to do routing. The network layer checks the IP address (logical address) of the packets and forwards them to the next router. It also translates each IP address to a MAC address. Large networks do not route data efficiently if only MAC addressing was used because MAC addresses are not logically grouped together.

IP addressing is another type of addressing that provides grouping of computers and network devices. Instead of traveling everywhere, data packets will go through selected groups of computers with certain routes with specific IP addresses in order to reach their destination. Because of the reduced traffic, routing becomes efficient. Internet Assigned Numbers Authority (IANA) developed the IP addressing structure. American Registry for Internet Numbers (ARIN) assigns the IP addresses to individuals, companies, and government.

With IPv4 protocol, each IP address is composed of 32 bits of binary code (4 bytes). There are dots between the bytes, and if it is formatted in decimal, it is called dotted decimal notation. For example, a computer that has an IP address of 172.16.4.35 in dotted decimal can also be represented in binary as 10101100.0 0010000.00000100.00100011. IP addresses are classified as classful and classless addressing.

Classful IP Addressing

Even though classful IP addressing is used rarely, it is still important to study it in order to comprehend subnetting and other modern networking practices. The IP addresses are divided into five classes: A, B, C, D, and E. Classes A, B, and C were assigned to individuals, companies, schools, universities, and governments. Classes D and E were assigned for research and multicasting. Multicasting is the process where one computer communicates with selected number of computers. Playing online games is an example of multicasting.

Class A Networks

All class A addresses start with binary 0 and they support very large networks. The first byte in the IP address is allocated to network (N) and the rest of the three bytes are for hosts (H) as follows in the textbox:

N(1 byte) . H(1 byte) . H(1 byte) . H(1 byte)
N = 00000001 = First class A address
N = 01111110 = 126 last class A address
N = 01111111 = 127 loopback address
Class A's network byte range is: 1–126
Maximum number of networks: $2^x - 2 = 2^7 - 2 = 126$
    x = the number of bits available for the network
Maximum number of hosts/network: $2^y - 2 = 2^{24} - 2 = 16,777,214$
    y = the number of bits available for hosts

The address 127.0.0.1 is the loopback address that is used for diagnostics to see whether IP is installed correctly in a computer. The troubleshooting tool called "packet Internet groper" (ping) checks the diagnostics. After entering "ping 127.0.0.1" in the command prompt of the computer, the replies need to be checked as shown in Fig. 11.15. If the packets are sent and received successfully, "4 packets sent and 4 packets received" will be the response.

```
C:\Users\owner>ping 127.0.0.1

Pinging 127.0.0.1 with 32 bytes of data:
Reply from 127.0.0.1: bytes=32 time<1ms TTL=128
Reply from 127.0.0.1: bytes=32 time<1ms TTL=128
Reply from 127.0.0.1: bytes=32 time<1ms TTL=128
Reply from 127.0.0.1: bytes=32 time<1ms TTL=128

Ping statistics for 127.0.0.1:
    Packets: Sent = 4, Received = 4, Lost = 0 (0% loss),
Approximate round trip times in milli-seconds:
    Minimum = 0ms, Maximum = 0ms, Average = 0ms
```

FIGURE 11.15 Ping example for PC diagnostic.

Class B Networks

Class B networks support medium- to large-size networks. All class B addresses start with binary bits 10. The first two bytes are allocated to network (N) and the next two bytes are for hosts (H) as shown in the following textbox:

N(1 byte) . N(1byte) . H(1 byte) . H(1 byte)
N = 10000000 = 128 first class B address
N = 10111111 = 191 last class B address
Class B's first network byte range is: 128–191
Maximum number of networks: $2^x - 2 = 2^{14} - 2 = 16382$
    x = the number of bits available for the network
Maximum number of hosts/network: $2^y - 2 = 2^{16} - 2 = 16382$
    y = number of bits available for hosts

Class C Networks

Class C networks support the small networks with a maximum 254 computers per network. All class B addresses start with binary bits 110. The first three bytes are allocated to network (N) and the fourth byte is for hosts (H) as shown in the textbox:

N(1 byte) . N(1 byte) . N(1 byte) . H(1 byte)
N = 11000000 = 192 first class C address
N = 11011111 = 223 last class C address
Class B's first network byte range is: 192–223
Maximum number of networks: $2^x - 2 = 2^{21} - 2 = 2,097,150$
    x = the number of bits available for the network
Maximum number of hosts/network: $2^y - 2 = 2^8 - 2 = 254$
    y = the number of bits available for hosts

Classless IP Addressing

Operating systems today use classful addressing in order to assign subnet masks to individual computers. As classful addressing was used historically until 2000, primarily classless addressing is used today. In classless addressing, companies and individuals get assigned IP addresses without considering any groups of classes. Classless IP addresses have fixed network sizes.

Subnetting

Subnetting is the technique of borrowing bits from the hosts to allocate more network addresses. It is an efficient way of designing networks and it optimizes network traffic. Manipulating the default subnet masks that are shown below in order to obtain more networks is called subnetting. In classes A, B, and C, subnet mask represents how many bits of an IP address are determined for network identification or host identification as follows:

Class A   N   H   H   H

Default subnet mask: 255 . 0 . 0 . 0

         or: 11111111. 00000000 . 00000000.00000000

24 bits (zeros) for hosts. We can allocate $2^{24} - 2 = 16,777,214$ hosts per network.

Class B   N   N   H   H

Default subnet mask: 255 . 255 . 0 . 0

         or: 11111111. 11111111 . 00000000.00000000

16 bits (zeros) for hosts. We can allocate $2^{16} - 2 = 16,382$ hosts per network.

Class C   N   N   N   H

Default subnet mask: 255 . 255 . 255 . 0

         or: 11111111. 11111111. 11111111.00000000

8 bits (zeros) for hosts.   We can allocate $2^{8} - 2 = 254$ hosts per network.

Example for Subnetting a Class C Network

Problem

The IP address of a network allocated to a company is 199.1.10.0 and the subnet mask provided is 255.255.255.224. Instead of having one large network, it is recommended that subnets for each floor of the company be created and accommodated. Answer the following questions:

1. How many subnets can be allocated?
2. How many hosts can be grouped together for each subnet?
3. Assign subnet IP addresses to each subnet and to each host within each subnet.

Solution

1. Convert 224 to binary: 224 = 11100000 where 1s symbolizes network bits borrowed from hosts and 0s are the number of bits available for the hosts.
   Use the formula $2^x - 2 = $ Number of subnets
   $x = $ The number of 1s borrowed from hosts = 3
   Get $2^3 - 2 = 6$ subnets

2. Use the formula $2^x - 2 = $ Number of hosts/subnet
   $x = $ Number of zeros for hosts = 5
   Get $2^5 - 2 = 30$ hosts per subnet

3. Major network IP address is 199.1.10.0.
   Subnet mask is 11111111.11111111.11111111.11**1**00000

   Find the multiplier or subnet identifier first. From left to right in the subnet mask, it is the weighting factor of the last **1** available for subnetting. Therefore the multiplier is $2^5 = 32$. Using the multiplier, the six subnets will have 32, 64, 96, 128, 160, and 192 as the last bytes of their subnet IP addresses. Subnet IP addresses can be assigned to the six floors as follows:

199.1.10.32	6th floor
199.1.10.64	5th floor
199.1.10.96	4th floor
199.1.10.128	3rd floor
199.1.10.160	2nd floor
199.1.10.192	1st floor

4. The next procedure is to assign host IP addresses for each subnet. As a rule of thumb, the first 30 numbers that are 199.1.10.1 to 199.1.10.31 are not used for hosts. As 30 hosts per subnet are calculated, keep assigning 30 addresses by adding 1 to the last byte starting with 199.1.10.33. Broadcast IP address in each subnet is the last number before the start of the next subnet IP address. Host IP addresses can be assigned to the six floors as follows:

   Subnet IP address: 199.1.10.32

   Host IP addresses are: 199.1.10.33–199.1.10.62
   Broadcast address is: 199.1.10.63
   Subnet IP address: 199.1.10.64

   Host IP addresses are: 199.1.10.65–199.1.10.94
   Broadcast address is: 199.1.10.95
   Subnet IP address: 199.1.10.96

   Host IP addresses are: 199.1.10.97–199.1.10.126
   Broadcast address is: 199.1.10.127
   Subnet IP address: 199.1.10.128

   Hosts IP address are: 199.1.10.129–199.1.10.158
   Broadcast address is: 199.1.10.159
   Subnet IP address: 199.1.10.160

   Host IP addresses are: 199.1.10.161–199.1.10.190
   Broadcast address is: 199.1.10.191
   Subnet IP address: 199.1.10.192

   Host IP addresses are: 199.1.10.193–199.1.10.222
   Broadcast address is: 199.1.10.223

Classless Interdomain Routing Notation

The number of masked bits in the IP address are designated or known as *classless interdomain routing* (CIDR) notation. Using the previous example, the subnet mask is 11111111.11111111.11111111.11100000. The number of 1s in the mask is 27 and the CIDR notation is 199.1.10.0/27. Examples of CIDR and subnetting are

- Problem 1: 192.168.4.30/28 host IP address or not?
  Answer 1: 28 means 28 1s in the subnet mask. The subnet mask is 11111111.11111111 .11111111.11110000 = 255.255.255.240. The shortcut to calculate the subnet identifier is: 254 (constant) − 240 (the last byte in the subnet mask) = 16. As a result, the subnets are 192.168.4.16, 192.168.4.32, etc. Because .30 falls in between .16 and .32 and it is not the broadcast address of .31, 192.168.4.30 is a host IP address.
- Problem 2: Design a network where the company needs 14 subnets based on the IP address 199.1.10.0.
  Answer 2: The first step is to find the subnet mask. Using $2^x - 2$ = Number of subnets = 14, we get $x = 4$. Therefore we need to borrow 4 bits from hosts. This means 11110000 = 240 is the last byte of the subnet mask. Therefore, the subnet mask is 255.255.255.240.

Router and IOS Commands

Routers filter traffic by creating collision and broadcast domains. Network administrators configure routers using Cisco internetwork operating system (IOS). IOS provides command line interface (CLI) to enter router commands.

There are three ways to access the routers: console port (Console), auxiliary (AUX) port, and virtual terminals (VTY). These ports are located at the back of the router as shown in Fig. 11.16. Console cable connects the router to PC, using RJ45 connector at the console port and DB9 connector at the PC as shown in Fig. 11.17. Through the AUX port remote the network administrator accesses the router by dialing into the router through a MODEM. As a third method, a router can be accessed by telnetting through five VTY terminals. The Trivial FTP (TFTP) server can also provide info to configure the routers.

Router Configuration Modes

There are several basic modes that a router uses for different purposes. Refer to the Appendix titled "Using Router Commands for Troubleshooting" for detailed information.

Network administrators can enter the following modes, in order to perform different functions:

- "user exec" mode: The administrator checks router status and sees whether the connections are working. The prompt is "RouterName >" and you can enter this

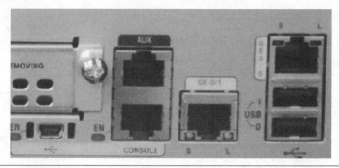

FIGURE 11.16   Ports of the router.

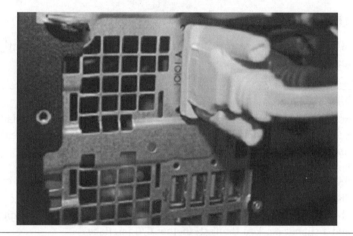

FIGURE 11.17   DB-9 Connector and console cable.

mode by hitting the return key and enter logout to exit. This mode is used to perform basic tests and display system information.
- "privileged exec" mode: The administrator can copy, erase, and set up router settings. The prompt is "RouterName#" and it can be entered from user exec mode by typing "enable" and enter "disable" to exit. This command is used to configure router operating parameters and show commands. Troubleshooting and debugging tasks are also performed in this mode.
- "global configuration" mode: The administrator can set clock, hostname, enable password, and secret passwords. The prompt is "RouterName(config)#" and it can be entered from privileged exec mode by typing "configure terminal" and exit with "Ctrl-Z" or "end." Use this mode to set router parameters.
- "interface configuration" mode: The administrator configures the Ethernet and serial interfaces and sets the IP addresses. The prompt is "RouterName(config-if)#" and it can be entered from global configuration mode by typing "interface [interface name]" and exit with "end" or "Ctrl-Z." Use this mode to set parameters of the router for serial and Ethernet interfaces.

- "line config" mode: The administrator configures console, VTY or AUX lines, and passwords. The prompt is "RouterName(config-line)#" and it can be entered from global configuration mode by typing "line console 0," "line vty 0 4," or "line aux 0." To exit enter "exit." Use this mode to configure terminal line parameters.
- "router configuration" mode: The administrator configures routing protocols. The prompt is "RouterName(config-router)#" and it can be entered from global configuration mode by typing "router [routing protocol name]" and exit by entering "exit." Use this mode for IP routing protocol configuration.

Router Commands

The following are the router commands that are used frequently in configuring and troubleshooting routers:

Router>enable ---> Enters the user exec mode
Router#show interface serial0/0/0 ---> Shows the status of the serial0/0/0 interface
Serial0/0/0 is administratively down; line protocol is down (disabled)
Router#config terminal ---> Enters the global config mode
Enter configuration commands, one per line. End with "Ctrl-Z"
Router(config)#hostname Chicago ---> Changes the router name to Chicago
Chicago(config)#interface FastEthernet 0/0 ---> Enters the interface config. mode
Chicago(config-if)#ip address 192.10.10.1 255.255.255.0 ---> Assigns an IP address
and a subnet mask to the FastEthernet 0/0 interface
(*Courtesy CISCO a registered trademark.*)

"?" gives all the commands under the mode you are in.
"exit" in the command line takes you back one level.
"end" or pressing "Ctrl-Z" takes you back to "routername #" prompt.
"Show running-config" command shows the present running-config file.
"Show interfaces" command shows all the configured interfaces.
"show ip route" shows the routing IP address table.
"Copy run startup-config" copies the running config file to the startup-config file.
"Show history" command brings up previous commands.
"Show startup-config" command shows the present startup-config.
"Show version" command displays the IOS version.
"Show cdp neighbor" shows the neighboring devices connected to the router.
"Show hosts" shows all the host configured on the router.
"↑" brings up the recent commands.
"Hostname [Name]" sets the router name.
(*Courtesy CISCO a registered trademark.*)

In order to configure the clocking serial interface s0/0/0 (IP address is 172.22.3.1) of the router A, enter the following commands:

```
RouterA#config t
RouterA(config)#interface s0/0/0
RouterA(config)#clockrate 64000
RouterA(config-if)#ip address 172.22.3.1 255.255.0.0
RouterA(config-if)#no shutdown
(Courtesy CISCO a registered trademark.)
```

If you would like to configure the IP address 192.168.1.1 with the default class C subnet mask for FastEthernet interface 0/0, type in the router the following commands:

```
Router#config t
Router(config)#interface FastEthernet 0/0
Router(config-if)#ip address 192.168.1.1 255.255.255.0
Router(config-if)#no shutdown
(Courtesy CISCO a registered trademark.)
```

Router Components

A router has four major components:

- RAM/dynamic RAM (DRAM): It is the working memory of the router, volatile, and it contains the "running-config" file.
- FLASH memory: It has fully functional IOS commands. Whenever the router is booted, Flash memory contents stay the same as before. Flash is an erasable programmable read-only memory (EPROM).
- Read-only memory (ROM): It has limited IOS commands. ROM stores the program to initialize the router's components.
- Nonvolatile RAM (NVRAM): It contains the "startup-configuration" file. When the router is booted, NVRAM contents stay the same as before. NVRAM contains the name of the router and the protocol configurations. When the configuration of the router is changed, the "running-config" file is copied from RAM into the "startup-config" file in NVRAM so that NVRAM is always updated.

Routing Protocols

Routing protocols are used for the routers to build routing tables so that routers can learn which network and subnetwork each PC is connected to. As a result, routers can route the packets in the direction where the destination PCs are located, which is called *path determination*. The number of routers that a data packet has to go through in order to reach the destination network is called *hop count*.

Metrics is the method that is used to decide which path the packets should travel. Measures of metrics are hop count, reliability, bandwidth, load balancing, and maximum transfer unit (MTU). Distributing the traffic evenly across different channels is called *load balancing*. MTUs are segments of data created by the transport layer to provide reliable communication.

Routing protocols are categorized into two: Static and dynamic routing. In static routing, the network administrator has to configure each path from a router to other networks. It is costly yet very reliable and is used for small networks. In dynamic routing, routers learn from each other by updating their routing tables automatically. As a result, packets know where to travel dynamically. It is less expensive than static routing, not as reliable, and is used for large networks.

In dynamic routing there are two subcategories: interior gateway protocols (IGPs) and exterior gateway protocols (EGP). Autonomous system (AS) consists of a group of routers under one organization. IGP is the protocol used within an AS. EGP is used between autonomous systems.

Interior Gateway Protocols

An example of interior gateway protocols (IGP) is the routing protocols used by the routers in a LAN of a company or organization. The subcategories of protocols under IGP are distance vector (DV) protocols and link state (LS) protocols.

Distance Vector Routing Protocol

DV protocols
- Routing information protocol (RIP)
- Routing information protocol version 2 (RIPv2)
- Enhanced interior gateway routing protocol (EIGRP)

LS protocols
- Open shortest path first (OSPF) is the common routing protocol.

The characteristics of DV routing include the following:

- Router sends the entire routing table to other routers.
- Router sends the tables periodically.
- Router sends the routing tables to directly connected routers only.
- It is slow for the network to converge.

Every 30 to 90 seconds (s) the router sends the updated table to the directly connected routers. Convergence occurs when all routers on the network have the most recent routing information. The problem with DV routing is that the information is not updated on all the routing tables at the same time. As a result of the delays, convergence problems could occur on the network.

Implementing the concepts of maximum hop count, split horizon, split horizon with poison reverse, and hold-down timers are methods to minimize problems with the DV protocol. Hop count is the number of routers the packets have to go through before they reach their destination. After the packets reaching a maximum hop count, they will get dropped. In RIP protocol, maximum hop count is 15. The sixteenth router that packets try to get through will drop them by providing a message that time to live (TTL) has been exceeded. The concept of maximum hop count will prevent infinite loops from happening in the network.

The split horizon method limits a router to receive information only from one router.

Split horizon with poison reverse method does the same as split horizon, but it also responds to the routers that it is not supposed to receive updates from. The network administrator will configure the routers that will implement these methods.

Hold-down timers set a period of time in a router during which the router does not accept any updates. Consequently the router will not forward false information to the network. The network will experience better convergence through implementing these four methods that minimize problems.

Link State Routing Protocol

The characteristics of LS protocol include the following:

- Router sends only updates in the routing table to the other routers.
- Router sends updates only when changes occur such as a network going down or a PC getting added to the network.
- Router sends updates to all other routers in the network at the same time.
- It is fast for the network to converge.

Exterior Gateway Protocols

The primary example of exterior gateway protocol (EGP) is border gateway protocol (BGP). BGP provides loop-free path selection between autonomous systems. It is used by ISP and major companies. Routing decisions are not based on metrics, but they are made according to network policies for different routing pathways.

Switches and VLANs

The main function of the network switch is to minimize traffic and improve network efficiency and performance. As a result, the switch creates separate collision domains and a collision would affect only the computers within a domain and would not cause traffic elsewhere. The network administrator can also configure virtual LANs (VLANs) that minimize traffic.

Through a VLAN, a group of computers can be logically connected together even though they are geographically separated by different floors and buildings in a company, school, or government facility as shown in Fig. 11.18. VLANs can logically segment the networks according to the group tasks/functions, departments, or project teams.

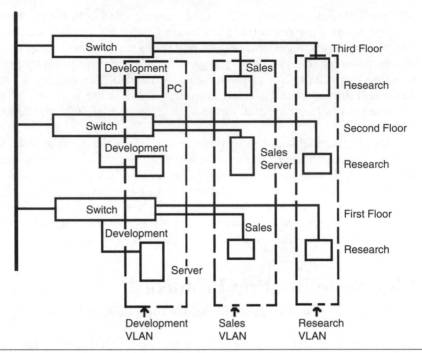

FIGURE 11.18   Virtual LANs (VLANs).

The benefits of VLANs can be classified as follows:

- High performance: Data traffic will be minimized due to the multiple logical networks and network performance will increase.
- Cost reduction: Because of the creation of high-performance networks through VLANs, maintenance needs and cost will be reduced.
- Improved security: Separate VLANs will have their own security policies. As a result, information will be kept confidential within the VLAN.
- Efficient IT staff: Because management and maintenance needs are reduced, IT will perform more efficiently.
- Broadcast storm: Broadcast storms occur when several broadcasts overload the networks and bandwidth is minimized. Because of LAN segmentation, broadcast storms will not propagate through all networks.
- Project management: Separating projects through VLANs would make the project management easier, faster, and more efficient.

A VLAN trunk is a crossover or a straight-through cable that carries multiple VLAN information between two Ethernet switch interfaces or between the interfaces of a switch and a router. VLAN trunking protocol (VTP) manages the VLAN switch configurations between the switches. It learns about the VLANs and stores them in the VLAN database file that is called vlan.dat.

Before VTP is configured, VLANs need to be created. For example, we would like to assign VLANs named Development, Sales, and Research to VLAN-ids 10, 20, 30 consecutively. The command "vlan [VLAN-id]" creates a VLAN and the command "name [VLAN-name]" creates the VLAN. The following commands need to be entered in the switch configuration after the # signs in order to create the three VLANs:

```
Switch(config)#vlan 10
Switch(config-vlan)#name Development
Switch(config-vlan)#vlan 20
Switch(config-vlan)#name Sales
Switch(config-vlan)#vlan 30
Switch(config-vlan)#name Research
Switch(config-vlan)#vlan 99
Switch(config-vlan)#name Management&Native
(Courtesy CISCO a registered trademark.)
```

Management&Native VLAN is usually assigned to 99, and it manages the IP address and subnet mask of the switch. Next, the VLANs need to be assigned to different access ports of the switch. For example, we would like to assign VLAN 10 to FastEthernet port 0/6, VLAN 20 to FastEthernet port 0/11, and VLAN 30 to FastEthernet port 0/18. The "switchport mode access" is the command to configure the port as an access instead of a trunking port. A VLAN is assigned to a certain port by the "switchport access vlan [VLAN-id]" command. The following commands need to be entered in the switch to complete the configuration:

```
Switch(config-if)#interface FastEthernet 0/6
Switch(config-if)#switchport mode access
Switch(config-if)#switchport access VLAN 10
Switch(config-if)#interface FastEthernet 0/11
Switch(config-if)#switchport mode access
Switch(config-if)#switchport access VLAN 20
Switch(config-if)#interface FastEthernet 0/18
Switch(config-if)#switchport mode access
Switch(config-if)#switchport access VLAN 30
(Courtesy CISCO a registered trademark.)
```

Troubleshooting Networks

Network performances and operations are always monitored for optimal performance and accuracy. As soon as a network is established, it is tested to make sure that the network is operational and documented. Network documentation is important because it provides information on the design of the network and its properties.

It is essential to measure the performance of the network and create a baseline to diagnose and repair any future issues.

There are four tables/diagrams necessary to include in order to provide complete documentation of the network. It is important to understand the fundamental information that is needed to construct each of these four tables/diagrams and logically record the network design. The documentation should also be available not only from a cloud but also a hardcopy to expedite repairs. The four tables/diagrams include the following:

1. Network topology
2. Table outline of the configuration
3. End-system configuration
4. Baseline of the network

Network Topology

The diagram of a network is referred to as the topology, which is basically a diagram of how each device in the network is connected to each other. In the diagram each network should be represented with all interconnecting devices and cables. A topology can be very detailed, but it should at least have the following information: identification of symbols for each device, whether the device connections are serial or Ethernet interface types, IP addresses, subnet masks, and WAN protocols. Troubleshooting any type of network connectivity is almost impossible without a network topology. There are two types of topologies:

- Physical topology: It identifies all device types, the model, and manufacturer of each device, operating systems and the versions, type of cable and its identifier, specifications of the cables, the types of connectors, and all the cabling endpoints used.
- Logical topology: It shows how the data is transferred on the network. Symbols are used on the topologies to identify the elements such as hubs, hosts, routers, servers, and virtual private networks (VPNs). Diagrams show connection types, data-link protocols, device identifiers, virtual circuits, interface identifiers, IP addresses with all subnet masks, routing protocols, site-to-site VPNs, static routes, and WAN technologies used.

Table Outline of the Configuration and End-System Configuration Table

The outline of the table configuration should include all the hardware, software, and wireless access points/bridges used in the network recorded in a clear and logical order on a spreadsheet or database. The spreadsheet or database is used to help locate and identify any faults of the network in a timely manner.

The table of the end-system configuration has the baseline information of end-system devices such as servers and workstations. The baselines will be inclusive of all the devices in the network. A configured network table can have a negative impact on the network performance. The following is the minimal amount of information that should be in an end-system configuration:

- Name of device
- Operating system and version
- IP addresses
- Subnet mask
- Default gateway, DNS server, Windows (WINS) server addresses
- Bandwidth network applications used

Baseline of the Network

Some problems with the network can be easily identified whereas problems such as excessive traffic or errors will not cause the network to fail but rather slow its performance. The performance of the network is documented in the network baseline document. To create the baseline document, collect the performance data during normal operations in order to reveal the true potential of the network.

Baseline reports show the physical connectivity, normal utilization, and average throughput of data transmission. One of the most important uses of baselines is for future upgrades to the network; another is to ensure peak performance optimization of the network. The baselines are the actual measure of the performance of the network in normal usage. Baselines can determine what thresholds settings should be for each device within the network. In the process of collecting the baseline data, problems with the network utilization and capacity, the network's true nature as well as most common errors can be recorded.

The baseline measurements directly affect future troubleshooting methods. It is very important to follow the steps and document the results very clearly. The following are the steps to take to initiate baseline development:

- Determine data to be collected. Checking CPU and interface utilization are good starting measures to analyze the data.
- Devices and ports need to be identified. Measurements of performance data should include all connected device ports, servers, users, and every device used in the regular operations of the network. The document should be detailed accurately. For example, it is necessary to show how a server is connected to which cable at what port.
- Determine the duration of the baseline. A rule of thumb for baseline duration is to conduct it during regular usage and its time frame should last from 2 to 6 weeks. Duration times should be regularly scheduled on hourly, daily, weekly, monthly, and yearly basis in order to accurately report usage of the network. Usage patterns will develop showing underutilization of system overloads.

Consistent and regular time analysis will show when the networks are optimized and when they are not.

Network Documentation

To document the network is time-consuming, but the information that is gathered is invaluable. Once the document is created, it must be updated to ensure accuracy. Changes in any device must be reflected in the topology and end-system documents. The following are the basic steps to create proper document:

1. Log in to the undocumented device.
2. Obtain important information of the device.
3. Enter network configuration information.
   A. If the device is important, add it to the topology diagram.
   B. Record important information to the network topology.
   C. Recheck for more information for the device.
4. Check other connected devices.
   A. If any undocumented devices exist, return to the second step.
   B. If there are no undocumented devices, then exit.

Useful commands used in network documentation are listed as follows:

Ping	Tests the connectivity with adjacent devices. MAC address autodiscovery process can be initiated by pinging to other PCs in the network.
Show cdp neighbor detail	Shows the information of directly connected adjacent devices.
Show ip interface brief	Shows IP addresses of all interfaces on a router or switch and the port up and down status.
Show ip route	Shows routing table in a router to discover the directly adjacent devices and the routing protocols that have been configured.
Telnet	Logs in at a distance to a device.
(*Courtesy CISCO a registered trademark.*)	

Network Troubleshooting through OSI Layers

Layers 4 through 7 of the OSI model relate to software problems. Physical and data link layers deal with hardware and software, and layers 1 through 3 deal with hardware problems.

Troubleshooting Physical Layer

The physical layer deals with tangible properties that data is transmitted on such as antennas, cards, and wires. The problems that occur can affect the whole network causing major delays in work and communications. Using the OSI model repairs and corrections to problems can be solved efficiently to minimize delays. Symptoms of usual physical layers troubles include the following:

- Congestion: Network congestion and high CPU usage can also cause issues.
- Error messages: LED displays an error message and/or there is a color indicator. Refer to the manual to be able to identify the issue.
- High collision counts: These counts occur usually in Ethernet networks using shared media hubs, and it is a problem in switch networks. A faulty cable or interface can cause collision problems. Collisions that overwhelm a system are usually over 5 percent of the total count of collisions.
- Loss of connection: The most obvious symptom that is due to lack of a link between devices. A simple ping test can determine whether the connection loss is due to a loose connection.

Troubleshooting the physical layer includes the following:

- Verifying to see whether all connections are installed properly in the device and the power source.
- Inspecting for faulty cables. Sometimes cables can be bent, crushed, or damaged. Then these cables need to be replaced immediately. Turn off all power sources before replacing cables.
- Ensuring that there is no corrosion at the connectors and ports.
- Checking to see whether the crossovers, straight-through, serial, or other types of cables are installed correctly for all devices.
- Organizing wiring closet by tagging all cables and wires and document.
- Making sure the interface configurations are correct for switches, including VLAN, duplex setting, and speed.
- Using "show" commands to observe input and output errors.

Troubleshooting Data Link Layer

The data link layer receives bits of data (1s and 0s) from the physical layer and forms frames which then transmit them to the network layer. Symptoms of data link layer troubles include the following:

- No connectivity from data link to network layer: Check to see whether amber LED link lights are on for the switches. These lights indicate trouble.
- Performance levels below baseline: Spanning tree protocol (STP) minimizes traffic problems by allowing or disallowing traffic at the switch's ports according to the utilization of the ports. Spanning tree topology when poorly designed

causes frames to take other paths than expected. As a result, frames are lost or misdirected.
- Broadcasting errors: These errors occur due to STP loops, improper configurations for applications, large number of devices, and PCs connected in a broadcast domain.
- Error messages: These messages are due to routing protocols not configured properly.

Troubleshooting the data link layer includes the following:

- Enabling STP by entering "spanning-tree VLAN-ID" command in the switch
- High CPU utilizations by routers, system log messages that show continuous learning of MAC addresses are ways to discover the STP loops
- Making sure that the interfaces are configured correctly
- Disconnecting the ports of the switch, which use the STP protocol
- Reconnecting the ports and seeing whether the switches perform normal STP operations without loops
- Exploring the network diagram to find out the redundant paths and reconfiguring the STP protocol

Troubleshooting Network Layer

A symptom of network layer problems is failure of the partial or entire network below baseline performance. Troubleshooting the network layer problems includes the following:

- Examining static/dynamic protocols: Confirm that all new PCs are added or any is removed from the network topology and review whether the routing protocols are updated.
- Testing connectivity: Ensure power is on and all cables, ports, and Internet are connected properly.
- Verifying neighbors: Confirm that routing protocols are applied to all neighboring devices.
- Checking routing tables: Make sure new routes are added and unused routes are removed from the routing tables.

Tools for Troubleshooting

Tools are used to gather and analyze symptoms as well as provide monitoring and reporting functions. New tools and technology are being developed on a daily basis for both software and hardware testing. When using any new tool or technology, refer to the manufacturer's manuals for specifications and proper use. To date, most of the new developments can be categorized into groups. Tools for troubleshooting include the following:

- Cable analyzers: Handheld devices are used to test copper and fiber cables. These devices include PC-based software to enable an upload to a PC to provide crucial data to generate accurate up-to date reports. Deluxe models will also include troubleshooting diagnostic features, such as the measure of the distance to the performance defect, corrective actions identification, and graphic display of crosstalk and impedance behaviors.

- Cable testers: These are the handheld devices that detect broken wires, crossover wiring, and improperly paired and shorted connections that are used to transmit data. Deluxe versions of cable testers include time-domain reflectometers (TDRs). A TDR sends signals along the cable and then calculates the time to receive the signal back to pinpoint the distance of a break or short in a cable. Optical TDRs (OTDRs) are used to test fiber-optic cables.
- Digital multimeters: These multimeters check power-supply voltage levels going into and out of devices, and also measure current, resistance, and voltage levels of electrical circuits.
- Protocol analyzers: These analyzers display network traffic as segments, packets, and frames that are helpful for analysis.

Network Troubleshooting Example Using a Flowchart

A good technique in troubleshooting network problems is to check the hardware problems first and then test for software problems next. Utilizing the seven layers of the OSI models, hardware problems relate to physical, data link, and network layers. Software problems relate to transport, session, presentation, and application layers. According to the flowchart shown in Fig. 11.19, the following are the major issues to explore in sequential order: Check hardware-related problems, TCP/IP settings, and client applications.

Check Hardware-Related Problems

When the PC cannot access the Internet, check hardware-related problems by taking the following steps:

1. Check the indicator lights at the Ethernet ports. If there is no blinking light, the problem could be that the Ethernet cable is not plugged in correctly.
2. After connecting the cable, if the problem still exists, check the lights on the other side of the cable connected to the hub, router, or switch. If the lights are on at the network device, check the NIC card by following the next step. If the lights are off at the network device, check whether the power is on. If the power is off, turn it on, and if it does not turn on, troubleshoot the switch or router. If the power is on, test the cables with a cable tester. If the cables are not good, replace them with new ones. If the cables are fine, proceed with step 3. Make sure straight-through and/or crossover cables are used properly.
3. Check to see whether the NIC card is properly inserted at the right expansion slot, without any damage and if the indicator light is on. If the NIC card is properly inserted, proceed with step 4. If not, reinsert the NIC card at the proper slot and proceed with step 4.

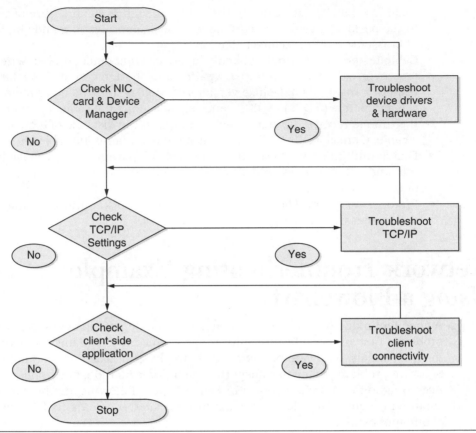

FIGURE 11.19 Troubleshooting network flowchart.

4. Check the network adapter or NIC card at the Device Manager software. Uninstall and reinstall the device drivers with new updates using the NIC card's manufacturer's website.
5. After updating the device drivers, if the Internet is still not accessible, replace the NIC card with a new one. The NIC card should be compatible with the network usually at 100-Mbps speed. Check the indicator light for the NIC card to see whether it is on. If it is on and the Device Manager shows that the device is working properly, connect to Internet. If you cannot access the Internet, proceed with the next section.

Check TCP/IP Settings

In order to test the connectivity, TCP/IP settings need to be checked. The following are the steps to take to achieve Internet connection:

1. Release the current IP address and receive a new one. In command prompt enter "ipconfig/release" and "ipconfig/renew" to receive a new IP address for your PC.

2. Check whether there are any problems by entering "ipconfig/all." You should observe IP address, subnet mask, default gateway, and MAC address of the PC. If "autoconfiguration" IP address starts with 169.254, the PC cannot access the Dynamic Host Configuration Protocol (DHCP) server. Because the DHCP server cannot be reached, the PC will not receive an IP address and will not be able to connect to Internet.

3. In command prompt interface, enter the loopback address by the "ping 127.0.0.1" command. This is the diagnostics test to see whether TCP/IP settings including IP addresses are working properly. If the results of the ping do not show connectivity, TCP/IP settings need to be set manually as follows:

   A. Enter "ipconfig/all" to see the TCP/IP settings and to check that DHCP is enabled. Record the IP address, subnet mask, and gateway IP address.

   B. In Windows 7 environment, do the following sequentially:

   > Control panel "Network and internet,"
   > "View network status and tasks," "Change adapter settings."
   > Right-click "Wireless network connection" or "Local area connection,"
   > "Properties," Highlight and double-click "Internet protocol v4,"
   > "Properties," "Use the following address."
   > Type the IP address, subnet mask, default gateway address you recorded.
   > Close the "Properties" box.
   > Close the "Network connection" box.

4. Check connectivity now. If the problem still persists, recheck the connections of the router to the Internet as follows:

   A. Turn off the digital subscriber line DSL, cable modulator-demodulator MODEM, and the wireless router.

   B. Turn the DSL or cable MODEM back on. Wait until the lights come on.

   C. Turn the router back on.

5. Check and see whether the router works. If it is down, it might be an ISP problem. Then disconnect the router and reconnect it.

6. If there is still no connectivity, call the ISP and they will help troubleshoot your network. If the network is still not accessible, proceed with the next section.

Check Client Applications

Firewall settings, secured connections, e-mail protocol settings, FTP, and voice over IP (VoIP) can be the problems related to client applications. Your router's security settings including firewall should allow communication with the Internet. A firewall is a software and hardware combination that either allows or blocks the information coming from the Internet to pass through your computer. It prevents malicious software or hackers from accessing your system.

Using Windows 7 OS, clicking on the following items in order will ensure the right firewall settings:

> Click "Start," "Control panel," "System and security," windows firewall, "Change notification settings,"
> Turn windows firewall on. Click "Notify me when windows firewall blocks a new program," "OK,"
> "Use recommended settings," "Restore defaults," "Yes."

In order to block programs on specific ports of the computer, open "Advanced settings" under "Firewall" settings. Click "Windows Firewall with Advanced Security" where you will be able to enable/disable inbound/outbound rules and create new rules.

Proxy servers are used by ISPs to speed up the Internet access. It also protects the networks like firewall and forces the employees of the corporation to follow company rules. If you cannot connect to the Internet using https (secured connections), using the wrong proxy server might be the problem. The network administrator should be able to assign the correct proxy server and provide secure connections to the employee.

In order to establish e-mail connections, by checking the ISP's website you can find the names of the incoming and outgoing e-mail servers and protocols being used. For example, if POP protocol for incoming mail and SMTP protocol for the outgoing mail are used, make sure you verify your settings accordingly.

FTP sites can be directly accessed by typing a uniform resource locator (URL) that starts with ftp instead of http. If the FTP site is not accessible, take the following steps:

> Add ports 20, 21 to Windows Firewall.
> Ping the FTP server to check connectivity.
> Check with the network administrator for your permissions to access the FTP site.

Through VoIP, voice is converted to digital data. VoIP is also connected to plain old telephone service (POTS) so that customers who do not have VoIP will still be able to communicate. Using the website of the VoIP provider, make sure you configure the VoIP service by obtaining an IP address from the DHCP server.

Troubleshooting Example for Routers

Figure 11.20 demonstrates a topology where network connectivity needs to be tested between two computers PC0 and PC1 that are connected through two routers representing a WAN connection. Figure 11.21 shows the IP addresses, subnet masks for the PCs, and the ports/interfaces of the routers R0 and R1. In order to have connectivity between PC0 and PC1, all the IP addresses, subnet masks, and routing

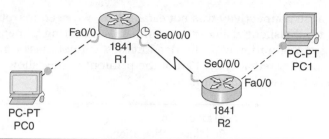

FIGURE 11.20 Network topology example.

Device	Interface	IP Address	S. M.	Default Gateway
R1	faØ/Ø	192.168.1.1	255.255.255.0	N/A
	serialØ/Ø/Ø	192.168.2.1	255.255.255.0	N/A
R2	faØØ	192.168.3.1	255.255.255.0	N/A
	serialØ/Ø/Ø	192.168.2.2	255.255.255.0	N/A
PCØ	NIC	192.168.1.2	255.255.0.0	172.16.3.1
PC1	NIC	192.168.3.2	255.255.0.0	172.16.1.1

FIGURE 11.21 IP addresses and subnet masks for Fig. 11.20.

protocols need to be configured for the routers. For simplicity, routing information protocol (RIP) has been selected to implement in both routers.

RIP requires the following commands to be entered in each router:

```
Entered in global config mode:
RouterName(config)#router rip
Entered in router config mode:
RouterName(config-router)#network [major network address]
```
(Courtesy of CISCO a registered trademark.)

First, we check connectivity by pinging from IP address of PC0 to PC1, which is 192.168.3.2 as follows:

```
PC0>ping 192.168.3.2
Pinging 192.168.3.2 with 32 bytes of data:
Reply from 192.168.1.1: Destination host unreachable.
Reply from 192.168.1.1: Destination host unreachable.
Reply from 192.168.1.1: Destination host unreachable.
Reply from 192.168.1.1: Destination host unreachable.
Ping statistics for 192.168.3.2:
Packets: Sent = 4, Received = 0, Lost = 4 (100% loss).
```

Because connectivity was not established, we need to troubleshoot the network. We can troubleshoot, using show commands. Entering "show run" will show the running configuration with the IP addresses, subnet masks, and routing protocols that are configured already. Part of the printout of the "show run" command for "Router1(R1)" is shown as follows:

```
R1#show run
Building configuration...
Current configuration : 612 bytes
version 12.4
no service timestamps log datetime msec
no service timestamps debug datetime msec
no service password-encryption
hostname R1
spanning-tree mode pvst
interface FastEthernet0/0
ip address 192.168.1.1 255.255.255.0
duplex auto
interface FastEthernet0/1
no ip address
duplex auto
speed auto
shutdown
interface Serial0/0/0
ip address 192.168.2.1 255.255.255.0
clock rate 64000
interface Serial0/0/1
no ip address
shutdown
interface Vlan1
no ip address
shutdown
ip classless
line con 0
line vty 0 4
login
end
```

(*Courtesy of CISCO a registered trademark.*)

The configuration for R1 shows that the FastEthernet 0/0, serial 0/0/0 interfaces, and subnet masks are configured. Also, observe that FastEthernet 0/1 and serial 0/0/1 are not configured. These results are expected as FastEthernet 0/1 and serial 0/0/1 interfaces are not connected to the network. Routing protocols were not configured. RIP needs to be entered and it can be the reason that PC0 did not ping to PC1. Before we configure RIP protocol in R1, let us check the running configuration of Router2 (R2). The following is the printout of the "show run" command at R2:

```
R2#show run
Building configuration...
interface FastEthernet0/0
ip address 192.168.3.1 255.255.255.0
interface FastEthernet0/1
no ip address
duplex auto
speed auto
shutdown
interface Serial0/0/0
ip address 192.168.2.2 255.255.255.0
interface Vlan1
no ip address
shutdown
router rip
network 192.168.2.0
network 192.168.3.0
ip classless
end
```

(Courtesy of CISCO a registered trademark.)

It is important to notice that "router rip" with the network addresses are entered in the configuration, which means RIP does not need to be configured for R2. Because "router rip" is not in the R2's running configuration, go back to R1 and configure RIP by entering the following commands:

```
R1#en
R1#conf t
Enter configuration commands, one per line.  End with CNTL/Z.
R1(config)#router rip
R1(config-router)#network 192.168.2.0
R1(config-router)#network 192.168.1.0
R1(config-router)#
```

(Courtesy of CISCO a registered trademark.)

After entering the RIP protocol commands in R1 we should be able to send packets from PC0 to PC1. Now we ping from PC0 to PC1's IP address as follows:

```
PC0>ping 192.168.3.2
Pinging 192.168.3.2 with 32 bytes of data:
Reply from 192.168.3.2: bytes=32 time = 60 ms TTL = 126
Reply from 192.168.3.2: bytes=32 time = 60 ms TTL = 126
Reply from 192.168.3.2: bytes=32 time = 60 ms TTL = 126
Reply from 192.168.3.2: bytes=32 time = 60 ms TTL = 126
Ping statistics for 192.168.3.2:
    Packets: Sent = 4, Received = 4, Lost = 0 (0% loss),
Approximate round trip times in milliseconds:
    Minimum = 60 ms, Maximum = 60 ms, Average = 60 ms
```

As observed, all four packets are sent and received successfully. The network does not need any more troubleshooting. Connectivity is established between PC0 and PC1 that belong to two different LANs separated by a long-distance WAN identified by two routers.

Preventive Maintenance

For the routers and switches, the network administrator needs to make sure that all IOS software gets updated. Also, in order to keep up with the demands of technology, more RAM memory cards need to be added frequently to the routers and switches.

Inspecting connectors, cables, wireline, wireless network devices, and PCs should be part of preventive maintenance. Checking interface errors, daily "event logs" for network devices, servers, cables, and connectors is crucial for the health of the network. Fixing errors should follow up the regular inspection. Wiring closets need to be maintained properly as well.

The following are the routine tasks to be performed on the network computers:

- Download and install antivirus programs. Update them with the most recent virus definitions. Install programs to clean up worms, spyware, and Trojan horses.
- Turn on the Windows Firewall.
- Update Windows settings and install service packs properly.
- Speed up the computer performance by defragmenting the hard drive.
- Change the administrator password frequently.
- Use "Chkdsk" utility in order to repair file system errors.
- Remove the unwanted programs by clicking on the "Add and Remove Programs" window.
- Clean the hard drive and keep about a quarter more of the drive free.

- Back up data by accessing "Back up your computer" in control panel.
- Click "View event logs" under "Administrative Tools" in order to observe whether there has been any unauthorized attempt to access the computer.
- Document the resolved issues and the issues to be resolved within a certain timeframe.
- Manage folders and files efficiently.
- Monitor CPU and memory utilization.
- Dust the PCs periodically, especially keyboards.
- Keep the print heads of the printers clean.
- Keep stock of network devices and components.

As a result of preventive network maintenance, the following items are achieved:

- Increased life of the network devices and components
- Efficient use of employees
- Improved quality and safety of networks
- Reduced network downtime
- Lower maintenance and repair costs

Self-Examination

Select the best answer:

1. In a server-based network
   A. Each station acts like a server
   B. There is no centralized station to manage network operations
   C. A server stores all the network-shared files
   D. There is no communication between server and client
2. In which topology are the computers connected to a central hub?
   A. Ring
   B. Star
   C. Bus
   D. Mesh
3. In which topology are all the computers connected to each other without a network device?
   A. Ring
   B. Star
   C. Bus
   D. Mesh
4. Which cable does not have metal shielding and is most susceptible to noise?
   A. Coaxial
   B. Optical fibers
   C. STP
   D. UTP

5. Coaxial cables have the following properties:
   A. Pulse spreading occurs.
   B. Dielectric material exists between the center and the shielding.
   C. Light propagates in one direction only.
   D. Infrared light is the major way of transmission.
6. The best way light travels is in
   A. Step index
   B. Multimode
   C. Single-mode step index
   D. Multimode graded index
7. In which cable does the signal travel up to a speed of 10 Mbps and up to 185 m?
   A. 100 Base 5
   B. 1000 Base T
   C. 10 Base 2
   D. 10 Base 5
8. Which protocol describes best the phenomenon where colliding PCs transmit according to back-off algorithm?
   A. Logical link control (LLC)
   B. Token ring
   C. CSMA/CD
   D. Wireless
9. Hubs are used at which layer of the OSI model?
   A. Data link
   B. Physical
   C. Network
   D. Transport
10. Dividing data into smaller pieces for transmission is called
    A. Encapsulation
    B. Multicasting
    C. Contention
    D. Segmentation
11. The following network device increases bandwidth and creates collision domains, but it does not create broadcast domains:
    A. Hub
    B. Bridge
    C. Switch
    D. Router
12. Which network device is a network layer device and creates broadcast domains?
    A. Hub
    B. Bridge
    C. Switch
    D. Router

13. Troubleshooting of a network is very easy to implement without a network topology.
    A. True
    B. False
14. For troubleshooting, the minimum amount of information in the "table of the end-system configuration" that should be included is
    A. Router and switch commands
    B. Type of connectors
    C. Operating system and version
    D. Network cable types
15. In Fig. 11.20 PC0 pings 192.168.2.2 that is serial 0/0/0 port of R2. The ping is unsuccessful. Assuming routing protocols and IP addresses are configured properly, what could be the reason of the connection failure?
    A. The cable between PC1 and R2 is not connected.
    B. Serial cable is not connected.
    C. PC1's IP address is not correct.
    D. R2's FastEthernet 0/0 IP address is not correct.
16. For troubleshooting purposes, "baseline report" for the network shows the following:
    A. Normal utilization of data transmission
    B. Average throughput of data transmission
    C. Physical connectivity of the network
    D. All of the above
17. Troubleshooting the physical layer includes
    A. Enabling STP
    B. Finding the redundant paths through exploring network topology
    C. Confirming routing protocols are applied to all neighbors
    D. Organizing the wiring closet by tagging all cables and wires, then documenting them
18. The following item is included as part of the physical layer troubleshooting:
    A. Examining routing protocols.
    B. Amber LED link lights are on for the switches.
    C. Checking to see whether crossover, straight-through, and serial cables are installed correctly.
19. For troubleshooting, protocol analyzers perform the following task(s):
    A. Check the voltage, resistance, and current levels of the cables.
    B. Test copper and fiber cables.
    C. Display crosstalk and impedance behaviors.
    D. Display network traffic including packets and frames.
20. A good rule of thumb for troubleshooting networks is to check for software problems first and then test for hardware problems.
    A. True
    B. False

Questions and Problems

1. Name the most important network architectures.
2. What are the advantages and disadvantages of bus topology?
3. Describe one of the steps of troubleshooting network layer problems.
4. Why exploring the network topology to find out the redundant data paths and reconfiguring the STP protocol is important?
5. According to the flowchart in Fig. 11.19, what are the major issues to explore in sequential order?
6. What does the blinking light at the Ethernet port of the PC indicate?
7. For troubleshooting, what do you check that is related to the cable connection between the PC and the router?
8. What is the procedure to check a network adapter and/or a NIC card?
9. After renewing the IP address, to check connectivity, what do you observe when you enter "ipconfig/all" in the command prompt of the PC?
10. State the purpose of DHCP server?
11. Explain what you observe after entering "ping 127.0.0.1" command in the command prompt of the PC.
12. Which five problems relate to client applications?
13. State the purposes of installing a firewall for the router.
14. What is the purpose of a proxy server?
15. How do you check POP and SMTP protocols in order to troubleshoot the e-mail operations?
16. State the steps to take in order to troubleshoot the FTP connectivity.
17. In Fig. 11.20, why is RIP used?
18. In the textbox for R1 running configuration, what is the reason that FastEthernet 0/1 and serial 0/0/1 interfaces do not have IP addresses?
19. State the characteristics of STP cables?
20. Which of the three types of fiber-optic cables is the best to use and why? Explain your answer.
21. Which type of fiber is the next best to use and why?
22. What are the network devices used at the physical, data link, and network layer's of the OSI model?
23. Describe MAC and IP addresses and their purposes.
24. Explain the advantages and disadvantages of using switches as network devices.
25. Why is a routing protocol needed to configure a router?
26. The IP address of a network is 199.1.10.0 and the subnet mask is 255.255.255.224. Up to how many subnets can be allocated in the system? Show your calculations.
27. In question 26, how many hosts can be grouped together for each subnet? Show your calculations.
28. Explain the tasks that can be performed in the "global configuration" mode of a router.
29. What are the tasks that can be performed in the "interface configuration" mode of a router?
30. State the purpose(s) of using VLANs in switches.

Chapter 12

Troubleshooting Embedded Microprocessor Systems

Embedded Systems

Embedded systems are computer systems that are designed to implement specified applications. Unlike typical computers that are flexible and can generally perform a wide range of applications, the embedded system typically contains a microcontroller or processor; it is efficiently designed and often mass-produced for monitoring and control operations. They are used for numerous applications and range from simplistic embedded microcontroller to sophisticated arrays of computers and networks.

Some typical applications of embedded systems range from consumer and home products such as appliances, dishwashers, refrigerators, and climate controls, to commercial applications such as vehicle controls, aviation, global positioning systems (GPSs), medical monitoring and diagnostic equipment, and automotive controls and braking systems. They also can be found in entertainment and personal applications such as video games, recording devices, DVDs, digital cameras, and smart phones. Chapter 9 on microprocessors covers how programmable devices enable easier redesign and provide the ability to update and upgrade performance without changing the circuit or its components. One example in Chap. 9 shows an array of logic gates replaced by a microcontroller and a simple program. A microcontroller is a special case of a microprocessor where the memory and some I/O ports are built into the chip, instead of being added on outside of the system.

In a computer system, the motherboard is a circuit board that contains the central processing unit (CPU), memory, and connections for peripherals that are also called *input/output* (I/O) devices. "Daughterboards" are attached to the motherboard and they contain such things as expanded memory, circuitry for video, or interfaces to the Internet. In a system that uses a microcontroller, most of the functions of the motherboard are on one chip. In Fig. 12.1, you see a Zilog® microcontroller that is optimized for motor control. It contains all the features of

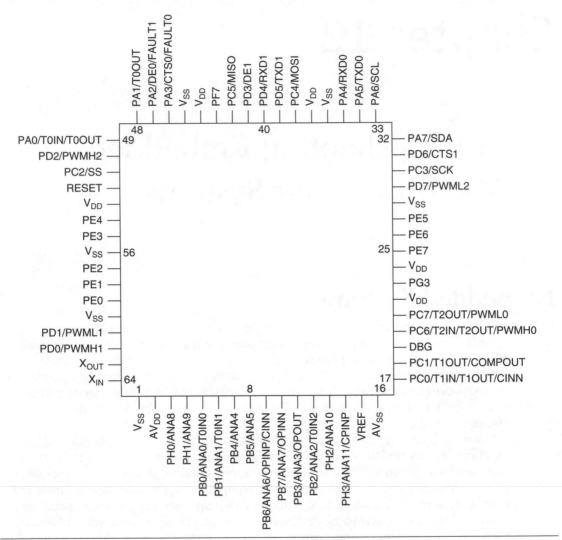

FIGURE 12.1 Zilog microcontroller. (*Courtesy of Zilog Corp.*)

a microcontroller and it is similar to a microprocessor. There are no pins for an address or data bus shown because the memory and I/O connections are actually inside the chip and not on a daughterboard.

Some of the pins that are identified in Fig. 12.1 are multiplexed. Multiplexing means that a pin like PC0/T1IN/T1OUT/CINN (pin 17) is part of an I/O port and it has three alternate functions. At any given moment, the pin could be connected to any of four different circuits inside the microcontroller. Depending on the control signals used, it could be the least significant bit of port C (PC0), an input to timer T1, an output from timer T1, or a negative input to a comparator (CINN).

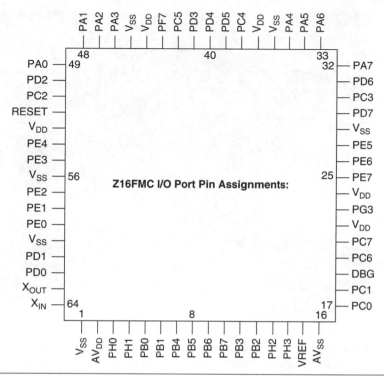

FIGURE 12.2 Z16FMC I/O port pin assignments. (*Courtesy of Zilog Corp.*)

If you just look at the pins that are used for general-purpose input/output (GPIO), with the alternate functions removed, a simplified picture of the chip looks like Fig. 12.2. You can see ports PA, PB, PC, PD, PE, and PH. Each one of these is an 8-bit port as shown in Fig. 9.4. Inside the chip, 8-bit registers called *data direction registers* (DDR) are loaded with bits that program each bit to be an input or an output. Ports F (PF7) and G (PG3) at pins 23 and 43 are 1-bit I/O pins and are not multiplexed to perform any other functions.

The Z16FMC integrated circuit (IC) chip has 64 pins and it is not a dual in-line package (DIP) with 32 pins on each side. The Z16FMC is a square block with surface-mount (flat-pack) pins. These pins are not plugged into a socket but instead are soldered-down flat onto flat conductors on the surface of a printed circuit board. The Z16FMC flat pack has 64 terminals around the outside of a square package and 16 on each side.

In Fig. 12.3 pin 1 is identified with an indented circle or dimple at the corner of the chip. Pins are counted upward from 1 to 64 going counterclockwise, so the pins at the bottom of the picture would start at 1 in the corner and going to the right, would count up to 16, continuing around the chip and ending at 64 as shown on the diagram. Some manufacturers use other methods of identifying pin 1, and the pins are counted the same way, counterclockwise from pin 1.

FIGURE 12.3 64-pin flat pack.

The Z16FMC IC is specialized for motor control, but it has features that could make it adaptable for many other uses. In addition to containing memory and I/O, it has several other features that are usually added on as separate chips or cards within a system. These features include the following:

- Flash memory: Z16FMC contains 128 kilobytes (kB) of the same kind of nonvolatile memory found in a memory stick/thumb drive. Flash memory can be programmed by the user, and if the power fails, it will remember the stored data. This allows the system to survive power interruptions without loss of code or data.
- 10-bit analog-to-digital converter (ADC): Z16FMC can be connected directly to sensors that have voltage outputs proportional to temperature, pressure, light intensity, or other measurements of real-world variables.
- An analog comparator and an operational amplifier: The comparator can respond to a signal that reaches a certain threshold without requiring a program coded to do the comparison. If a signal from some transducer is too weak for the ADC's input, the signal can be amplified by the operational-amplifier (op-amp) before it is applied to the input of that converter.
- Universal asynchronous receiver-transmitters (UART): This feature allows serial data to be received or transmitted by doing parallel-to-serial and serial-to-parallel conversion.
- Infrared (IR) encoder-decoders (ENDECs): ENDEC can be used with the UART to send and receive serial data using infrared between microcontroller and other equipment. For example, smart phones with transfer capability use IR protocols and they could also be used for distributed processing where a large number of processors have to be coordinated wirelessly.

- Direct memory access (DMA) controller: DMA controller allows transfer of data into or out of memory by a hardware signal, without the use of a program to carry out the transfer of data.
- Pulse-width modulator (PWM): PWM provides a signal with varying duty cycles to switch. This feature provides control of the three-phase electric motors by varying amounts of power delivered to the motor.
- Timers: Timers are used for timing and counting event durations. In combination with the PWM, timers would allow the motor speed to vary in specific patterns. An application of timers is the walking gait of a two-legged robot climbing a flight of stairs.
- Priority interrupt controllers: These controllers allow a hardware signal to force specialized interrupt software to run if a sensor detects an exceptional condition. For instance, if the robot climbing the stairs lost its balance, a gravity sensor could kick in an emergency program to make it grab for a railing before it falls.
- Crystal oscillator: This internal circuit can provide a frequency from 32 kilohertz (kHz) to 20 megahertz (MHz). It could output frequencies all the way from an audio tone (beep) for a speaker, to a radio wave frequency for wireless transmission via an antenna.
- On-chip debugger: This feature allows programming of the chip and analysis of software performance without an external emulator or analysis unit.

All these features are on the Z16FMC chip. In a PC all these features would require additional chips, several cards, boards, or a large number of circuits on the motherboard. Not all microcontrollers are designed to have the same features that Z16FMC contains. For any specific microcontroller, you will need to download or order the data sheet from the manufacturer. Read the data sheet to see what capabilities the chip has and what the timing diagrams for various outputs look like, before attempting to analyze a system with an embedded processor. Just like the microprocessor testing, microcontrollers may use a DMA to identify problems with the hardware or software.

Data Sheet Information for the Troubleshooter and System Documentation

Approximately only 10 percent of the data sheet provides relevant information for troubleshooting. Data sheets are used mainly by designers and programmers to build and configure a system around the chip. As a troubleshooter, you already have a system that is designed and you only need a part of the information on the data sheet such as

- Timing diagrams: These diagrams show a timeline of what happens at the outputs as a result of certain inputs.
- Waveforms: These show analog signals that are not just On or Off, as is the case with the logic outputs of timing diagrams. For instance, DACs will have these kinds of output waveforms.

- Pin configuration diagram(s): For the package used in your design, this diagram tells you the function of the pins.
- Electrical characteristics: Electrical measurements associated with normal operation which includes many timing diagrams and waveforms.
- The on-chip debugger provides essential information about how Z16FMC's features are used to analyze problems.

Designers and programmers need to document how the microcontroller functions within the system. When a problem occurs with the outlying ports of the system, the documentation will be essential for troubleshooting and debugging.

The system designers may not have to use all the features of a chip. For example, if a Z16FMC chip is used in a communications panel, its motor control functions are not used, but the UART and IR ENDECs functions are needed primarily. If the microcontroller is embedded in a microwave oven, its pulse-width modulation feature may be used to adjust the power level of the microwave cooking intensity, instead of the speed of a motor. In a smart phone or video game design, the microcontroller might need an additional amount of on-board graphics and video random access memory (RAM).

Troubleshooting Comparison of Microcontroller-to Microprocessor-Based Systems

Troubleshooting a microcontroller-based system would be similar to troubleshooting a microprocessor-based system, with a few exceptions as follows:

- As Z16FMCs memory is included inside the chip, there would be no external address bus and data bus connections to a memory card, and no cables or connections to check between these two parts of the system. As a result, when a memory fault occurs, the microcontroller chip has to be replaced.
- As stated above there are no address and data bus connections to the chip's on-chip I/O ports. For example, if port A fails to work on all pins, the microcontroller chip must be replaced as there are no loose cables or connectors to replace.
- If the microcontroller has a built-in clock oscillator, you may not even be able to see the clock-pulse waveform on any of its pins. A clock oscillator cannot be replaced when it fails; instead you would have to replace the microcontroller chip.
- Adding memory may not be an option. This is not likely to matter unless there is a software update for the system that is not compatible with all versions of the chip. For example, the Z16FMC exists in three package styles: Z16FMC32, Z16FMC64, and Z16FMC28. The first two contain 32 kB and 64 kB of flash memory, whereas the third contains 128 kB. If the updated software loads into, for example 100 kB of memory, there is no way to add external memory to a system that contains a Z16FMC64. The CPU itself will have to be replaced with the Z16FMC28 chip, one that has twice as much flash memory.

Timing Diagrams for a Microprocessor

Timing diagrams for microprocessor or microcontroller systems are generally more complicated to read than for simpler digital systems. All timing diagrams are essentially graphs of voltage versus time, like a trace on an oscilloscope or logic analyzer. Figure 12.4 shows an example of a modulus-16 (mod-16) counter called a ripple counter. The only input shown on this diagram is the clock (CLK). For simplicity the preset and clear inputs are not shown and are presumed to be inactive. The four outputs Q_0, Q_1, Q_2, and Q_3 can be seen, with each output below the previously numbered one. The lower part of Fig. 12.4 shows the binary numbers at the outputs; reading from top to bottom, each binary number from 0000 to 1111 appears from left to right following the rising edge of each clock pulse.

Now, compare the simple timing diagram from the counter of Fig. 12.4 with one for the Z80180 8-bit (data) microprocessor with the 19-bit address bus shown in Fig. 12.5. Read the diagram from left to right just like the counter figure. In this case, there are five control signals M1, MREQ, IORQ, RD, and WR, and four operations are depicted. First, the program counter of the CPU fetches in an instruction from memory. Following that is an output-port operation, presumably because the instruction caused

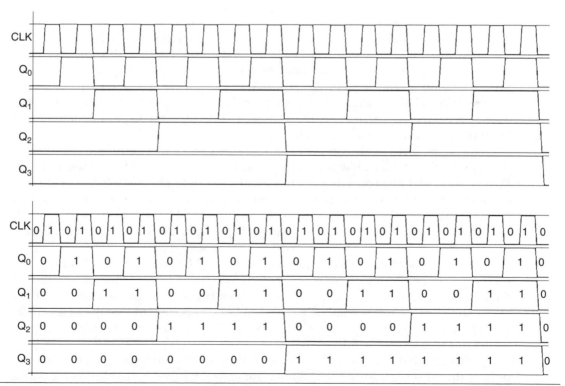

FIGURE 12.4   Timing diagram for a mod-16 counter.

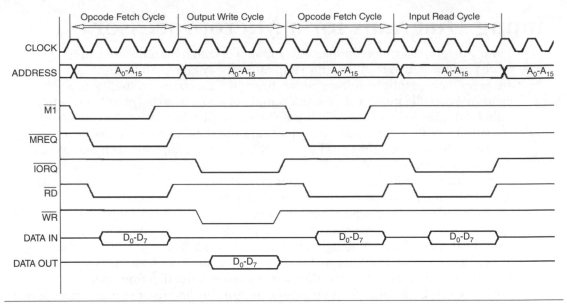

FIGURE 12.5 Timing diagram for an 8-bit microprocessor.

an output. Then the program counter fetches another instruction in machine code from memory and causes an input-port operation in the following cycle.

Notice that all the control signals; M1, MREQ, IORQ, RD, and WR are active-low. At the beginning, M1 is low, indicating an instruction is being fetched by the program counter. That signal triggers the ADDRESS to change. MREQ indicates that the memory is being used, and that is where the instruction is coming from. The only other input that is active is RD, indicating that the data bus will be taking DATA IN to the computer's instruction decoder.

The signal lines on the timing diagram represent all 19 lines of the ADDRESS bus. Where the high and low lines crisscross, it means all 19 address lines of a 19-wire bus have just had a new set of 1s and 0s loaded. The address is a 19-bit number that comes from the program counter in the CPU. As the counter counts up, instructions are fetched from consecutive-numbered addresses so the next time M1 becomes active, another instruction will probably be fetched from the next address in memory.

On the DATA IN line of the diagram, when the 8 bits of the instruction in memory are fetched, there are really eight pins or lines carrying the bits of the operation code for the instruction (opcode). The DATA IN is inactive until the fetch cycle, so the place where the line pops open into D_0–D_7 means that an 8-bit number has appeared on those pins and will be decoded inside the CPU to carry out the action of the next instruction.

Following that fetch cycle, there is an output cycle. During those four clock pulses, the IORQ and write (WR) signals are active, indicating that I/O is requested to write DATA OUT from the CPU to an outside device. The ADDRESS changes to the address of an output port. This number is unrelated to the address that the opcode was fetched from. Data D_0–D_7 appears again, but the data pins are working in the outbound direction instead of inward, so this data appears on the line called DATA OUT. DATA IN and DATA OUT are actually measured on the same eight pins of the processor. For DATA OUT, data is flowing out of the CPU to a port instead of flowing into the CPU from a memory device.

Another instruction is fetched in during the next four clock pulses. M1, MREQ, and RD are all active just like they were in the first four clock pulses at the beginning of the diagram. It looks exactly like the last opcode fetch cycle, but the ADDRESS is a different number, because the program counter has counted up.

Finally, an input operation takes place in the last cycle of four clock pulses. The DATA IN shows another burst of 8-bit data appears on the data bus lines. It looks like the instruction fetch, except MREQ and M1 are not active, so the data is not an opcode coming from memory. Instead, IORQ and read (RD) are active, which shows the direction of data is DATA IN. Just like you read and take "data in" through your eyes, the CPU terminology uses the same term RD for "taking in data."

The point here is that a timing diagram for a microprocessor or microcontroller cannot be like one for a 4-bit counter, because too many pins are involved. Because a diagram with each pin's state would make the figure too crowded, the ADDRESS, DATA IN, and DATA OUT lines on the diagram are an attempt to represent many signals all at once.

Timing Diagrams for an Embedded Processor or Microcontroller

A standard microprocessor can be built into a controller product as its embedded processor. A controller is simply a digital or analog device that controls a process or an aspect of its environment. Microcontrollers have more peripherals on the chip. A design that uses a microcontroller instead of a microprocessor will be more compact and have fewer failure points.

Timing diagrams for microcontrollers are very different from those of microprocessors. Many microcontroller applications do not use any external address and data bus connections. The microcontroller timing diagrams of Fig. 12.6 are very different from the timing diagrams for the Z80180 in Fig. 12.5.

In Fig. 12.6, during clock pulses 1 to 17, microcontrollers with four built-in GPIO ports are shown. The first four operations are inputs from ports A, B, C, and D. The times between the inputs, instructions are being fetched and executed. Because the program memory is inside the CPU, no address or data signals are visible that can identify the status of instruction fetches from memory. During clock pulses to 17 signals RD and IORQ are active, but WR is not. This indicates that I/O is being used and that data is being taken in to the CPU from the outside world or read into the CPU from the ports.

During clock pulses 18 to 33, the next four operations are outputs to ports A, B, C, and D, consecutively. Now signals IORQ and WR are active, but RD is not.

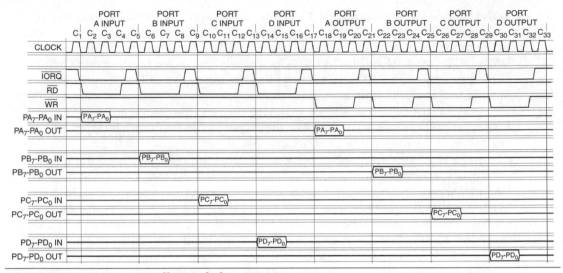

FIGURE 12.6   Microcontroller with four GPIO ports.

This indicates that I/O is being used, but data is traveling outward or being written from the CPU to the ports. This data is processed to be used in the outside world.

These timing and waveform diagrams are intended to depict what is going on inside the microprocessor or microcontroller products. They are of limited value when compared to the actual measurements that a troubleshooter might see when trying to repair or diagnose a system with these components.

Waveforms from Real Test Instruments

Most modern test equipment is programmable and they multiplex their pins to play many different roles. The pin you are looking at for a data signal may be a part-time control signal. Data bus lines that are carrying output port signals at one moment may be address bus lines a moment later, all coordinated by changes in control signals. Depending on whether there is a "memory request," "I/O request," "address latch enable," "read," or "write," the same pin might be performing one of the six different functions.

An added degree of complexity comes from what the pins do when they are not being used at all. Multiplexing in most of these chips is done using three states of logic, namely:

1. Logic 0 or low: Where the pin is electrically pulled down to a low potential voltage near zero.
2. Logic 1 or high: Where the pin is electrically pulled up to a higher-potential voltage, perhaps 3 volt (V), 5 V, 12 V, or whatever level positive voltage is used in the system for powering the chip.

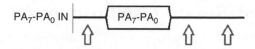

FIGURE 12.7   Port A input for the microcontroller.

3. A high-impedance (high-Z) state: Where the pin is not being pulled into either a low- or high-potential voltage. Instead, it has a very high resistance or open-collector situation where none of the transistors attached to that pin are in a conducting state. High-Z causes the most trouble comparing signals on a real chip's terminals with the signals represented on the timing diagrams in the data sheet.

The real waveforms observed on an oscilloscope or logic analyzer will be different from the theoretical diagrams used in Figs. 12.5 and 12.6.

Figure 12.7 shows the part of Fig. 12.6 where the input from port A takes place. The parts marked with the arrows are high-Z intervals between input operations. If a buffer gate array from outside were attached so it would pass data into the eight pins of port A, it would be disabled at this time and no input signals (high or low) would be presented to these pins.

Figure 12.8 depicts a real oscilloscope trace you will see instead of the theoretical Fig. 12.7. The features of Fig. 12.7 differ from reality in three ways:

1. The instrument is really only looking at one input pin at a time. In this case, it is pin 7 of input port A shown in Fig. 12.8.
2. That pin is really only high or low at the time it is active. In between the arrows, the pin is pulled low during the time the input is taking place.
3. The flat parts of Fig. 12.7, between active input events on that pin, are in reality not flat. What is shown in Fig. 12.8 is electrical noise. The noise is so far from flat that some of the voltages in the waveform are high enough to be legitimate 1s and others are low enough to be legitimate 0s.

If the gate is "broken" when it is supposed to drive the pin 7 low, it can still register a high or low signal. The signal is due to the noise present at the last moment when the input pin is sampled. This will cause intermittent behavior that may almost resemble normal operation, except that the outputs will not behave the way they would if the system were receiving its normal inputs.

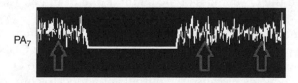

FIGURE 12.8   Oscilloscope trace of the waveform for bit 7 at PA_7.

Recurrent-Sweep Oscilloscopes

The earliest type of oscilloscope would sweep a spot of light from left to right for an interval of time. The process is repeated from the left again, at a rate (frequency) set with a control knob by the user.

A varying voltage applied to the vertical input will move the spot of light up and down a distance proportional to the number and polarity of volts at the input as it moved steadily across the screen. That would produce a trace identical to a graph of voltage versus time. The input's sensitivity would be controlled with a gain knob for an amplifier, to adjust the size of the graph. As the spot where the light would remain glowing for a while, you could see the whole trace from left to right as a wavy line across the screen of the scope.

This kind of oscilloscope is useful to examine repeating patterns. A good example is a sine-wave voltage from an alternate-current (ac) electric generator, which is exactly the same sequence of voltages, repeated at an unvarying rate. You would twiddle the time-base adjusting knob until the sweep rate of the oscilloscope is as close as possible to the repetition rate of the voltage being measured. Then, it would only drift to the left or right slowly, so you could examine the waveform for glitches or noise. This instrument is also useful in medicine for looking at the electrical signals in a heartbeat, which repeats at a fairly steady rate. Recurrent-sweep oscilloscopes are not useful for complex waveforms such as a serial data stream or an unsteady heartbeat.

Triggered-Sweep Oscilloscopes

Shortly after the recurrent-sweep oscilloscope was developed, a triggering circuit was added to the horizontal sweep circuit. The "trigger" would allow a new sweep to take place "prematurely," before the oscilloscope reached the normal end of its sweep. By adjusting the voltage threshold to trigger a new sweep, the scope could produce a waveform "locked" to a repeating event. As a result, you no longer have to "fiddle" with a sweep control to get a stable waveform on the scope.

At first, this triggering feature was at the same input that produced the vertical deflection of the wave pattern, and it was adjusted to be "sensitive" to a voltage at the triggering level nearby to the time of the normal end of the sweep. Later, a separate control input for triggering was added to the scope design, and it could be triggered by a level or edge that was separate from the input. Additional channels displaying multiple lines made it possible, for instance, to display more than one trace. If the falling edge of Q_3 is used for the trigger on a two-channel scope, the CLOCK and the Q_2 outputs of the counter circuit in Fig. 12.4 can be displayed on the screen.

With the setup described above, the correct adjustment would produce a new trace once every 16 clock cycles. The channel with the CLOCK input would show 16 square wave cycles, whereas the channel attached to Q_2 would show two full square wave cycles. Because the outputs of a mod-16 counter repeat identically every 16 clock cycles, this pattern would be stable and would hold still on the screen,

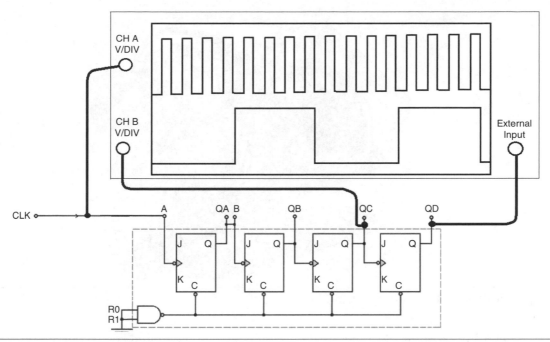

FIGURE 12.9   Clock and Q_2 displays of a mod-16 counter.

looking like a static graph of the timing diagram for that counter. Figure 12.9 shows part of the timing diagram for the counter. With only two inputs and two channels, you would have less than half of displays shown in Fig. 12.4.

Logic Analyzers

To troubleshoot a data or an address bus, it would be useful to have oscilloscopes that display multiple channels such as 8, 16, or 24. If you are going to examine the signals of digital system such as a microcontroller, graphing a lot of detail between the bottom and top of the waveform may not be as critical. Digital waveforms are generally square waves; either on or off, they have generally flat lines between transitions from one state to another.

A logic analyzer is essentially a special-purpose computer running a program that analyzes the data. It has the ability to represent the signals realistically like an oscilloscope, or symbolically, to appear like timing diagrams rather than oscilloscope traces. A triggering event starts the input connectors to sample data from many inputs for a specific interval of time, and then displays everything that is sampled.

Software analyzer programs exist for desktop and laptop computers that permit them to use the same "pods" that connect an analyzer to a circuit being tested, and display similar results on a more portable machine. The pods in Fig. 12.10 connect eight lines each to an input on the analyzer or to a PC programmed to act like an analyzer. Figure 12.10 shows clips that grab the sides of IC chips on the board and

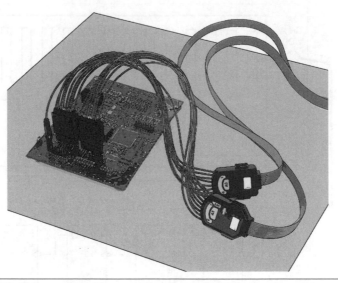

FIGURE 12.10 Pod figure.

allow the reading of 16 signals at once, plus a common ground connection for a current return path.

The biggest difference between analyzer programs that run on a PC and the actual logic analyzer is that the PC software versions run more slowly and cannot take as many samples in the same amount of time. However, the PC or laptop may cost less than an analyzer.

Digital Storage Oscilloscopes

Somewhere in between the dual-trace triggered oscilloscope and the logic analyzer is the digital storage oscilloscope (DSO). A storage oscilloscope is triggered by an electrical event at its trigger input, and it begins sampling its inputs, storing the samples in memory. At a preset time, the scope ends its sampling and displays the waveform(s) on the display, as a "snapshot" of what happened in the interval of time after it was triggered.

Unlike the dual-trace scope in Fig. 12.9, this machine in Fig. 12.11 is digital, converting each sample voltage from analog into digital and storing the consecutive samples in digital form. Also unlike the triggered-sweep scope, it does not scan the screen repeatedly, which would show the waveform changing in time over many repetitions. It only "plays back" what happened during that one interval of time unless it is programmed to re-trigger and display a set samples of another waveform.

Although its display looks much like that of the triggered-sweep oscilloscope, DSO does a software-driven presentation of the samples it collects. It can also apply analysis and timing information to the display which would be impossible for an analog oscilloscope. The sampling rate and analog-to-digital conversion (ADC)

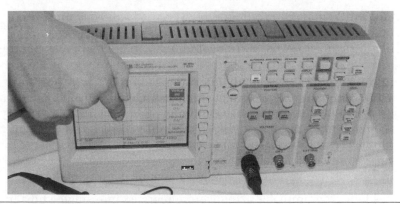

FIGURE 12.11 Digital storage oscilloscope (DSO).

rates limit the amount of samples DSO can take in a microsecond or a nanosecond. This sampling/conversion rate will not produce a reasonable replica of the voltage waveform applied unless the frequency of the sampling is at least twice as large as the cycle repetition rate of the input waveform. At least two samples are needed for a data waveform that must be high and low during one cycle.

Signature Analyzers

In the early days of microcomputers, a way to test digital signals from an input was to sample the digital 1s and 0s into a machine that would store data during a sampling period of 16 clock cycles. It would use a triggering event like the triggered oscilloscope, but instead of displaying a square wave on a screen, the set of samples would be shifted into a 16-bit shift register and displayed as four hexadecimal numbers.

The test instrument shown in Fig. 12.12 would be used to take sample measurements at various points on a microprocessor board running a repetitive program (a program loop). The output or "signature" would be compared with the "signature" at the same point on a known good board running the test loop. On earlier and simpler microprocessor systems, this was a fairly good piece of test equipment for tracing faults at the board level.

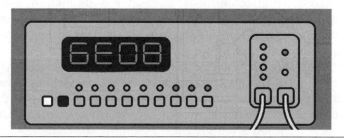

FIGURE 12.12 Signature analyzer.

Other Programmable Devices

Microprocessors and microcontrollers are not the only types of programmable devices that could be embedded in digital systems. Chapter 9 covers software running on a processor that could replace single-purpose hardware made of gates. For most problems, single-purpose hardware made of gates is generally faster than software solutions. It is not necessary for the hardware to be built gate by gate and assembled by wiring it together; there is also a programming solution.

PALs

Gates in a programmable array were integrated into single chips in the late 1970s. These chips called *programmable array logic* (PAL) devices contained NOT, AND, and OR gates with true and inverted inputs connected to fuses. These fuses were burned out to program the connected and unconnected inputs to the gates. The same fuse-burning technology was used in early programmable read-only memory (PROM) to store 1s and 0s. In both cases programming 1s and 0s into the devices was called "burning-in" the data.

Any logic-gate truth table can be implemented as a Boolean algebra logic equation with a number of terms multiplied in products of AND logic added together in a sum of OR logic. This logic equation is called *sum of products* (SOP). In a PAL, a simplified logic cell might look like Fig. 12.13.

In Fig. 12.13a, an unprogrammed SOP cell is shown. There are 16 fuses connecting each of four AND gates to two inputs and their inverted forms. The outputs of the AND gates go to the inputs of an OR gate. The OR gate can combine the sum of the four products (the outputs of the two input AND gates), each product having two inputs each.

In Fig. 12.13b, the cell has been programmed to copy the truth table of a two-input XOR gate: A XOR B that underutilizes the capability of that cell. Only two of the AND gates are used, so they are the only ones programmed. The others are tied LOW at the inputs, which means they output 0 and contribute nothing to the logic sum at the output of the OR gate.

The two gates at the top of Fig. 12.13b have fuses burned-out, and these inputs are pulled up by the resistors to HIGH logic states. As a result, the first AND gate (on top) will have a high output only when A and not-B are high; the second AND gate will have a high output when B and not-A are high. At the OR gate, the output goes high if A and not-B or B and not-A are high. This is the condition for the EXCLUSIVE-OR (XOR) gate. Therefore, in Fig. 12.13b, the whole cell acts like an XOR gate with two inputs.

PAL assembly language is used by the programmer to blow out the four fuses connected to the top two AND gates. This language is programmed using the SOP Boolean equation as its input. The program interpreter turns that into a pattern of burned and unburned fuses. When the program is "burned" into the chip, the voltage is temporarily raised to a higher level than the conventional 5-V power level (the programming voltage) and then returned to 5 V afterward.

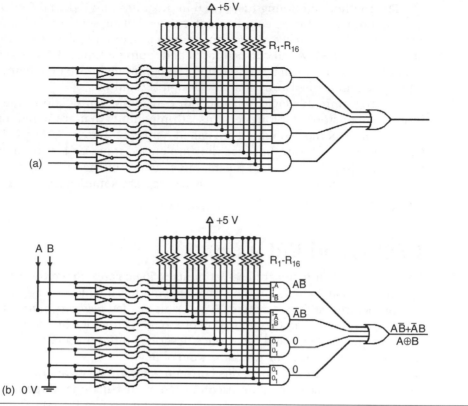

FIGURE 12.13   SOP logic cell.

Larger logic cells with more inputs and gates can be used to implement more complicated logic equations. Because every truth table for static logic can be converted into a SOP equation, the only restriction on the complexity of PAL is how many gates, resistors, and fuses can fit on the chip.

CPLDs

Complex programmable logic devices (CPLDs) combine the PAL features of SOP circuits with programmable inputs with some sequential logic and memory storage logic devices. CPLDs are at a level of complexity between PALs and the field-programmable gate arrays (FPGAs). The only major difference between CPLDs and FPGAs is that the CPLDs use SOP like a PAL and FPGAs use the method of lookup tables (LUTs). Each CPLD is programmed differently by the designer. From a troubleshooting standpoint, how CPLDs are programmed has no significance in locating problems in the systems.

The method of burning fuses used in programming PAL and CPLD devices has been replaced by the complementary metal-oxide semiconductor (CMOS) switches. CMOS switches are controlled by a program in flash memory, which is as permanent as PROM and can be erased and reprogrammed. PALs and CPLDs with the new technology can be reused and reprogrammed instead of being programmed only once and thrown away if a new design is required.

In programmable logic, bits used for controlling connections and gate functions are stored for long-term use. Components that are wired together when the system is being assembled are called *hardware*. Bits in RAM used to run a CPU are called *software*, because they are erased if power is turned off or a new program is loaded. The term for the bits that are "burned in" like in the case of the fuse-link technology is *firmware* because they are somewhere between hardware and software.

FPGAs and FPLAs

Gate arrays are prefabricated assortments of logic gates. Since the 1980s the number of gates that can be built on the same-size chip has doubled every 18 months. Field-programmable gate arrays (FPGAs), also called *field-programmable logic arrays* (FPLAs), originated as a combination of PROM technology and logic gates on one chip. The terms *FPGA* and *FPLA* once described different products but are now interchangeable terms for the same kind of device.

In FPGA the interconnections between gates are reconfigurable, which means that they can be connected or disconnected by instructions stored in a memory device on the chip. Early designs were programmed like PROMs. Control of the interconnections to internal bus lines inside the chip are now configured using a hardware description language (HDL). HDL is specifically designed for logic design on programmable logic arrays. An example of an HDL designed specifically for FPGAs is IEEE 1364 Verilog.

As with the PAL example mentioned earlier, PROM has been replaced with flash memory in recent FPGA designs. It has the permanence of read-only memory (ROM) but can be rewritten like RAM. In an FPGA, not only the connections but the gates themselves are programmable. Using LUTs, many designs allow a single logic cell to imitate any gate with up to four inputs and one output. The interconnections between the logic cells and the behavior of each logic cell are all programmable using the HDL language.

Troubleshooting FPGAs/FPLAs

Troubleshooting an FPGA system may differ from troubleshooting a microprocessor, because an array of gates does not necessarily require a clock pulse. It may simply react to its inputs by producing an output. For example, a combination of NAND or NOR gates are used to produce a specific truth table of outputs for any given inputs. FPGAs are used in these systems that would be troubleshot like boards containing basic logic gates. The systems are then tested to see whether their outputs match the truth table constructed and designed from the gates.

Because FPGAs can include sequential components such as flip-flops and registers, a clock for sequencing these components may be part of the design. Testing the machine containing the gate array will be a lot more like testing a microprocessor or microcontroller system. Troubleshooting may involve substituting an emulator into the socket for the gate array to see whether its peripherals are operating normally or whether there is a fault in the FPGA.

FPGAs may also contain arithmetic units such as adders and multipliers, to perform arithmetic operations without a computer. Operations such as these are commonly needed for digital signal processing (DSP), and they will probably be found in systems where the FPGA is part of a telecommunications board. DSP applications also include analog signals, which are converted inside many FPGA chips by internal DAC converters. Troubleshooting analog signals such as audio may also be a part of the system analysis on these machines.

Figure 12.14 shows a FPGA containing 64 logic cells. It contains address and data bus connections for programming the controlling information for gates and connections. This FPGA is near to the complexity of a microprocessor. Many field-programmable logic devices (FPLDs) such as FPGAs require an external configuration ROM. With an external memory device, these FPLDs resemble microprocessors. With multiple modes of operation, the terminals in Fig. 12.14 have multiple uses, like the terminals in microcontrollers and microprocessors.

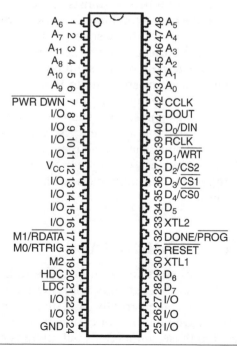

FIGURE 12.14   FPGA containing 64 logic cells.

From the designer's point of view, what the chip does inside is very different from a microcontroller/microprocessor. From the troubleshooter/signal tracer's point of view, there is almost no difference. Troubleshooting a system containing an FPLD is a matter of looking at the control signals and timing diagrams in the data sheet for comparison with the actual behavior of the chip. This procedure will ensure that the inputs are producing the expected outputs in a way that conforms to the documentation. There will still be high, low, and high-Z states at the outputs. Problems will still be more common at interconnecting plugs, sockets, and places in the system with moving parts. Power failure and/or low voltage will still cause intermittent or unreliable performance.

Because FPGAs include embedded microprocessor CPU(s), separation of gate arrays and microcontrollers is becoming harder. With the passage of time, the difference between troubleshooting FPGAs and microcontrollers is getting smaller.

Raspberry Pi

Raspberry Pi is an ultrasmall computer system with a 2¼- × 3½-inch (in) circuit board. It has a system with an ARM 700-MHz microprocessor, VideoCore technology, 256 megabytes (MB) of RAM (can be expanded to 512 MB), and a secure digital (SD) card. This is a system called "system on a chip" (SoC).

The ARM microprocessor is used in smart phones, cable televisions, and tablet computers. Fewer numbers of transistors, power efficiency, costs, and heat reduction are part of the characteristics of an ARM processor. This ultrasmall processor makes it possible for the Raspberry Pi to be used in embedded systems efficiently. VideoCore technology uses a multimedia architecture that provides video—2D and 3D graphics and high-resolution camera support at low power.

ARM allows faster clocking speed up to 1 gigahertz (GHz), which is called *overclocking*. As a result, the computer runs faster. Because Raspberry Pi does not have any hard drive, long-term storage and the booting-up process are accomplished by an SD card. A Wi-Fi adaptor or universal serial bus (USB) port is used to provide network connection. Because there is no real-time clock, the operating system such as Linux asks the user the clock information. Raspberry Pi is used in applications such as tablet computer, home appliances, space, satellite, robot, lighting, and even coffee machine controls.

Preventive Maintenance

Systems with microprocessors and with microcontrollers may differ in how many of the peripheral daughterboard components are embedded inside the CPU as well as how many are connected externally. A system such as a program memory inside the CPU chip would have no cables or connectors outside the microcontroller that need to be cleaned. If those connections fail, there is no way to replace anything but the whole CPU chip itself.

In a microcontroller-based system such as Z16FMC's motor controller, there are motherboards holding the CPU and its support chips and daughterboards containing components such as the gate drivers, sensors, and power metal-oxide semiconductor field-effect transistors (MOSFETs). These systems will have cables, plugs, and board-edge connectors that should have periodic checking and cleaning during routinely scheduled downtime.

Display panels with touch-screen technology are replacing keyboards. As they become more common, there will be no keyboards and switch panels that need cleaning and testing. There will still be devices connected to the main machine being plugged into sockets. Small computer system interface (SCSI) connectors, 50-pin Centronics connectors, and 25-pin RS-232 sockets with parallel signals have been replaced by USB sockets that are simpler to clean and maintain. USB sockets are more reliable because there are only four contacts and a lot fewer failure points.

Where a mouse is used as a controller, the mechanical version that contains a ball periodically requires cleaning the ball and its sensors. An optical mouse is more reliable than a mechanical mouse for two reasons. First, it has no parts that would actually bring contaminants into its interior. Second, because no moving, sliding, or rolling parts can get stuck. The undersurface of the optical mouse can still pick up contaminants that will block light and it needs periodic cleaning. In an abrasive environment the underside of the optical mouse will get scratched eventually and become less reliable.

For preventive maintenance, a detailed informative log should be kept and followed in order to

- Build a database identifying the most frequent faults.
- Set up a schedule of testing that examines the most frequent failure points.
- Set a cleaning schedule.
- Enter all new parts and when they were replaced in the log.
- Identify motors and actuators that require periodic cleaning and lubrication. Even if that is not a digital or microcontroller issue, it should be included in the database of scheduled maintenance.

There are fewer connection points with microcontrollers due to subsystems that are embedded on one chip. Maintenance schedules for systems with embedded microcontrollers, FPGAs, CPLDs, and PALs may require less downtime than the maintenance for microprocessor-based systems doing the same task. If the system operates the same electromechanical peripheral devices, these devices still need the same maintenance regardless of how reliability improves in the microcontroller.

Self-Examination

1. The microcontroller does not contain the following:
   A. RAM
   B. I/O devices
   C. Power supply
   D. ROM or flash

2. Compared to a microprocessor-based system, a microcontroller-based system would probably have fewer
   A. Daughterboards
   B. Motherboards
   C. Breadboards
   D. Featherboards
3. Which kind of systems would likely have fewer connectors that might fail?
   A. Microprocessor-based systems
   B. Microcontroller-based systems
   C. Both A and B equally
4. Multiplexed data pins on a microcontroller are limited to
   A. Only memory write operations
   B. Only output port operations
   C. Only input port operations
   D. Only memory read operations
   E. Memory or I/O operations in one direction at a time
5. Most IC packages have a special mark on one corner of the chip that identifies
   A. V_{CC} and ground pins
   B. Whether the chip is a DIP, flat-pack, pin-grid array, or ball-grid array
   C. Pin number 1
6. The DDR register in a microcontroller controls
   A. I/O port direction for each bit on the data bus
   B. Memory-in and memory-out direction for each bit on the data bus
   C. Which bits on the address bus are for memory and which are for I/O
7. A DSP system used to process audio signals from a microphone and send them to a speaker with noise reduction would be possible on one chip if the microcontroller had the following feature(s):
   A. Serial input
   B. Serial output
   C. ADC and DAC
   D. RS-232 Port
8. Serial data is transmitted when a multi-bit character inside the CPU is shifted out through a single pin 1 bit at a time. The feature (or chip) that does this in a digital microcontroller is called a(n)
   A. ADC
   B. DAC
   C. PWM
   D. UART
9. A DMA controller controls
   A. Data going into (or out of) a memory device
   B. Data going into (or out of) the CPU from an I/O port
   C. Analog data into or out of an analog I/O port

10. An ENDEC deals with signals that are
    A. Analog
    B. Parallel
    C. Infra-red
    D. Pneumatic

11. A PWM that switches power to a dc motor changes from a 50 percent duty cycle to a 75 percent duty cycle. The motor speed will
    A. Increase
    B. Decrease
    C. Remain the same

12. A microcontroller controlling a 100-ton industrial crane is running a program to transfer a locomotive into a repair barn after it has arrived at the railhead. Half-way along, a fire is detected in the power room. A fire alarm signal starts emergency code running that sets the engine down gently, before the power cables burn through and the engine stops suddenly. This kind of emergency code would be called a(n):
    A. DMA transfer
    B. Interrupt
    C. On-chip debugger
    D. Multiplex

13. What kind of information would you not look for in a data sheet for the microcontroller CPU?
    A. Timing diagrams
    B. Electrical specifications
    C. Package types and pin layouts
    D. Prices of replacement parts

14. If you wanted to see signals on a great multitude of pins on a microcontroller, the best instrument would probably be a
    A. Storage scope
    B. Recurrent-sweep scope
    C. Multichannel analyzer
    D. Triggered sweep dual-channel scope

15. You are examining a timing diagram for a microcontroller and something called L_0–L_{15} looks like this: ———<⎯⎯>—<⎯⎯⎯> This pattern indicates a
    A. Single-pin switching serial data
    B. Bus-carrying multiple signals
    C. Three different levels of voltage at the same time on a single pin
    D. Ambiguous tristate logic

16. The flat part in the middle of the "sausage links" in the preceding diagram stands for
    A. Signals L_0–L_{15} have no usable data.
    B. Signals L_0–L_{15} are in a high-Z state.
    C. Signals L_0–L_{15} are not being driven HIGH or LOW by the CPU.
    D. All of these.

17. The Z16FMC microcontroller can be programmed using
    A. The in-system debugger
    B. The ADC input
    C. The PWM
    D. The ENDEC
18. You are looking at a timing diagram for a microcontroller and you see a line marked Q_{OUT}, which looks like this: $\overline{T_1\,T_2}$|. Based on the fact that the name Q_{OUT} does not have a bar or line over it, at which time is the signal Q_{OUT} active?
    A. T1
    B. T2
    C. Not possible to determine given the available data
19. Which is the simplest of the gate arrays named here?
    A. CPLD
    B. FPGA
    C. PAL
    D. FPLA
20. Which of the following is likely to be fastest in responding to its inputs?
    A. Microprocessor
    B. FPGA
    C. Microcontroller
    D. PAL

Questions and Problems

1. When troubleshooting a system with an embedded processor, what documentation do you need besides the data sheet?
2. There was a question in the last section about a multichannel logic analyzer. What tool for microprocessor/microcontroller troubleshooting would be even better for checking out the whole system?
3. Would adding an external memory card to expand a Z16FMC32 microcontroller be possible? Why or why not?
4. In what ways would troubleshooting be simpler in a system based on a microcontroller than in a system based on a microprocessor?
5. Control devices such as switches, dials, keyboards, and mouse inputs are being replaced by more reliable controls with fewer moving parts. Name some of these control devices.
6. RAM memory "forgets" what has been stored in it when power fails or has been turned off. For machine operation, a type of memory that does not forget the main program code is needed. What are some of the memory technologies that can be used for this purpose?
7. Chips with downward-pointing pins plugged into sockets are being replaced by newer connecting technologies. What are some of these technologies, and are they more reliable than chips plugged into sockets? Why or why not?
8. A PAL is used to replace a whole board full of gates. How does this affect the reliability of the system? Why?

9. What features embedded in the Z16FMC would be useful in a router for networking data in a computer system?

10. Go online and find some products used for preventive maintenance, which will help you set up a database. Compare and contrast two of these products.

11. Go online and find a data sheet for one of the following components listed here. Discuss what features the data sheet promotes as being important about this unit.
    A. A microprocessor
    B. A microcontroller
    C. A FPGA
    D. A CPLD

12. Firmware can be used to control gate and interconnect configuration in a programmable gate array. What types of memory are used for firmware storage?

13. A DSO can display a waveform for a one-time sequence of digital events, whereas an ordinary triggered sweep scope requires a repeating series of events to produce a stable waveform for analysis. Why is this and what is the difference between the two types of oscilloscope?

14. A signature analyzer will display a sequence of bits at a test point. How are they displayed, and how can the "signature" be used for troubleshooting?

15. Suppose you do not have any system documentation for a nonworking microcontroller-based machine, but you have documentation for the systems that are working. How would you use comparison method to diagnose the problem? Which measuring instruments do you think would be most useful and why?

16. How would you make a bunch of microprocessor-controlled mini-helicopters fly in formation if they could not be connected to one another by wires? Suggest a feature of the Z16FMC that could do this.

17. Look up how flash memory works, and explain why it might lose its firmware in an industrial location with very high levels of ionizing radiation.

18. Suggest a different type of firmware storage that might be better in a high-radiation environment like that of the previous question, and explain why it would be better.

19. There are many producers of diagnostic software for computers. Look up this subject on the Internet and describe one product that might be useful for your laptop.

20. IR communication through fiber-optic cables in an electrically noisy industrial environment is preferable to wireless or Ethernet network cables. Why?

21. Noise on an ac line can interrupt normal operation of digital equipment if voltage surges or noise are affecting the signals delivered by the power supplies. How would you test for this and what instruments would you use?

22. If you had the right "pods" and the right software, you could turn an ordinary laptop into a multichannel analyzer for most chips. Compared with a professionally built analyzer, what are the advantages and disadvantages of this setup?

23. On boot-up, instead of the normal touch-screen interface, the screen shows a clown-face and a rude message and does not respond to any attempts to operate or reset it. Do you think this is a software or hardware problem and why?

24. A system is spontaneously rebooting for no apparent reason. The fan is audible at start-up but is making a loud humming noise and does not appear to be spinning. What do you suspect and how could you check for it?

25. A microcontroller-based display screen outside a store uses red, green, and blue light-emitting diodes (LEDs) to display a traveling message. Recently, the sign has begun blinking partially on and off with only the red color LEDs staying lit. The green LEDs have faded out and the blue ones have not been lighting at all. Now even the green lights do not show up and the processor has begun running intermittently. The processor loses messages and requires reprogramming with the bargains of the day. Discuss what you would look for first, and why, including what equipment you would use to test for the problem (Hint: Check the operating electrical characteristics of red, green, and blue LEDs.)

Appendix A

Motor Troubleshooting Guide

| | Motor Type | | | | AC Polyphase (Two or Three Phase) | Brush Type (Universal, Series, Shunt, or Compound) |
| | AC Single Phase | | | | | |
Trouble	Split Phase	Capacitor Start	Permanent Split Capacitor	Shaded Pole		
Will not start.	1, 2, 3, 5	1, 2, 3, 4, 5	1, 2, 4, 7, 17	1, 2, 7, 16, 17	1, 2, 9	1, 2, 12, 13
Will not always start, even with no load, but will run in either direction when started manually.	3, 5	3, 4, 5	4, 9		9	
Starts, but heats rapidly.	6, 8	6, 8	4, 8	8	8	8
Starts, but runs too hot.	8	8	4, 8	8	8	8
Will not start, but will run in either direction when started manually—overheats.	3, 5, 8	3, 4, 5, 8	4, 8, 9		8, 9	
Sluggish—sparks severely at brushes.						10, 11, 12, 13, 14

(Continued)

| Trouble | Motor Type | | | | AC Polyphase (Two or Three Phase) | Brush Type (Universal, Series, Shunt, or Compound) |
| | AC Single Phase | | | | | |
	Split Phase	Capacitor Start	Permanent Split Capacitor	Shaded Pole		
Abnormally high speed—sparks severely at the brushes.						15
Reduction in power—motor gets too hot.	8, 16, 17	8, 16, 17	8, 16, 17	8, 16, 17	8, 16, 17	13, 16, 17
Motor blows fuse, or will not stop when switch is turned to off position.	8, 18	8, 18	8, 18	8, 18	8, 18	18, 19
Jerky operation— severe vibration.						10, 11, 12, 13, 19

1. Open in connection to line.	11. Dirty commutator or commutator is out of round.
2. Open circuit in motor winding.	12. Worn brushes and/or annealed brush springs.
3. Contacts of centrifugal switch not closed.	13. Open circuit or short circuit in the armature winding.
4. Defective capacitor.	14. Oil-soaked brushes.
5. Starting winding open.	15. Open circuit in the shunt winding.
6. Centrifugal starting switch not opening.	16. Sticky or tight bearings.
7. Motor overloaded.	17. Interference between stationary and rotating members.
8. Winding short-circuited or grounded.	18. Grounded near switch end of winding.
9. One or more windings open.	19. Shorted or grounded armature winding.
10. High mica between commutator bars.	

SOURCE: Courtesy Bodine Electric Co.

Appendix B

Motor Control Troubleshooting Guide

Trouble	Cause	Remedy
OVERLOAD RELAYS Tripping	1. Sustained overload.	1. Check for grounds, shorts, or excessive motor currents and correct cause.
	2. Loose connection on load wires.	2. Clean and tighten.
	3. Incorrect heater.	3. Heater should be replaced with one of correct size.
MAGNETIC AND MECHANICAL PARTS Noisy Magnet	1. Broken shading coil.	1. Replace magnet and armature.
	2. Magnet faces not mating.	2. Replace magnet and armature.
	3. Dirt or rust on magnet faces.	3. Clean.
	4. Low voltage.	4. Check system voltage and voltage dips during starting.
Failure to Pick Up and Seal	1. Low voltage.	1. Check system voltage and voltage dips during starting.
	2. Coil open or shorted.	2. Replace.
	3. Wrong coil.	3. Replace.
	4. Mechanical obstruction.	4. With power off, check for free movement of contact and armature assembly.
Failure to Drop Out	1. Gummy substance on pole faces.	1. Clean pole faces.
	2. Voltage not removed.	2. Check coil circuit.
	3. Worn or rusted parts causing binding.	3. Replace parts.
	4. Residual magnetism due to lack of air gap in magnet path.	4. Replace magnet and armature.
PNEUMATIC TIMERS	1. Foreign matter in valve.	1. Replace timing head completely or return timer to factory for repair and adjustment.

(Continued)

Trouble	Cause	Remedy
Erratic Timing Contacts Do Not Operate	1. Maladjustment of actuating screw.	1. Adjust as per instruction in service bulletin.
	2. Worn or broken parts in snap switch.	2. Replace snap switch.
LIMIT SWITCHES Broken Parts	1. Overtravel of actuator.	1. Use resilient actuator or operate within tolerances of the device.
MANUAL STARTERS Failure to Reset	1. Latching mechanism worn or broken.	1. Replace starter.
COMPENSATORS (MANUAL) Welding of Contacts on Starting Side	1. Inching, jogging, and operating handle slowly.	1. Excessive inching and jogging not recommended (caution operator); move handle swiftly and surely to start position.
Welding of Contacts on Running Side	1. Moving handle slowly to run position.	1. Move handle swiftly and surely to run position as motor approaches full speed.
	2. Lack of sufficient spring pressure.	2. Replace contacts and contact springs.
Damaged or Burned Transformer	1. Repeating inching and jogging.	1. Excessive inching and jogging not recommended (caution operator).
	2. Holding handle in start position for long periods.	2. Hold handle in start position only until motor approaches full speed.
CONTACTS Contact Chatter	1. Broken shading coil.	1. Replace magnet and armature.
	2. Poor contact in control circuit.	2. Replace the contact device or use holding circuit interlock (three-wire control).
	3. Low voltage.	3. Correct voltage condition. Check momentary voltage dip during starting.
Welding or Freezing	1. Abnormal inrush of current.	1. Check for grounds, shorts, or excessive motor load current; or use larger contactor.
	2. Rapid jogging.	2. Install larger device rated for jogging service.
	3. Insufficient tip pressure.	3. Replace contacts and springs, check contact carrier for deformation or damage.
	4. Low voltage preventing magnet from sealing.	4. Correct voltage condition. Check momentary voltage dip during starting.
	5. Foreign matter preventing contacts from closing.	5. Clean contacts with Freon; contactors, starters, and control accessories used with very small current or low voltage should be cleaned with Freon.
	6. Short circuit.	6. Remove short or fault and check to be sure fuse or breaker size is correct.

(Continued)

Trouble	Cause	Remedy
Short Tip Life or Overheating of Tips	1. Filing or dressing.	1. Do not file silver tips. Rough spots or discoloration will not harm tips or impair their efficiency.
	2. Interrupting excessively high currents.	2. Install larger device or check for grounds, shorts, or excessive motor currents.
	3. Excessive jogging.	3. Install larger device rated for jogging.
	4. Weak tip pressure.	4. Replace contacts and springs, check contact carrier for deformation or damage.
	5. Dirt or foreign matter on contact surface.	5. Clean contacts with Freon.
	6. Short circuits.	6. Remove short or fault and check to be sure fuse or breaker size is correct.
	7. Loose connection.	7. Clean and tighten.
	8. Sustained overload.	8. Check for excessive motor load current or install larger device.
COILS Open Circuit Roasted Coil	1. Mechanical damage.	1. Handle and store coils carefully.
	1. Overvoltage or high ambient temperature.	1. Check application circuit and correct.
	2. Incorrect coil.	2. Install correct coil.
	3. Shorted turns caused by mechanical damage or corrosion.	3. Replace coil.
	4. Undervoltage, failure of magnet to seal in.	4. Correct system voltage.
	5. Dirt or rust on pole faces increasing air gap.	5. Clean pole faces.

SOURCE: Courtesy Square D Co.

Appendix C

Radio and Stereo Troubleshooting Guide

Symptom	Section	Probable Cause
Weak sound, very distorted output.	Output amplifiers/drivers.	Replace output, check circuit.
No output or extremely low output.	Driver circuits.	Check drivers, outputs.
Hum, low volume.	Power supply.	Filter.
Poor selectivity/sensitivity.	Tuner/alignment.	Check converter, align receiver.
Noisy, rattling vibration.	Speakers.	Check connections, speaker, etc.
Motorboating, squealing.	Tuner/power supply.	Filter, tuner, check connections.
Intermittent operation.	All circuits.	Check connections, amplifiers, etc.
Station drift.	Local oscillator.	Check oscillator/leaky capacitors.
Volume varies.	AGC.	Check AGC circuit.
One channel dead and/or highly distorted with low volume.	Outputs/driver circuits, etc.	Check outputs, coupling, bias, circuits, AM/FM switch.
Good FM, but low volume, distorted AM.	AM circuits.	Outputs, etc.
Poor treble/bass fidelity.	Preamp crossover system.	Check circuits.
Noisy volume or balance control.	Control potentiometer.	Clean or replace control.
Faulty balance control.	Balance control circuit.	Check circuit.

Appendix D

DVD Player Troubleshooting

Symptoms	Remedy Actions
No power	1. Check whether the power cord has been properly connected to the power outlet. 2. Check whether the main power has been switched on. (DVD player's power supply transformer could be burned out.)
Screen is "frozen" or completely black, or Power button does not work	1. Try turning the power off and wait 3 seconds before turning it on again. 2. Make sure the TV and VCR (if connected) are set to the correct channel or line.
Does not play disc	1. No disc; Load a disc into the tray. 2. Disc has been loaded upside down. Place the disc with the label side up. 3. Disc region code does not match region code of the player. 4. Disc type is not the correct type to play. 5. Disc is damaged or dirty. Clean the disc or try another. 6. Moisture may be condensed inside the unit. Remove the disc and leave the power on for 1 or 2 hours.

(Continued)

Symptoms	Remedy Actions
No sound	1. Check whether the TV and amplifier (if connected) is turned on and set correctly. 2. Check whether the TV, DVD player, and amplifier are connected properly. 3. Ensure the TV, DVD player, and amplifier volumes are not set to Mute. 4. There will be no sound output during Reverse Play/Pause/Step/Slow/Search. 5. If the DTS soundtrack is enabled in the DVD Disc Menu, but you are using Stereo Analog audio, the sound will not work. Try switching the soundtrack to Dolby 5.1 or PCM mode in the Disc Menu.
Volume too high	1. Set TV to medium loudness; adjust volume with remote.
No surround sound	1. If digital output has been applied, enter the Setup menu, Player, and select proper setting under the Digital Audio section. 2. Turn on the active loudspeakers.
No sound or noise while playing MP3 disc	1. When program files or other data files are mixed together with MP3 files in the same disc, they may play the non-MP3 with noise or no sound. 2. Check the disc for MP3 files. 3. Skip the file or try another file.
Alternative audio soundtrack or subtitle cannot be selected	1. An alternative language cannot be selected with discs that contain only one language. 2. Some discs are programmed such that the Audio or Subtitle button cannot be used to select an alternative audio soundtrack or subtitle. Try selecting it from the DVD menu.
Angle cannot be changed	1. This function is dependent on software availability. Even if a disc has a number of angles recorded, these angles may be recorded at specific scenes only.
Cannot SKIP or SEARCH	1. Some discs are programmed such that they do not allow users to Skip or Search forward at some sections, especially at the WARNING section in the beginning. 2. Single-chapter discs cannot apply the Skip function.
No picture	1. Ensure all power cords are connected. 2. Try pressing the Enter/Play key. 3. Ensure the TV is turned on and the input is set properly. 4. Ensure the TV source is set properly on the TV. (This is probably done on the screen; check your TV manual.)

(Continued)

Symptoms	Remedy Actions
Picture noise/distortion	1. Disc is dirty or damaged. Clean the disc or try another disc. 2. Reset the color system of the player or the TV set. 3. Try a direct connection from the DVD player to the TV set, instead of via other components.
Picture not full screen	1. Enter Setup and then Player menu to select the correct screen format. 2. Select the screen format from the DVD Disc Menu or press Zoom key.
TV displays only static or "snow" on some channels	1. Perform Channel Scan (select Setup, then TV) to skip channels that do not broadcast in your area. (You cannot perform Channel Scan if you have a cable box.)
TV displays only static or "snow" on every channel	1. Ensure connections are correct. 2. Ensure the TV is set to proper input source. 3. Select Setup, then TV to perform the Channel Scan.
TV shows picture on only one channel	1. Ensure your VCR and/or cable box are tuned to the output channel (3 or 4) of the cable or satellite box.
Unable to access all TV channels from my cable box	1. You can only tune in to your cable channels on Channel 3 and you may have to use the remote for your cable box to select program channels.
Viewing resolution is poor in dim movie/TV scenes	1. Apart from the RCA connections, try to disconnect any other video output cable from the connectors (e.g., S-Video) on the DVD player. And/or, if there are other RCA inputs on the TV, you may try to connect the video and audio output to them.
Remote does not work	1. Ensure no objects obstruct the line of sight between the remote and player. 2. The remote may need new batteries.
Cannot enter menu selection	1. Use the Arrow keys to select option and then activate by pressing the Enter key.

SOURCE: Courtesy COLUMBIA ISA Audio/Video, 2013.

Appendix E

Antenna Television Troubleshooting

Symptom	Possible Cause or Remedy
System cannot locate satellite	• Check power to modem • Ensure that all cables are connected • Turn system off and recycle modem power • Ensure that cables are connected to dish and mounting apparatus • Navigate the antenna slowly • Check condition of cables and connectors • Check motor, GPS system • Check for damage to system due to weather or lighting
Reception has ghosts and smears	• Check whether cables or connectors are bad • Reposition the antenna, possible multipath reception disturbance • Check impedance for mismatch • Substitute different television for possible bad television • Check whether wrong directional antenna • Check the amplifiers and splitters for defects or poor connections • Check the television for wrong settings
Poor picture quality or noise and interference	• Rotate the antenna to another direction • Change the height of the antenna • Check the television for wrong settings • Check or replace all components of the antenna systems such as the antenna, connectors, preamplifier, amplifier, cable, splitters, filters, couplers, etc. • Check local area for mechanical, high-power, high-frequency interference • Check power quality, proper installation of system • Wrong wires and cables: replace them and test system • Check for damage to antenna and system due to weather or lightning

Appendix F

Generator Troubleshooting Guide

Symptom	Cause/Remedy
Stops operating in warm temperatures.	1. Check regulator.
	2. Check cooling fan.
Produces low voltage.	1. Check all meters.
	2. Inspect all connections and wires.
	3. Check diodes and surge suppressor.
	4. Check regulator.
	5. Check generator windings and armature.
No output.	1. Possible blown fuse.
	2. Verify connections.
	3. Inspect all wires for shorts or opens.
	4. Check regulator, diodes, or surge suppressor.
Voltage fluctuation.	1. Check tachometer.
	2. Inspect governor.
	3. Examine for loose connections.
	4. Check regulator.
	5. Test diodes, exciter, and windings.
Excessive high-voltage output.	1. Test meters.
	2. Check all connections.
	3. Test regulator.
	4. Check diodes and polarity.

Appendix G

General Troubleshooting Guide

Suspected Problem	Suggested Techniques*
Fuse	1, 2, 13
Filter (capacitor)	2, 4, 11, 16
Transistor	2, 4, 6, 7, 10
Diode	2, 3, 6, 10
Circuit board	1, 2, 5, 6, 7, 8, 12
Integrated circuit	1, 2, 3, 6
Tube	1, 2, 3, 5, 15
Coil	2, 3, 13, 16
Transformer	1, 2, 3
Resistor	1, 2
Switch	1, 2
Lamp	2, 3
Wire/cable	1, 2, 17
Motor winding	13, 14, 17
Armature	14
Electronic circuit	1, 2, 5, 6, 8, 9, 10, 12, 16
*Options (correspond to numbers in table)	

1. Voltage measurement.	10. Diode/transistor tester.
2. Resistance measurement.	11. Spark test.
3. Substitution.	12. Resoldering.
4. Bridging.	13. Test lamp.
5. Tapping.	14. Growler.
6. Heat/freeze.	15. Tube tester.
7. Transistor cutoff.	16. Other component testers.
8. Signal tracing.	17. Megohmmeter.
9. Oscilloscope.	

Appendix H

Testing Ultracapacitors Guide

Maxwell Technologies tests its ultracapacitors, using a straightforward constant current charge/discharge cycle. A constant current discharge test may be useful for customer evaluation of the product prior to application testing.

All ultracapacitors are stored discharged for safety. We recommend completely discharging any capacitors that will not be installed into equipment, and installing a shorting strap or wire across the terminals.

Below is a list of equipment required to perform a typical constant current discharge test:

- Bidirectional power supply (supply/load)
- Or, separate power supply and programmable load (constant current capable)
- Voltage versus time measurement and recording device (digital scope or other data acquisition)
- Current versus time measurement and recording device (optional if you can trust the power supply and load settings)

Before testing, connect data acquisition equipment to the device terminals, and set recording speeds as fast and reasonably as possible (100 milliseconds (ms) preferred, the faster, the more accurate the calculations).

Setup

- Set the power supply to the appropriate voltage and current limits, and turn the supply output off.
- The current limit can be anything at or less than the maximum rated current for the cell. Maxwell typically tests its cell at the 20 to 50 percent of the full rated current, depending on cell size. When repetitive high current testing is performed, cooling air should be provided.
- The voltage limit is the maximum cell voltage, times the number of cells in series. A single cell should be limited to 2.5 volts (V). Six cells in series, for example, can be operated at any voltage up to 15 V (6×2.5 V = 15 V).

- Connect the ultracapacitor to the power supply (having preset the current and voltage limits).
- Cooling air may be required to keep the ultracapacitor within operating temperature limits, depending on the test current and duration.
- Connect the voltage and current measuring/recording devices.

Charge

- With the power supply preset and the ultracapacitor connected, turn the supply output on.
- Charge the ultracapacitor at the appropriate current to the appropriate voltage.

Discharge

- Note: If using a separate programmable load instead of an integrated bidirectional power supply, disconnect the charging power supply prior to discharging. (Don't simply turn it off or change its set points, as many supplies will sink current when not regulating.)
- Set the load to the appropriate constant current, and discharge to 0.1 V or as low as the load can be controlled.
- *Immediately* remove the load once the minimum voltage is reached, allowing the device's voltage to "bounce" back.
- (The discharge can actually be stopped at any voltage. Maxwell, depending on equipment, measures some units discharged to 0.1 V and others discharged to one-half of the initial voltage. Values of capacitance will be slightly higher when discharged to one-half of the initial voltage rather than 0.1 V.)
- Measure the following parameters:

V_w = initial working voltage V_{min} = minimum voltage under load

I_d = discharge content V_f = voltage 5x after removal of load

t_d = time to discharge from initial voltage to minimum voltage

Capacitance calculation:

$$\text{Capacitance} = (I_d \times t_d)/(V_w - V_f)\ 5(I_d \times t_d)/V_d$$

This change in voltage $(V_w - V_f)$ is used because it eliminates the voltage drop due to the equivalent series resistance.

Equivalent series resistance (ESR) (at "dc") calculation:

$$\text{ESR} = V_f - V_{min}/I_d$$

(An LCR meter or bridge can be used to measure ESR at higher frequencies. The ESR at frequencies up to 100 hertz (Hz) will typically be 50 to 60 percent of the "dc" ESR. The capacitance will be much lower, due to the structure of the electrode.) (Note that calculations for capacitance and resistance can also be done on the charge cycle, using analogous measurement points.)

Safety Considerations

As in all electrical testing, you as the investigator should take appropriate cautions in the design and execution of the test. Proper precautions for the appropriate voltage should be observed. Any interconnections should be sized for the maximum anticipated current and insulated for the appropriate voltage. If repeated testing is performed, cooling air may be required to keep the test unit within its operating temperature range.

(Courtesy of Maxwell Technologies.)

Appendix I

Troubleshooting Acronyms

AC	Alternating Current
ACK	Acknowledgment
ADC	Analog-to-Digital Conversion
af	Audio-Frequency
AFC	Automatic Frequency Control
AGC	Automatic Gain Control
AM	Amplitude Modulation
ARIN	American Registry of Internet Numbers
ARP	Address Resolution Protocol
AS	Autonomous System
ASCII	American Standard Code for Information Interchange
AUX	Auxiliary Port
BGP	Border Gateway Protocol
BJT's	Bipolar Junction Transistors
BMET	Biomedical Engineering Technology
BMP	Bitmap
BNC	Bayonet Neill–Concelman
C	Chest
CB	Citizens' Band
CD	Compact Disk
CFL	Compact Fluorescent
CIDR	Classless Interdomain Routing
CLI	Command Line Interface
CLRN	Clear
CM	Common-Mode
CMOS	Complementary Metal-Oxide Semiconductor
CNC	Computer Numerically Controlled
Console	Console Port
CPLDs	Complex Programmable Logic Devices
CPU	Central Processing Unit
CRT	Cathode-Ray Tube
CSMA/CD	Carrier Sense Multiple Access/Collision Detect
DAC	Digital-to-Analog Converter
dB	Decibel
DC	Direct Current

DICOM	Digital Imaging and Communications in Medicine
DIP	Dual In-Line Package
DLCs	Double-Layer Capacitors
DLP	Digital Light Processing
DMA	Direct Memory Access
DMM	Digital Multimeter
DNS	Domain Name Service
DOD	Department of Defense
DOS	Digital Storage Oscilloscope
DSP	Digital Signal Processing
DSS	Digital Storage Scope
DVM	Digital Voltmeter
ECG	Electrocardiogram
EEG	Electroencephalogram
EGP	Exterior Gateway Protocols
EIGRP	Enhanced Interior Gateway Routing Protocol
EMF	Electromotive Force
EMG	Electromyography
EMT	Electrical Metallic Tubing
ENDECs	Encoder-Decoders
ENIAC	Electronic Numerical Integrator and Computer
ESD	Electro-Static Discharge
ESR	Equivalent Series Resistance
F	Farad
FCS	Frame Checking Sequence
FDDI	Fiber Distributed Data Interface
FET	Field-Effect Transistor
FF	Flip-Flops
FM	Frequency Modulation
FM MUX	Frequency Multiplexing
FPGAs	Field-Programmable Gate Arrays
FPLAs	Field-Programmable Logic Arrays
FPLDs	Field-Programmable Logic Devices
ft^2	Square Feet
FTP	File Transfer Protocol
g	Grams
GFI	Ground Fault Interrupter
GPIO	General Purpose Input/Output
H	Hosts
HDL	Hardware Description Language
HDLC	High-Level Data Link Control
HDTV	High-Definition Televisions
High-Z	High-Impedance
HSSI	High-Speed Serial Interface
HTTP	Hypertext Transfer Protocol
i.f.	Intermediate Frequency
I/O	Input/Output
IANA	Internet Assigned Numbers Authority
IC	Integrated Circuit
ICE	In-Circuit Emulator
ICMP	Internet Control Message Protocol

IGFET	Insulated-Gate Field-Effect Transistor
IGP	Interior Gateway Protocols
IORQ	Input/Output Request
IOS	Internetwork Operating System
IP	Internet Protocol
JFET	Junction Field-Effect Transistor
L	Left
LA	Left Arm
LAN	Local Area Network
LCD	Liquid Crystal Display
LCD	Lowest Common Denominator
LED	Light-Emitting Diode
LL	Left Leg
LLC	Logical Link Control
LS	Link State
LSB	Least Significant Bit
LSI	Large-Scale Integration
LUTs	Lookup-Tables
m	Meter
MAC	Media Access Control
µF	Microfarads
MOSFETs	Metal-Oxide Semiconductor Field-Effect Transistors
MPEG	Moving Picture Experts Group
MREQ	Memory Request
MRI	Magnetic Resonance Imaging
MSB	Most Significant Bit
MTU	Maximum Transfer Unit
N	Network
n	Negative
NEC	National Electrical Code
NEMA	National Electrical Manufacturers Association
NIC	Network Interface Card
NOS	Network Operating System
NVRAM	Nonvolatile Random Access Memory
OCR	Optical Character Recognition
Op-Amp	Operational-Amplifier
OSI	Open System Interconnect
OSPF	Open Shortest Path First
OTDR	Optical Time Domain Reflectometer
OUI	Organizational Unit Identifier
p	Positive
PAL	Programmable Array Logic
PC	Personal Computers
PCO	Port CO
pF	Picofarads
ping	Packet Internet Groper
PIPO	Parallel-In, Parallel-Out
PISO	Parallel-In, Serial-Out
PLC	Programmable Logic Controller
PM	Preventive Maintenance
POP	Post Office Protocol

POTS	Plain Old Telephone Service
PRN	Preset
PROM	Programmable Read-Only Memory
R	Right
RA	Right Arm
RAM	Random Access Memory
RARP	Reverse Address Resolution Protocol
RD	Read
rf	Radio-Frequency
RGB	Red, Green, Blue
RIP	Routing Information Protocol
RL	Right Leg
RPM	Revolutions Per Minute
SCR	Silicon Controlled Rectifier
SCSI	Small Computer Systems Interface
SDLC	Synchronous Data Link Control
SIPO	Serial-In, Parallel-Out
SISO	Serial-In, Serial-Out
SMDs	Surface-Mounted Devices
SMTP	Simple Mail Transfer Protocol
SOP	Sum of Products
SQL	Structured Query Language
STP	Shielded Twisted-Pair
STP	Spanning Tree Protocol
TCP	Transmission Control Protocol
TDR	Time-Domain Reflectometer
TELNET	Terminal Emulation Protocol
TFTP	Trivial File Transfer Protocol
TPS	Temporary Protected Status
TTL	Transistor-to-Transistor Logic
UART	Universal Asynchronous Receiver-Transmitter
UDP	User Datagram Protocol
UHDTV	Ultra-High-Definition Television
UHF	Ultra-High-Frequency
UL	Underwriters Laboratories
ULSI	Ultra-Large-Scale Integration
USB	Universal Serial Bus
UTP	Unshielded Twisted-Pair
V	Volts
var	Volt-Ampere Reactive
VCSEL	Vertical Cavity Surface Emitting Laser
VGA	Video Graphics Array
VHF	Very High-Frequency
VLSI	Very Large Scale Integration
VoIP	Voice Over Internet Protocol
VOM	Volt-Ohm-Milliamp Meter
VPN	Virtual Private Network
VSAT	Very Small Aperture Terminal
VTP	VLAN Trunking Protocol
VTY	Virtual Terminals

W	Watt
W/lb	Watts Per Pound
WAN	Wide-Area Network
WAV	Waveform Audio File
WINS	Windows
WR	Write
XNOR	EXCLUSIVE-NOR
XOR	EXCLUSIVE-OR

Appendix J

Using Router Commands for Troubleshooting

Global Configuration Mode to Interface Configuration Mode	router(config)#	router(config)# means in Global Config Mode
	router(config)# interface [interface name]	enter: interface [*interface name*]
	router(config-if)#	router(config-if)# means in Interface Config Mode
Global Configuration Mode to Privileged EXEC Mode	router(config)#	router(config)# means in Global Config Mode
	router(config)#exit	enter: exit
	router#	router# means Privileged EXEC Mode
Interface Configuration Mode to Global Configuration Mode	router(config-if)#	router(config-if)# means in Interface Config Mode
	router(config-if)# exit	enter: exit
	router(config)#	router(config)# means in Global Config Mode
Privileged EXEC Mode to Global Configuration Mode	router#	router# indicates in Privileged EXEC Mode
	router# configure terminal	enter: configure terminal
	router(config)#	router(config)# means in Global Config Mode
Privileged EXEC Mode to User EXEC Mode	router#	router# indicates in Privileged EXEC Mode
	router# disable	enter: disable
	router>	router> in User EXEC Mode
User EXEC Mode to Privileged EXEC Mode	router>	router> means in User EXEC Mode
	router> enable	enter: enable
	router#	router# means in Privileged EXEC Mode

Global Configuration Mode	router(config)#banner motd [@ message @]	enter: banner motd [@ message@] this allows you to post the "message of the day" between the @ symbols
	router(config)#enable secret [privileged EXEC mode password]	enter: enable secret [Privileged EXEC mode password] enters password in Privileged EXEC Mode
	router(config)#hostname [Name of Router]	enter: hostname [Name of Router] names the router
	router(config)#interface ethernet [Number]	enter: interface ethernet [Number] to configure the specific Ethernet interface
	router(config)#interface ethernet [Slot Number]	enter: interface [Slot Number] to configure slot and port number
	router(config)#interface [interface Type and Number]	enter: interface [interface Type and Number] to configure interface and type
	router(config)#interface serial [Number]	enter: interface serial [Number] to configure specific serial interface that you specify
	router(config)#interface [Subinterface Type and Number]	enter: interface [Subinterface Type and Number] to enter subinterface
	router(config)#line console [line #]	enter: line console [line #] to configure the Console Line #
	router(config)#line vty [# #]	enter: line vty [# #] to configure the Console Line 1st # to 2nd #
	router(config)#ip host [Name of Router] [IP Address of Router Interface]	enter: ip host [Name of Router] [IP Address of Router Interface] adds the IP address to the router name
	router(config)# no [command]	enter: no [command] will reverse or make it negative
Interface Configuration Mode	router(config-if)#clock rate 64000	enter: clock rate 64000 sets the clock rate at the DCE end of the serial interface
	router(config-if)#encapsulation [ENCAPSULATION TYPE]	enter: encapsulation [ENCAPSULATION TYPE] sets the encapsulation type for your interface
	router(config-if)#ip address[Interface address] [subnet mask of this Interface]	enter: ip address[Interface address] [subnet mask] Configures the IP address and subnet mask of the interface
	router(config-if)#no shutdown	enter: no shutdown to save the interface configurations
Line Configuration	router(config-line)#exit	enter: exit to return to the configuration mode
	router(config-line)#line vty [line numbers # #]	enter: line vty [line numbers 1st# 2nd#] changes to line configuration mode to set passwords
	router(config-line)#login	enter: login user logins to access the console line
	router(config-line)#password [to Console Line]	enter: password [to Console Line] specifies the password to the specific console line

Routing Protocols	router(config-router)#exit	enter: exit to return to Global Config Mode
	router(config-router)#network [IP address]	router(config-router)#network [IP address] defines the network that RIP will advertise
	router(config-router)#network [Network #]	enter: network [Network #] to specify the major networks RIP will advertise
	router(config-router)#no passive-interface [Interface Name]	enter: no passive-interface [Interface Name] enables the [Interface Name] to send and receive RIP updates
	router(config-router)#passive-interface default	enter: passive-interface default disables the forwarding advertisements on all interfaces
	router(config-router)#router rip	enter: router rip enables RIP routing on the router
Ethernet Interface Configuration	router(config-if)# interface e0/0	enter: interface e0/0 to access Ethernet interface
	router(config-if)# ip address [IP Address] [Subnet-Mask]	enter: ip address [IP Address] [Subnet-Mask] configures the IP address and subnet mask of the Ethernet interface
	router(config-if)# no shutdown	enter: no shutdown to save the interface configurations
Serial Interface Configurations	router(config-if)# interface s0/0/0	enter: interface s0/0/0 to access serial interface s0/0/0
	router(config-if)#ip address [IP Address] [Subnet-Mask]	enter: ip address [IP Address] [Subnet-Mask] configures IP address and subnet mask of the serial interface s0/0/0
	router(config-if)#clock rate 64000	enter: clock rate 64000 to configurate clock rate at Data Communication Equipment (DCE)
	router(config-if)#interface s0/0/1	enter: interface s0/0/1 to access serial interface s0/0/1
	router(config-if)#ip address [IP Address] [Subnet-Mask]	enter: ip address [IP Address] [Subnet-Mask] configures IP address and subnet mask of the serial interface s0/0/1 which is the Data Terminal Equipment (DTE) end of the serial cable
	router(config-if)# no shutdown	enter: no shutdown to save the interface configurations

Appendix K

Using Switch Commands for Troubleshooting

Global Config Mode to Interface Config Mode	switch(config)#	switch(config)# means in Global Mode
	switch(config)# interface [interface name]	enter: interface [*interface name*]
	switch(config-if)#	switch(config-if)# means in Interface Config Mode
Global Config Mode to Privileged EXEC Mode	switch(config)#	switch(config)# means in Global Mode
	switch(config)#exit	enter: exit
	switch#	switch# means in Privileged EXEC Mode
Interface Config Mode to Global Config Mode	switch(config-if)#	switch(config-if)# means in Interface Config Mode
	switch(config-if)# exit	enter: exit
	switch(config)#	switch(config)# means in Global Config Mode
Privileged EXEC Mode to Global Config Mode	switch#	switch# indicates in Privileged EXEC Mode
	switch# configure terminal	enter: configure terminal
	switch (config)#	switch(config)# means in Global Config Mode

Privileged EXEC Mode to User EXEC Mode	switch#	switch# indicates in Privileged EXEC Mode
	switch# disable	enter: disable
	switch>	switch> in User EXEC Mode
User EXEC Mode to Privileged EXEC Mode	switch>	switch> means in User EXEC Mode
	switch> enable	enter: enable
	switch#	switch# means in Privileged Exec Mode
Virtual Local Area Networks (VLAN) Commands		
Adding a VLAN	switch(config)#	switch(config)# indicates in Global Config Mode
	switch(config)#vlan [VLAN-ID]	enter: vlan [VLAN-ID]
	switch(config-vlan)# name [VLAN-Name]	enter: name [VLAN-Name]
	switch(config-vlan)# end	enter: end to return to Privileged EXEC Mode
Trunk Configuration	switch(config)#	switch(config)# indicates Global Config Mode
	switch(config)#interface [interface ID]	enter: interface [interface ID]
	switch(config-if)#no switchport trunk	enter: switchport mode trunk link switches to trunk
	switch(config-if)#switchport trunk native vlan [VLAN ID]	enter: switchport trunk native vlan [VLAN ID] to specify a native VLAN
	switch(config-if)#switchport trunk allowed vlan and vlan-list	enter: switchport trunk allowed vlan and vlan-list to add allowed VLANs to trunk
	switch(config-if)# end	enter: end to return to Privileged EXEC Mode
Trunk Modification	switch(config-if)# no switchport trunk allowed vlan	enter: no switchport trunk allowed vlan not to link switches to trunk
	switch(config-if)# no switchport trunk native vlan	enter: no switchport trunk native vlan to reset the native VLAN
	switch(config-if)# switchport mode access	enter: switchport mode access to return to static access mode port

Virtual Trunking Protocol (VTP) Commands	switch(config)#	switch(config)# indicates Global Config Mode
	switch(config)#vtp version [Number]	enter: vtp version [Number] configures the number
	switch(config)#vtp password [Password]	enter: vtp password [Password] configures password
	switch(config)#vtp domain [Domain Name]	enter: vtp domain [Domain Name] configures domain name
	switch(config)#vtp mode [Client|Server|Transparent]	enter: vtp mode [Client|Server|Transparent] used in global configurations

Appendix L

Troubleshooting Function Keys

Keys	Function of the Key or Command
Any key at the -More- prompt	Goes back to the EXEC prompt
Backspace	Character to the left is erased
Ctrl-A	Cursor is moved to the start of the command
Ctrl-B	Cursor is moved left only one character
Ctrl-C	Ends configuration mode and returns to privileged EXEC mode, ends the command prompt in the setup mode, also used to cancel a command or exit the configuration mode
Ctrl-D	Character at the cursor is erased
Ctrl-E	Cursor is moved to end of the command line
Ctrl-F	Cursor is moved right only one character
Ctrl-I	Shows the entered message on the command line or the system prompt
Ctrl-K	Characters at the cursor to the end of the command are erased
Ctrl-L	Shows the entered message on the command line or the system prompt
Ctrl-N	When using "Ctrl-P" to scroll down, use "Ctrl-N" to scroll up
Ctrl-P	Remembers the most recent commands, each time pressed again displays the previous commands
Ctrl-R	Shows the entered message on the command line or the system prompt, also used to avoid retyping the line due to any interruption
Ctrl-Shift-6 and then "x"	Ends any sequence, also used to end DNS lookups, pings, and traceroutes
Ctrl-U	Characters at the cursor to the beginning of command line are erased

Ctrl-W	Words to the left of the cursor are erased
Ctrl-X	Characters at the cursor to the beginning of command line are erased
Crtl-Z	Ends configuration mode and returns to privileged EXEC mode directly, used instead of exiting each mode
Enter Key at the -More- prompt	Shows the next line
Esc D	Characters at the cursor to the end of the word are erased. Note: To use Esc function, press and release the "Esc" key and then the press "D" key
Esc F	Cursor is moved forward one word to the right. Note: To use Esc function, press and release the "Esc" key and then press "F" key, then another key
Left-arrow key	Cursor is moved left only one character
Right-arrow key	Cursor is moved right only one character
Spacebar at the -More- prompt	Shows the next screen
Tab	Command name completion, also used to see other commands
Up-arrow key	Remembers the most recent commands, each time pressed again displays the previous commands

Answers to Self-Examinations

Chapter 1

1. E
2. D
3. A
4. C
5. B
6. D
7. B
8. C
9. A
10. D
11. B
12. A
13. C
14. A
15. E
16. B
17. B
18. B
19. B
20. D
21. A
22. B
23. C
24. C
25. C
26. A
27. C
28. C
29. B
30. A

Chapter 2

1. C
2. D
3. E
4. B
5. E
6. D
7. D
8. E
9. E
10. D

Chapter 3

1. D
2. D
3. C
4. B
5. A
6. D
7. E
8. A
9. D
10. A
11. C
12. A
13. A
14. C
15. D
16. C
17. D
18. A
19. A
20. A

Chapter 4

1. C
2. A
3. B
4. A
5. C
6. A
7. C
8. C
9. B
10. B
11. A
12. C

13.	D
14.	B
15.	B
16.	A
17.	B
18.	C
19.	B
20.	C
21.	C
22.	A
23.	D
24.	D
25.	B
26.	D
27.	A
28.	A
29.	C
30.	C

Chapter 5

1.	E
2.	D
3.	D
4.	C
5.	E
6.	C
7.	A
8.	D
9.	D
10.	E
11.	A
12.	C
13.	D
14.	C
15.	C
16.	D
17.	D
18.	B
19.	B
20.	C

Chapter 6

1.	B
2.	D
3.	D
4.	B
5.	A
6.	B
7.	E
8.	C
9.	C
10.	A
11.	E
12.	C
13.	C
14.	B
15.	B
16.	C
17.	D
18.	B
19.	C
20.	A
21.	D
22.	B
23.	D
24.	A
25.	B
26.	B
27.	C
28.	B
29.	B
30.	C
31.	B
32.	D
33.	B
34.	D
35.	B
36.	C
37.	C
38.	B
39.	D
40.	C

41.	A
42.	B
43.	C

Chapter 7

1.	B
2.	C
3.	B
4.	A
5.	C
6.	B
7.	B
8.	C
9.	D
10.	B
11.	C
12.	B
13.	B
14.	A
15.	C
16.	A
17.	B
18.	D
19.	A
20.	B

Chapter 8

1.	B
2.	A
3.	B
4.	D
5.	A
6.	B
7.	A
8.	A
9.	A
10.	C
11.	C
12.	C

13. B
14. C
15. C
16. A
17. A
18. B
19. B
20. D

Chapter 9

1. A
2. C
3. D
4. A
5. B
6. A
7. B
8. C
9. B
10. B
11. A
12. B
13. B
14. C
15. B
16. B
17. C
18. C
19. D
20. A

Chapter 10

1. B
2. A
3. D
4. B
5. D
6. B
7. D
8. D
9. C
10. B
11. C
12. D
13. B
14. A
15. B
16. D
17. B
18. B
19. A
20. C

Chapter 11

1. C
2. B
3. D
4. D
5. B
6. A
7. C
8. C
9. B

10. D
11. C
12. D
13. B
14. C
15. B
16. D
17. D
18. C
19. D
20. B

Chapter 12

1. C
2. A
3. B
4. D
5. C
6. A
7. C
8. D
9. A
10. C
11. A
12. B
13. D
14. C
15. B
16. B
17. A
18. A
19. C
20. D

Index

447